Software-Assisted Method Development in High Performance Liquid Chromatography

Other Related Titles from World Scientific

Analytical Applications of Ionic Liquids
edited by Mihkel Koel
ISBN: 978-1-78634-071-9

Fast Liquid Chromatography–Mass Spectrometry Methods in Food and Environmental Analysis
edited by Oscar Núñez, Héctor Gallart-Ayala, Claudia PB Martins and Paolo Lucci
ISBN: 978-1-78326-493-3

High Performance Liquid Chromatography Fingerprinting Technology of the Commonly-Used Traditional Chinese Medicine Herbs
by Baochang Cai, Seng Poon Ong and Xunhong Liu
ISBN: 978-981-4291-09-5

Problems of Instrumental Analytical Chemistry: A Hands-On Guide
by JM Andrade-Garda, A Carlosena-Zubieta, MP Gómez-Carracedo, MA Maestro-Saavedra, MC Prieto-Blanco and RM Soto-Ferreiro
ISBN: 978-1-78634-179-2
ISBN: 978-1-78634-180-8 (pbk)

Software-Assisted Method Development in High Performance Liquid Chromatography

Editors

Szabolcs Fekete
University of Geneva, Switzerland

Imre Molnár
Institute of Applied Chromatography, Berlin, Germany

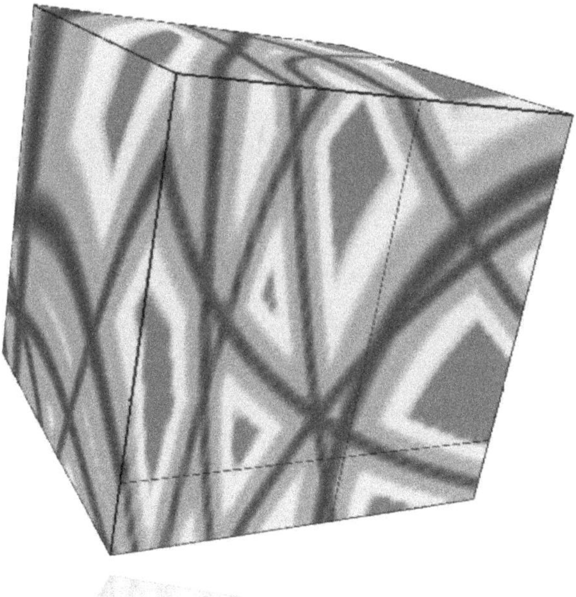

World Scientific

NEW JERSEY • LONDON • SINGAPORE • BEIJING • SHANGHAI • HONG KONG • TAIPEI • CHENNAI • TOKYO

Published by

World Scientific Publishing Europe Ltd.

57 Shelton Street, Covent Garden, London WC2H 9HE

Head office: 5 Toh Tuck Link, Singapore 596224

USA office: 27 Warren Street, Suite 401-402, Hackensack, NJ 07601

Library of Congress Cataloging-in-Publication Data

Names: Fekete, Szabolcs (Chemist), editor. | Molnár, Imre, 1943– editor.
Title: Software-assisted method development in high performance liquid chromatography /
edited by Szabolcs Fekete (University of Geneva, Switzerland),
Imre Molnár (Molnár-Institute for Applied Chromatography, Germany).
Description: New Jersey : World Scientific, 2018. | Includes bibliographical references.
Identifiers: LCCN 2018008579 | ISBN 9781786345455 (hc : alk. paper)
Subjects: LCSH: Liquid chromatography. | Chromatographic analysis--Data processing.
Classification: LCC QD79.C454 S64 2018 | DDC 543/.84028553--dc23
LC record available at https://lccn.loc.gov/2018008579

British Library Cataloguing-in-Publication Data

A catalogue record for this book is available from the British Library.

For any available supplementary material, please visit
http://www.worldscientific.com/worldscibooks/10.1142/Q0161#t=suppl

Desk Editors: Suraj Kumar/Jennifer Brough/Shi Ying Koe

Typeset by Stallion Press
Email: enquiries@stallionpress.com

Preface

High-performance liquid chromatography (HPLC) coupled with several detectors is now considered as the workhorse in several domains for the analysis of compounds of different size and polarity present in various matrices. Today, this method has strongly evolved to meet some of the requirements from different areas in terms of (i) high throughput or elevated resolution, through the use of innovative stationary phases and instruments, (ii) selectivity by using alternative modes of separation and (iii) sensitivity, thanks to the efficient coupling of HPLC with different detectors and more particularly with mass spectrometers (MS). Of course, these methods need development with appropriate tools in order to save time and improve the quality as well as the knowledge of the separation parameters. Moreover, for quantitative analyses, these methods have to be validated according to published guidelines. For this purpose, today there are different softwares available based on well-known retention models to assist the development of HPLC methods by varying simultaneously different important parameters. In this book, the current situation of HPLC retention–resolution modeling and its applications in the bio- and pharmaceutical industry are reviewed.

Chapter 1 gives an introduction to the method development in liquid chromatography (LC) assisted by software. A brief history of retention and resolution modeling is given and different software packages are presented. Chapter 2 is dedicated to the quality by design in HPLC, and more particularly to the use of DryLab software; Chapter 3 describes the software Chromsword for method development. In Chapter 4, intelligent systems are described to predict retention of analytes from their molecular properties in reversed-phase HPLC. The EluEx software is discussed for achieving resolution of the solutes defined by their chemical structure. Chapter 5 discusses the importance of statistical methods in the Quality by Design (QbD) approach to LC method development. Peak capacity and its optimization in isocratic and gradient elution modes are addressed in Chapter 6 followed in Chapter 7 by a description of a simple tool for teaching LC called "HPLC teaching simulator". The book ends with examples of applications of software-assisted method development for the analysis of small pharmaceuticals in Chapter 8, for the characterization of therapeutic proteins by reversed-phase chromatography in Chapter 9, by ion-exchange chromatography in Chapter 10, by hydrophobic interaction chromatography in Chapter 11 and, finally, by hydrophilic interaction chromatography in Chapter 12.

All these chapters are written by recognized experts in their fields, and therefore this book is recommended for every chromatographer who has as a main goal saving time, costs and efforts in method development and wants to understand the retention behavior of their compounds.

Jean-Luc Veuthey

About the Authors

Hermane T. Avohou is a biostatistician and Junior Scientist at the Laboratory of Pharmaceutical Analytical Chemistry of the University of Liège. Hermane has a Master degree in Biostatistics. He is currently interested in probabilistic modeling of spectroscopic signals with applications to quality control of medicines. His research is also focused on analytical quality by design strategy using Bayesian statistics.

Balazs Bobaly received his PhD in analytical chemistry from Budapest University of Technology and Economics, Hungary. He started work at the same institution and, since 2016, is a postdoctoral researcher at the University of Geneva, Switzerland. He has contributed ~20 articles and authored many book chapters. His research is focused on the liquid chromatographic characterization of therapeutic proteins using various (RP, IEX, SEC, HIC and HILIC) techniques. He is interested in the evaluation of new sample preparation and method development approaches as well as column technologies.

Bruno Boulanger is currently a Chief Scientific Officer of PharmaLex Statistical Solutions and Lecturer in Statistics at the School of Pharmacy, University of Liège since 2000. Bruno also has been a USP Pharmacopeia Expert since 2010 and a member of the Committee of Experts in Statistics. Bruno has 25 years of experience in pharmaceutical industry, working in Europe and in USA.

Benjamin Debrus is a Scientific Collaborator of the Laboratory of Pharmaceutical Analytical Chemistry of the University of Liège. He performed his PhD thesis in Pharmaceutical and Biomedical Sciences working on new methodologies for the development of chromatographic methods using design of experiments and quality-by-design tools. He followed this with postdoctoral research in Analytical Chemistry and Chemometrics at the University of Geneva focusing on multivariate data analyses used for the treatment of LC-MS and GC-MS data.

Szabolcs Fekete holds a PhD degree in analytical chemistry from the Technical University of Budapest, Hungary. He worked at the Chemical Works of Gedeon Richter Plc at the analytical R&D department for 10 years. Since 2011, he has been working as a scientific collaborator at the University of Geneva in Switzerland. He has contributed ~100 journal articles, authored many book chapters and edited handbooks. His main interests include liquid chromatography (RP, IEX, SEC, HIC, SFC and HILIC), column technology, method development, pharmaceutical and protein analysis and mass transfer processes.

Sergey V. Galushko is a specialist in analytical chemistry with more than 30 years experience in HPLC method development. He has more than 70 scientific publications in the analytical chemistry field. His research interests involve structure–retention relationship, computer modeling and optimization of separations of small and large molecules in liquid chromatography.

Dr. Galushko is the President and head of R&D of ChromSword (from 1998). ChromSword is a provider of method development service and specialized software for computer-assisted and automatic HPLC method development.

Davy Guillarme holds a PhD degree in analytical chemistry from the University of Lyon, France. He is now senior lecturer at the University of Geneva in Switzerland. He has authored 190 journal articles related to pharmaceutical analysis. His expertise includes HPLC, UHPLC, LC-MS, SFC and analysis of proteins and mAbs. He is an editorial advisory board member of several journals including *Journal of Chromatography A, Journal of Separation Science, LC-GC North America* and others.

Krisztián Horváth is an associate professor at the University of Pannonia (Veszprém, Hungary) and a member of the board of the Hungarian Society for Separation Sciences and the Analytical Division of Hungarian Chemical Society. He graduated with a degree in environmental engineering in 2002 and obtained his PhD in chemistry in 2007 from the University of Pannonia. His research interests include the study of retention behavior of small and large molecules in HPLC, and method development and optimization in 1D- and 2D liquid chromatography. He has contributed 30+ journal articles and also several textbooks.

Cédric Hubert is a Senior Scientist at the Laboratory of Pharmaceutical Analytical Chemistry of the University of Liège. Cédric has a Masters degree in Chemical Sciences and received his PhD in 2015 in Pharmaceutical and Biomedical Sciences. His main research interests concern the development and optimization of chromatographic methods using analytical quality by design strategy for bioanalysis and quality control for pharmaceutical industry. Cédric is also recognized for his expertise in the field of analytical method validation.

Philippe Hubert is Professor of analytical chemistry at the University of Liège. He received his PhD in 1994 in pharmaceutical analytical chemistry. He has published more than 150 peer reviewed articles with the SCI over 2000 times cited. Currently, his research focuses on separation sciences for the determination of active ingredients in various matrices, vibrational spectroscopy (NIR and Raman) in the framework of FDA's Process Analytical Technology and validation and chemometrics aspects including experimental design and quality by design.

Róbert Kormány is a chemist and special analyst in chromatography. He graduated from the University of Debrecen then completed his education with an additional degree in chromatography and PhD degree in chemical sciences from the Department of Inorganic and Analytical Chemistry of the Budapest University of Technology and Economics in the laboratory of Prof. Dr. Jenő Fekete. He works at the pharmaceutical company Egis and specializes in developing UHPLC methods for the separation of pharmaceutical compounds using computer modeling.

Pierre Lebrun is Director Statistics at Pharmalex Statistical Solutions. Pierre developed statistical methodologies to promote quality-by-design, with a strong emphasis on the use of Bayesian statistics during early characterization and validation stages. Pierre has a Masters degree in computer sciences and economy and in statistics from the University of Louvain-la-Neuve (Belgium). He holds a PhD in statistics from the University of Liège (Belgium), on the topic of Bayesian models and Design Space applied to the pharmaceutical industry.

Imre Molnár is the president of Molnár-Institute and has more than 35 years of experience in the field of HPLC. He specialized in pharmaceutical research and analysis and works with industrial and academic groups on research topics in pharmaceutical and biopolymer analysis.

He received his PhD in 1975 and spent the following two years as a postdoctoral fellow at the Department of Bioengineering at Yale University. Later, he began working on the development of DryLab software, which is now widely used in both the pharmaceutical industry and the life science community. He has contributed 50+ journal articles and authored book chapters in handbooks.

György Morovján graduated from the Department of Chemical Engineering, Technical University, Budapest. His PhD thesis related to analytical method development of pharmaceuticals from biological matrices. He has a general interest in separation science, especially liquid-phase separation methods, and analytical and preparative chromatographic method development. He is also professionally involved in intellectual property, mainly in the field of pharmaceuticals. He is a Hungarian and European Patent Attorney.

Norbert Rácz is an analytical chemist who obtained his BS and MS degree in chemical engineering at the Budapest University of Technology and Economics. He is currently working to obtain his PhD degree in chemistry under the supervision of Róbert Kormány. His research is focused on the development of new methods with the aid of modeling software. He works as an analyst at R&D Analytical Laboratory for APIs of Egis Pharmaceuticals PLC.

Hans-Jürgen Rieger has been working with Molnár-Institute since 1999 as a chemist specializing in software programming. He is the Vice-President and product manager of the company. He gives DryLab courses worldwide and continuously works on software development. He has co-authored several journal articles.

Oksana Rotkaja has been an application specialist at ChromSword Baltic from 2011. Oksana has BSc (2006), MSc (2012) degrees and is currently a PhD student in analytical chemistry at Latvian University, Riga.

Serge Rudaz is an Associate Professor at the University of Geneva. He is interested in metabolomics, (UHP) LC and CE coupled to MS, advances in sample preparation, analysis of pharmaceuticals and counterfeit medicines, biological matrices, clinical and preclinical studies, including metabolism and toxicological analysis. Serge Rudaz is an expert in various chemometric approaches, including experimental design, validation and regulation (ISO17025) as well as multivariate data analysis.

Irina Shishkina graduated from Kie State University in physic of spectroscopy and obtained her PhD at the Institute of Organic Chemistry of Ukrainian Academy of Science. She specialized in analytical chemistry of organic compounds and has more than 30 publications to her credit. Her current research and development activity are in the field of computer chromatography and automated method development.

Evalds Urtans has been a software development project manager and a researcher in ChromSword Baltic SIA from 2013. Evalds is a PhD candidate in Computer Science at Riga Technical University Latvia and holds an MSc in intellectual robotic systems speciality of Riga Technical University, Latvia and BSc in computer games development, University of Glamorgan, UK, 2009.

Jean-Luc Veuthey is a Professor at the School of Pharmaceutical Sciences, University of Geneva, Switzerland. He also acted as President of the School of Pharmaceutical Sciences, Vice-Dean of the Faculty of Sciences and finally Vice-Rector of the University of Geneva from 1998 to 2015. His research domains are development of separation techniques in pharmaceutical sciences, and more precisely study of the impact of sample preparation procedures in the analytical process; fundamental studies in liquid and supercritical chromatography; separation techniques coupled with mass spectrometry; and analysis of drugs and drugs of abuse in different matrices. He has published more than 300 articles in peer-reviewed journals.

Contents

List of Abbreviations and Symbols

2D	Two-dimensional
3D	Three-dimensional
α	Selectivity
ACN	Acetonitrile
ADC	Antibody–drug conjugate
ATP	Analytical target profile
Cb	Buffer concentration
CDS	Chromatography method development data system
CMP	Critical method parameters
CQA	Critical quality attribute
d_p	Particle size of the stationary phase
DoE	Design of experiments
DS	Design space
F	Flow rate of the mobile phase
FFD	Full factorial design
GMP	Good manufacturing practice
HIC	Hydrophobic interaction chromatography
HILIC	Hydrophilic interaction liquid chromatography
HPLC	High-performance liquid chromatography
ID	Inner diameter of the chromatographic column
IEX	Ion-exchange chromatography
k	Retention factor
L	Column length
LC	Liquid chromatography
LSS	Linear solvent strength

$\log D$	Logarithm of distribution coefficient between 1-octanol and water for ionizable compounds
$\log P (\log P_{ow})$	Logarithm of partition coefficient between 1-octanol and water for non-ionizable compounds
mAb	Monoclonal antibody
MeOH	Methanol
MODR	Method operable design region
MP	Mobile phase
NPLC	Normal-phase liquid chromatography
OFAT	One-factor-at-a-time
OoS	out of specification
OoT	Out of trend
PACP	Post-approval change process
PBD	Plackett–Burman partial factorial design
p*Ka*	Negative logarithm of the acid dissociation constant
QC	Quality control
QbD	Quality by design
QSRR	Quantitative structure retention relationships
RPLC	Reversed-phase liquid chromatography
R_s	Resolution
$R_{s,\mathrm{crit}}$	Critical resolution (lowest resolution among all resolutions)
SMR	Standard multivariate regression
SP	Stationary phase
SST	System suitability test
T	Temperature
tC	Ternary composition
tG	Gradient time
t_R	Retention time
UHPLC	Ultra-high pressure liquid chromatography
V_d	Dwell volume of the chromatographic system
$V_{\mathrm{ext.col.}}$	Extra-column volume of the chromatographic system

V_{inj} Injected volume (by the auto-sampler of the chromatographic system)

w Peak width

WP Working point (optimal condition in a design space where resolution criterion is fulfilled)

Chapter 1

Introduction: The First Steps of Method Development in Liquid Chromatography

Imre Molnár*,‡ and Szabolcs Fekete†

*Molnár-Institute, Institute for Applied Chromatography, Schneeglöckchenstrasse 47, D-10407 Berlin, Germany

†School of Pharmaceutical Sciences, University of Geneva, University of Lausanne, CMU — Rue Michel Servet 1, 1211 Geneva 4, Switzerland

‡imre.molnar@molnar-institute.com

1.1 Introduction

Modeling in high-performance liquid chromatography (HPLC) was started in 1975 by Csaba Horváth at Yale University, Haven, CT, USA. He purchased a PDP-11 computer and studied the theory of band spreading in HPLC with this computer [1]. The fundamentals of reversed-phase liquid chromatography (RPLC) were established based on compounds of relevance to the field of life science, such as catecholamines and their derivatives and metabolites [2]. The analysis time for 100 organic acids could be reduced from 48 h to less than 30 min using RPLC [3]. A few months later, the separation of amino acids and peptides was achieved first time on an octadecyl (C18) stationary phase [4]. The basic relationships were investigated and systematic work was carried out, typically one-factor-at-a-time (OFAT), to be able to understand the reason for peak movements in a chromatogram during the optimization process. The observed relationships created a new theory of solvophobic interactions that was reported in a series of papers which explained the significant differences observed between the solute retentions observed in water-rich and organic-modifier-rich mobile phases [5–7]. The power of water to enable retention on a C18 stationary phase — due to its high surface tension — can be reduced by adding methanol (MeOH) or acetonitrile (ACN), as is usually done in RP gradient elution. These generic gradients have the advantage of being able to elute almost any organic compound by a continuous increase in the content of the organic modifier in the aqueous mobile phase, thus leading to a new culture of gradient elution in HPLC.

In the 1980s, Snyder and Kirkland studied column properties by measuring ca. 1000 columns at DuPont to understand how solute diffusion in the pores influences band spreading. They included the van Deemter and the Knox equations into their models. In 1985 Lloyd Snyder (LC Resources) and Imre Molnár (Molnár-Institute) began to develop a software to assist method development for HPLC separations. The software was first an isocratic variant which could calculate tables of retention factor (k) ranges and resolution (R_s) values based only on a few experiments and could predict the optimal mobile phase composition to achieve the highest resolution between the compounds in the shortest possible time. Snyder named the software "DryLab I" (I for isocratic) [8]. It could set retention limits for a separation ($1 < k < 10$) and also introduced the principle of "equal band spacing" (EBS) based on the resolution of the least well-separated peak pair, called the "critical peak pair" ($R_{s,\mathrm{crit}}$). If the critical peak pair is separated with baseline resolution ($R_s > 1.5$), it can be presumed that all other peak pairs will also be atleast baseline separated.

A short time later, the development of a software version working in gradient elution mode was created, based on the theory developed by Snyder, under the name DryLab G (G for gradient) [9, 10]. Here, the initial and final mobile phase composition could be varied — with up to 10 gradient steps — and the influence of column dimensions (L, ID, d_p), flowrate and instrument factors (dwell volume, extra-column volume) could also be considered. It was the first attempt to calculate multifactorial influences on chromatographic separation in gradient elution.

In 1989, Snyder and Joseph Glajch provided an impressive collection of contributions from experienced method development professionals regarding their work. A set of 43 papers was the final result, with a premium selection of articles by well-recognized authors like Berridge, Deming, Billiet,

Galan, Snyder, Dolan, Jandera, Lankmayer, Schoenmakers, Massart, Valkó, Jinno and others who contributed important research to this field [8].

Unlike the initial interest on band spreading, it was only somewhat later that other variables like mobile phase pH, temperature, buffer composition, ternary eluent composition, ion-pair concentration, etc., were added to be able to study the influence of multifactorial changes on the position of peaks in a chromatogram, and even later that they were commercially available under the name DryLab Imp (Isocratic multiparameter version). The impact of column length, inner diameter, particle size, flow rate, system dwell volume and extra-column volume on the separation could also be calculated. DryLab® was therefore a multifactorial tool even during 1988–1990 in separation modeling and helped to understand how retention is changing and how to control selectivity changes of moving peaks to be able to comply with regulatory expectations.

To study the influence of two measured variables at the same time, 2D models were developed. These 2D resolution maps showed the separation of the critical peaks in the chromatogram as a function of two simultaneously changed experimental variables. The gradient time (*tG*) and the mobile phase temperature (*T*) were chosen as the two most relevant variables (*tG-T*-model) as a simple method for peak tracking. The concept of a "method operable design region" (MODR) or "design space" (DS) was laid down for HPLC here the first time. A more detailed summary of the contributions is compiled in Ref. [11].

After the great success of the *tG-T*-model, advocated by Snyder and Dolan, they introduced this concept in 2000 for column characterization [12].

It was in 2009 that it first became possible to study three measured variables at the same time and calculate the influence of an additional six/eight other parameters, such as flow-rate, column dimensions, instrument parameters and gradient conditions, in the so-called "cube". Launching this 3D critical resolution map [13], new avenues were opened not only in method development but also in robustness testing. Several HPLC instruments (e.g. Waters, Shimadzu, Thermo) can be controlled today with the DryLab® software for the automated processing of the necessary experiments to enable quicker separations in an automated fashion. Nowadays, experiments that require build-up of four resolution cubes (on four columns providing different selectivity) can be performed in one single day as a complete method screening and optimization protocol [14].

The latest version of the software (DryLab®4, version 4.3) enables "dry" modeled robustness testing. From the DS, as defined in a 3D resolution map, it is possible to obtain robustness information for the measured parameters, including *tG*, *T*, *tC* (ternary composition) and mobile phase *pH*. In addition, based on the models included in the software, the retention time of any compound can be calculated to account for the influence of additional variables, such as flow rate or initial and final mobile phase composition (expressed in %B) through the gradient. Consequently, the impact of changes in any of these six variables on the resolution can be assessed using simulated 2^6 or 3^6 type factorial designs. No additional experiments are necessary for performing the simulated robustness calculation [13]. The possible deviations from the nominal values just need to be defined and then the software makes the calculations for $2^6 = 64$ or $3^6 = 729$ conditions. With two additional gradient points (each gradient point corresponds to two additional variables), the variants sum up to $2^{10} =$ ca. 59,000 chromatograms, which are calculated and evaluated in less than 1 min. At the end, the software provides a "frequency distribution graph" showing how often a certain critical resolution value occurs under any combination of the possible parameters. This graph also shows the failure rate, i.e. number of experiments that could fall outside the required critical resolution in routine work. On the other hand, "regression coefficients" can also be obtained to show the effect of each variables, related to the selected deviation from the nominal value, for the critical resolution. The robustness feature allows to reduce the "out of specification" (OoS) results.

In this book, the current situation of HPLC retention–resolution modeling and its applications in the bio- and pharmaceutical industry are reviewed.

1.2 Modeling Alternatives

Currently, there are many other commercially available software packages on the market such as DryLab®4 modules (PeakTracking, 3D-Cube, Robustness Module, Know-ledge Management Module, Column Comparison

Module, Molnár-Institute, Germany), ChromSword® packages (Developer, AutoRobust, Scout, ChromSword, Latvia), ACD packages (LC simulator, ChromGenius, AutoChrom, ACD/Labs, Canada), Fusion (S-Matrix, USA) and Osiris (Datalys, France). Some of them mainly focus on the quantitation process using statistical approaches to find out if a method is not robust or to support method screening (Fusion). Other packages, like DryLab, explain why a method is not robust and how to change conditions to get back to the validated region (DS).

Some tools start from molecular structure and try to derive an approximation of the retention time at which a molecule would elute from the column (ChromSword). Other tools (e.g. EluEx) use log*D*, log*P* and p*Ka* values to approximate the mobile phase composition and pH range for a decent resolution. Other packages offer a mathematical statistics-oriented development approach (Fusion).

Presently, the trend is to look at robust conditions in a multivariable space in different ways. First of all, it is most important to understand peak movements in HPLC separations, which are based on a sufficiently wide range of eluent properties (pH, gradient time and program, flowrate, etc.), before time-consuming experiments based on *trial & error* can be investigated. Multifactorial modeling simplifies and speeds up the process of developing reliable chromatographic separations by allowing the user to model changes in the separation conditions using a computer. As an example, the analysis time of an old pharmacopeia method was cut down from 160 min to less than 3 min [15]. The most important advantage is, however, the better understanding of the scientific process of the separation and to prove the suitability of the HPLC method for communication of results to the regulatory agencies (FDA, EMA) in order to receive commercial authorization for drugs.

1.3 What is the Purpose of Method Development?

Method development in HPLC means the search for the optimal chromatographic operating conditions (type of mobile and stationary phase, temperature, gradient steepness, pH, ionic strength, etc.) resulting in the proper separation of a mixture into its constituents within a given analysis time frame [16]. Because of the high probability for peak overlap and the high dependence of the retention time on the employed chromatographic parameters, the method development process is often tedious and time-consuming (up to several weeks of work) [17]. It requires the knowledge and expertise of the analyst, but still involves a lot of trial-and-error processes. Computer-assisted method development however has the potential to speed up the process significantly, if adequate retention models exist. The gain in analysis time can be particularly significant for regulated laboratories (pharmaceutical, food and environmental analytical laboratories) to prove that the method is based on solid science. Before beginning method development, the chromatographers need to review what is known about the sample [18]. The goals of the separation should also be defined at this point. The chemical composition of the sample can provide valuable clues for the best choice of initial conditions for an HPLC separation. The analytical target profiles (ATP's) of the HPLC separation need to be specified clearly. Is it a quantitative or qualitative analysis? Do we need to determine the known main component(s) or the unknowns, like impurities, degradation products and excipients? Is it necessary to resolve all the sample components or just some of them? What level of accuracy and precision is required? What should be the limit for the analysis time? Many other questions have to be clarified before starting the experimental work. In most cases, it is sufficient to have baseline resolution.

Method development typically involves a scouting process and an optimization phase. During the optimization phase, accurate retention modeling is required to find the optimal separation conditions (errors as low as $\sim$1–2 % in retention time). On the other hand, during the scouting phase, including, e.g. the choice of the chromatographic technique and column stationary phase, prediction errors up to 10% could be tolerated. Quantitative structure retention relationships (QSRRs), which could potentially replace the initial exploratory experiments by prediction solely based on the structure of the molecule, are of interest to speed up this phase. However, much lower prediction errors can be achieved using analytical, empirical models established through fitting of a limited number of experimental retention data. Besides speeding up the method development process, the systematic experimentation using the state-of-the-art software packages also allow to meet quality by design (QbD) requirements in industrial laboratories by providing a tool to improve the robustness of a chromatographic method. Finally, computer-assisted method development reduces the solvent consumption by limiting the required number of experiments. Hence, it can be considered as a green strategy in liquid chromatography (LC) [19].

Note that for some applications, involving a high number of compounds (proteomics), peak capacity optimization can be considered as an additional alternative to model the critical resolution [20]. Achievable peak capacity per analysis time is then often the response function of the method development process.

1.4 How to Select the Most Important Method Variables?

The method variables are the main factors (e.g. gradient steepness, mobile phase temperature, pH, buffer concentration, stationary phase, etc.) that impact the separation of the compounds of interest to us. The method variables have to be carefully selected on the basis of the separation mode and the characteristics of the sample components. This book discusses the most important modes of LC today such as: RPLC, ion-exchange (IEX) chromatography, hydrophobic interaction chromatography (HIC) and hydrophilic interaction liquid chromatography (HILIC). Obviously, the different modes require the optimization of different variables due to the differences in retention mechanisms. These variables are discussed in detail in the corresponding chapters.

Among the different variables, the gradient time *tG* (gradient steepness) is of primary interest as it has a huge impact in any mode of chromatography, which are based on the linear solvent strength (LSS) theory. Mobile phase temperature is the second most important variable, as temperature affects the strength of interactions between solute and stationary phase and changes the viscosity, sample diffusivity and solubility. So far, the most popular start in HPLC method development is the so called *tG-T* design.

For ionic (charged or ionizable) samples, the mobile phase pH should be studied as a variable. The combination in ternary eluents (*tC*) (such as mixtures of AcN and MeOH) changes the separation selectivity and should be explored to figure out the best separation conditions. DryLab® offers for both combinations two different cubes (*tG-T*-pH and *tG-T-tC*) with 12 basic experiments. The concentrations of additives (salt, ion-pairing reagent, buffer, etc.) can also influence the separation and be modeled. Some of these factors impact the retention in a nonlinear fashion, therefore necessitating a study of their effects at three or more levels, resulting in a cube with 18 basic experiments, such as *tG-tC-pH* cube.

1.5 Who Should Read this Book?

This book is recommended for practicing chromatographers who are interested to save time, costs and efforts in method development and want to understand the retention behavior of their samples. Computer-assisted method development is also useful for studying conditions for method robustness and for the transfer/scale of methods and to automate the method development procedure and understand peak tracking.

The book reviews the most important chromatographic modeling software (and their features) which are available today and explains the screening and optimization procedures in a step-by-step manner for various modes of separations and for various samples. Several industrial examples illustrate the potential and benefits of computer-assisted method development from various fields of analysis in the following chapters. Another set of case studies is available in Ref. [14] and at www.molnar-institute.com/literature, with more than 200 scientific papers.

References

[1] C. Horváth, H.J. Lin, Band spreading in liquid chromatography general plate height equation and a method to individual plate height contributions, *J. Chromatogr.* 149 (1978) 43–70.

[2] I. Molnár, C. Horváth, Catecholamins and related compounds — Effects of substituents in reversed phase chromatography, *J. Chromatogr.* 145 (1978) 371–381.

[3] I. Molnár, C. Horváth, Rapid separation of urinary acids by high-performance liquid chromatography, *J. Chromatogr.* 143 (1977) 391–400.

[4] I. Molnár, C. Horváth, Separation of amino acids and peptides on nonpolar stationary phases in HPLC, *J. Chromatogr.* 142 (1977) 623–640.

[5] C. Horváth, W. Melander, I. Molnár, Solvophobic interactions in liquid chromatography with nonpolar stationary phases (solvophobic theory of reversed phase chromatography, Part I.), *J. Chromatogr.* 125 (1976) 129–156.

[6] C. Horváth, W. Melander, I. Molnár, Liquid chromatography of ionogenic substances with nonpolar stationary phases (Part II.), *Anal. Chem.* 49 (1977) 142–154.

[7] C. Horváth, W. Melander, I. Molnár, Liquid chromatography of ionogenic substances with nonpolar stationary phases (Part III.), *Anal. Chem.* 49 (1977) 2295–2305.

[8] L.R. Snyder, J.L. Glajch, *Computer-assisted method development for high performance liquid chromatography*, eds. J.L. Glajch and L.R. Snyder, Elsevier, 1990, ISBN 0-444-88748-2; *J. Chromatogr.* 485 (1989) 1–640.

[9] L.R. Snyder, *High Performance Liquid Chromatography. Advances and Perspectives*, Vol. 1, ed. C. Horváth, Academic Press, New York, 1980.

[10] J.W. Dolan, L.R. Snyder, M.A. Quarry, Computer simulation as a means of developing an optimized reversed-phase gradient-elution separation, *Chromatographia* 24 (1987) 261–276.

[11] I. Molnár, Computerized design of separation strategies by reversed-phase liquid chromatography: Development of DryLab software, *J. Chromatogr. A* 965 (2002) 175–194.

[12] J.W. Dolan, L.R. Snyder, T. Blanc, L. van Heukelem, Selectivity differences for C-18 reversed phase columns as a function of temperature and gradient steepness, *J. Chromatogr. A* 897 (2000) 77–116.

[13] I. Molnár, H.J. Rieger, K.E. Monks, Aspects of the "Design Space" in high pressure liquid chromatography method development, *J. Chromatogr. A* 1217 (2010) 3193–3200.

[14] I. Molnár, H.-J. Rieger, R. Kormány, Modeling of HPLC methods using QbD principles in HPLC, *Advances in Chromatography*, Vol. 53, eds. Eli Grushka and Nelu Grinberg CRC Press, Boca Raton, London, NewYork, 2017, pp. 331–350.

[15] A. Schmidt, I. Molnár, Using an innovative quality-by-design approach for development of stability-indicating method for ebastine in the API and pharmaceutical formulations, *J. Pharm. Biomed. Anal.* 78–79 (2013) 65–74.

[16] E. Tyteca, J.L. Veuthey, G. Desmet, D. Guillarme, S. Fekete, Computer assisted liquid chromatographic method development for the separation of therapeutic proteins, *Analyst* 141 (2016) 5488–5501.

[17] J.M. Davis, J.C. Giddings, Statistical theory of component overlap in multicomponent chromatograms, *Anal. Chem.* 55 (1983) 418–424.

[18] L.R. Snyder, J.J. Kirkland, J.W. Dolan, *Introduction to Modern Liquid Chromatography*, John Wiley & Sons, Inc., Hoboken, NJ, USA, 2010.

[19] J. Plotka, M. Tobiszewski, A.M. Sulej, M. Kupska, T. Górecki, J. Namiesnik, Green chromatography, *J. Chromatogr. A* 1307 (2011) 1–20.

[20] X. Wang, D.R. Stoll, A.P. Schellinger, P.W. Carr, Peak capacity optimization of peptide separations in reversed-phase gradient elution chromatography: Fixed column format, *Anal. Chem.* 78 (2006) 3406–3416.

Chapter 2

HPLC Method Development by QbD Compatible Resolution Modeling (DryLab4)

Szabolcs Fekete[*,§], Imre Molnár[†], Hans-Jürgen Rieger[†] and Róbert Kormány[‡]

[*]*School of Pharmaceutical Sciences, University of Geneva, University of Lausanne, CMU — Rue Michel Servet 1, 1211 Geneva 4, Switzerland*
[†]*Molnár-Institute, Institute for Applied Chromatography, Schneeglöckchenstrasse 47, D-10407 Berlin, Germany*
[‡]*Egis Pharmaceuticals PLC, Keresztúri út 30-38, H-1106 Budapest, Hungary*
[§]*szabolcs.fekete@unige.ch*

2.1 Introduction

High-performance liquid chromatography (HPLC) method development, robustness and quality by design (QbD) play an important role in the global economy, where pharmaceutical and chemical products are distributed worldwide and the method transfer process has probably been running for the same product in different countries and in different laboratories. Regulatory authorities (FDA, ICH, EMA, etc.) nowadays are promoting and requesting the application of QbD principles to ease the exchange of complex information about chromatographic selectivity and resolution of support method — and quality control (QC), including method development, transfer and robustness testing. By applying QbD approaches, a better understanding and fine tuning of the method can be performed to ensure the requested help is in place to support separation in the preparation of an analytical design space (DS) [1]. Modeling is an excellent way to apply analytical QbD development and QbD-based documentation.

In addition, the International Council for Harmonisation Q8(R2) guideline (ICH Q8(R2)) [2] made a clear movement toward elimination of trial and error in HPLC, adding more flexibility to support the systematic development of new products in industrial environments, to understand peak movements in HPLC based on solid science and less solely on statistical evaluations. The appearance of terms such as QbD and DS are an indication of this growing trend [3, 4], which also requires a high level of understanding of the basic rules of HPLC.

One of the steps in implementing QbD principles in HPLC method development is the elaboration of the analytical DS [3, 4]. A key benefit of defining a DS is a significant gain in flexibility, as an alteration of the working point (WP) within this space is not considered to be a "change" and therefore would not initiate a regulatory post-approval change process (PACP). A DS, as defined by the ICH Q8 (R2), is the "multi-dimensional combination and interaction of input variables that have been demonstrated to provide assurance of quality". In chromatographic terms, this means that all parameters (input variables) that have a strong influence on retention and selectivity (separation quality) should be studied in varied combination, thus defining a completely known multi-dimensional space. Among all the influencing factors, the most critical variables in the majority of HPLC separations are the gradient time (*tG*), mobile phase temperature (*T*), pH of the aqueous mobile phase (eluent A), the ternary composition (*tC*) of the organic modifier (eluent B) and the chemistry of the stationary phase. This is an important difference when compared with gas chromatography (GC), where the stationary-phase chemistry is the dominant term influencing selectivity and the mobile phase plays a less important role in selectivity. As indicated in the ICH Q8 (R2) toward product development, it is possible to either "establish an independent DS for one or more unit operations, or to establish a single DS that spans multiple operations".

2.2 The Basics of DryLab Software

The software DryLab was the first revolutionary HPLC method development and optimization software, that predicts chromatograms under a much wider range of experimental conditions than would ever be possible to perform in the laboratory. With this software, one can quickly and easily determine exactly how the separation would progress as the chromatographer simultaneously varies multiple method parameters, such as pH, temperature, buffer concentration and many other variables. Anybody developing HPLC methods who wish to optimize complex separations and economize resources spent developing and running methods can benefit from the many advantages offered by such a modeling software.

The beauty of the software is that by using data generated from only 2–12 input experiments, DryLab predicts resolution and retention times for millions of unique, virtual conditions (chromatograms). The first step of the modeling is to define the analytical target profile (ATP) and then let the software, working in a systematic way, suggest initial method conditions and final optimized WP. The chromatographer simply runs the limited number of input experiments required to build up the retention and resolution models, and then imports the experimental results to the software to further optimize the separation *in silico*.

DryLab uses real data to create color-coded maps plotting critical resolution as a function of one, two or three method parameters. In addition to visualizing the interaction of these parameters, one can also predict chromatograms for changes in other method conditions, such as column dimensions, flow rate, gradient elution, instrument parameters and many others. Each point within the map corresponds to a unique chromatogram, displayed directly below the resolution map, and the user can follow how the resolution changes as the method parameters are varied (adjusted). Figure 2.1 shows the main window of the software, displaying the resolution map and calculated chromatogram on the right-hand side and the method parameters and variables on the left-hand side (such as column data, gradient table, gradient time, temperature, critical peak pairs, run time, the volume of used mobile phase, etc.). Please note that the display

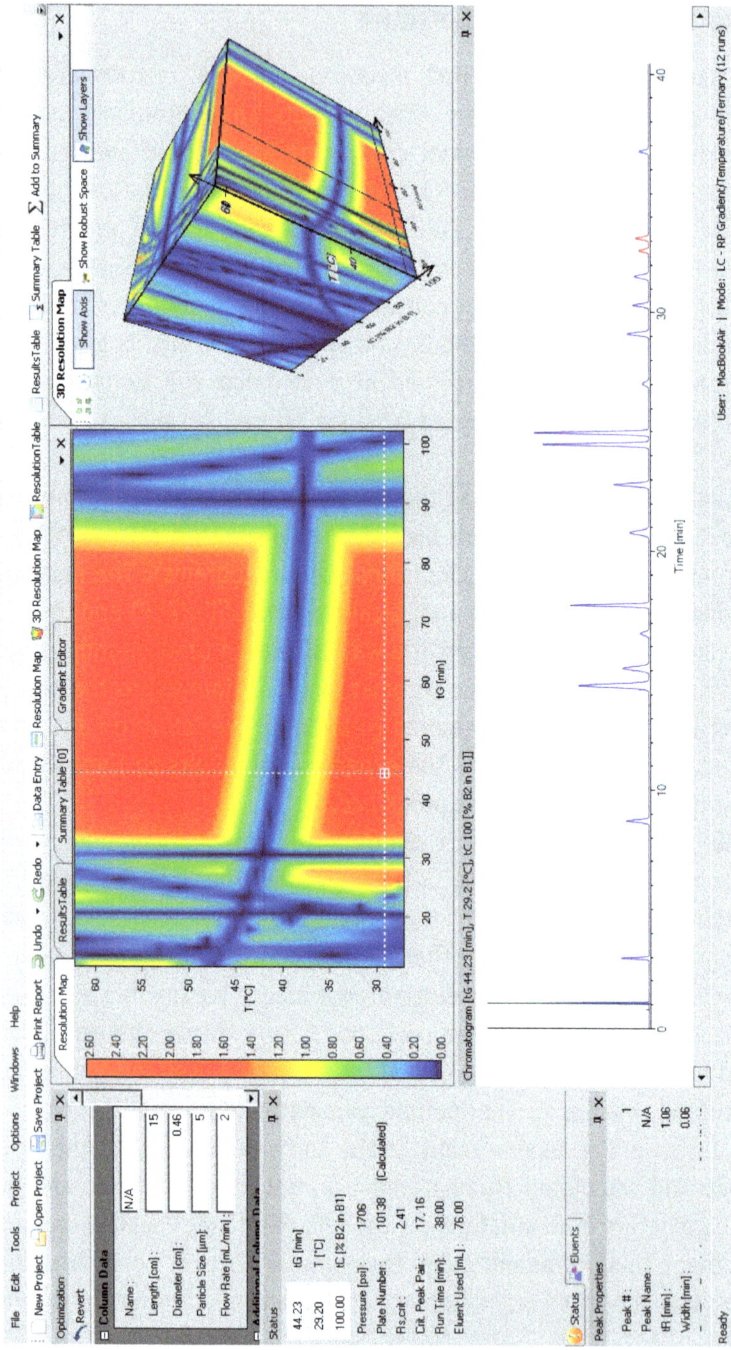

Figure 2.1: The main window (optimization) of the DryLab software.

of the main window can arbitrarily be changed by the user by simply dragging and dropping the different windows and changing their size.

The identification and assignment of peaks from a set of systematic experiments is an important first step in controlling the HPLC method development process. DryLab's "Peak Tracking" feature includes both peak areas and molecular masses and offers an efficient tool for preparing a peak table in an organized and systematic manner. In the peak table, the user can reorder and turn peak positions, separate double and triple peaks, and reduce complexity. It is color-coded to indicate the likelihood of correct peak identification, and the "Comparison Feature" compares the original experimental runs to the model to help further control for possible errors in peak tracking.

The "Gradient Editor" module is a powerful tool for optimizing the separation in gradient elution mode. While the input experiments must be run with simple linear gradients, once the retention model is built, it is possible to modify the gradient time, change the start and end %B, and add gradient steps (for multi-linear gradient separations). It is also possible to combine isocratic steps with gradient segments. The gradient conditions can be controlled manually or also automatically by allowing the software to find the best linear or step gradient. This "Gradient Editor" feature helps not only drastically reduce run times but also significantly increase resolution between peak pairs.

Another useful feature is the "Column Match" module which lets the chromatographer compare the selectivity of different columns included in a huge database. Taking into account various contributions such as hydrophobicity, steric selectivity, hydrogen bond acidity, hydrogen bond basicity and ion-exchange properties of the silanol groups at different pH values, "Column Match" supports the selection of an equivalent — or at least very similar — column. In the event that you want to discover hidden peaks, you can also select columns that are very different in their selectivity.

The "Robustness Module" tests the tolerance limits of the selected WP by computing the number of out-of-specification (OoS) results that can occur because of small fluctuations in method variables and parameters. A chromatogram is generated for every possible combination of errors and shows the range of resolution values that can be expected during routine application when the values of variables and parameters are not perfectly set (e.g. $\pm 1^{\circ}$C difference in column temperature between two instruments having either a still air or a forced air oven). Based on the number of successful experiments, one can choose a new WP to ensure safer results during routine application (this does not require new validation). Moreover, it is possible to evaluate which method parameters exert the highest influence on separation, a fact that is highly useful for setting up an efficient control strategy.

The "Knowledge Management" module is a reporting tool for documenting and archiving the history of the method development. It encourages a QbD approach to method development and ensures that the method conforms to these standards by providing a comprehensive method report, including a platform for the step-by-step justification of the method choices. The "Knowledge Management" provides an analytical method summary report to be signed and dated by the author and supervisor, making it GMP compliant.

The latest version of the software now provides a "Column Comparison" module which can be a useful tool to find alternative (replacement) columns. In this module, various 3D resolution maps can be compared (overlapped), which can help to study the measured points — in a DS — obtained on different stationary phases and find a common zone where the sample components are all separated with sufficient selectivity and resolution. The advantage of this approach is the mapping of the retention behavior of the compounds of interest (and not common test solutes) in an entire 3D DS, instead of some selected conditions (as suggested by earlier column tests).

2.3 Building up a Retention Model and Design Space in DryLab

2.3.1 *Data input*

For building up a computer model of an HPLC separation, the following information are needed:

— The variables of the design of experiment (DoE) such as gradient time (tG), temperature (T), pH, ternary composition (tC), buffer concentration (Cb), etc.

- Column Data: length (L), inner diameter (ID), particle size (d_p) and packing material.
- Instrument Data: Brand name, dwell volume (V_d), extra-column volume ($V_{ext.col.}$) and injected volume (V_{inj}).
- Eluent Data (if not included as method variables): pH and buffer concentration (Cb), ternary eluent composition (tC), organic modifier type, additives, temperature, gradient range and flow-rate.
- Chromatograms in AIA-format, or from individual brand data (like Shimadzu's*.lcd-format) or retention data as an Excel table.

In the judgment of a method, it is imperative to have the basic input chromatograms. They play an extremely important role as, many times, obvious problems are already visible here at this stage. This is the case if some of the chromatograms are obtained with limited quality or reproducibility, including baseline noises, obvious equilibration issues, bad column performance, etc. One should model methods only when the input chromatograms of the DoE are reproducible, i.e. the retention times of the sample components from repeated injections are within the tolerance limit of a few seconds.

2.3.2 *Design of experiments (DoE)*

There are many different designs available in HPLC method development, but only a few are really of practical value. The chromatographer is often faced with a great amount of "statistically significant" experiments, which may be challenging to interpret. As peaks are moving, these movements have to be understood in the first place as they are the reasons for the many OoS and out-of-trend (OoT) data in routine QC and in the numerous PACPs. The costs to rectify OoS and apply PACP are tremendous owing to the complexity of the separations and could potentially be attributed to the low level of understanding of the actual problems by the persons responsible for conducting the experiments/ analysis.

Sometimes it is suggested to carry out many number of runs in the framework of a "method scouting" exercise; however, we can easily lose overview of peak movements and get confused between the many chromatograms. It is advisable to reduce the complexity and limit the DoE to

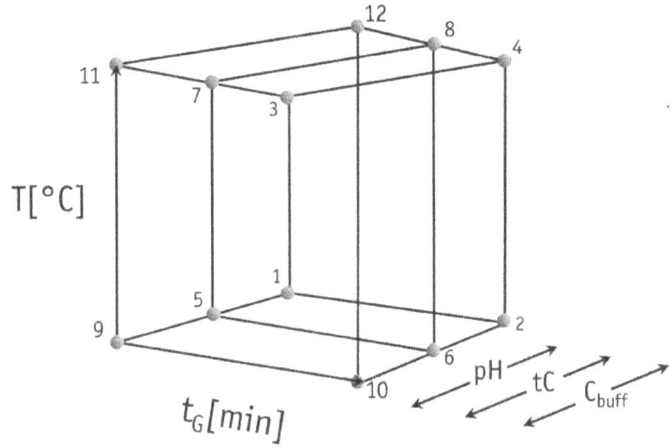

Figure 2.2: DoEs, used to obtain the 3D-models. The experiments 1, 2, 5, 6, 9 and 10 were carried out at the low temperature, (i.e. 30°C), and the experiments 3, 4, 7, 8, 11 and 12 at the high temperature (i.e. 60°C). The experiments 1, 3, 5, 7, 9 and 11 were carried out with a steep gradient (i.e. *tG* = 1.5 min), and the flat gradient experiments were 2, 4, 6, 8, 10 and 12 (i.e. *tG* = 4.5 min). The pH of the eluent A could be, i.e. pH = 4.4 with experiments 1, 2, 3, 4, it could be, pH 3.8 with experiments 5, 6, 7 and 8, and pH = 3.2 with experiments 9, 10, 11 and 12. Similarly, the ternary composition (*tC*) of eluent B (the ratio of ACN vs. MeOH) could be 100% ACN in run 1, 2, 3 and 4, it could be (ACN:MeOH) (50:50) (V:V) in run 5, 6, 7, and 8 and it could be 100% MeOH in runs 9, 10, 11 and 12. Similarly, the buffer or additive concentration could also be varied as the third variable. Running the experiments, we start with those at the low temperature T_1 (1, 2, 5, 6, 9 and 10) and, after heating up, we continue with the high-temperature runs at T_2 (3, 4, 7, 8, 11 and 12) (adapted from Ref. [1] with permission).

4–12 experiments and to create 2D or 3D resolution spaces based on two or three method variables [1]. The basic elements of a DoE are shown in Fig. 2.2. It is important to carry out 3 so-called *tG*-*T*-models (see Fig. 2.2 experiments 1-2-3-4, or 5-6-7-8, or 9-10-11-12) as in such a case the peak tracking process is less complex and is supported by a simple logic, as described in the legend of Fig. 2.2. Then, a third variable can also be added, typically pH, *tC* or *Cb*.

It is fairly simple to make peak tracking (identifying the peaks in the different runs) in a *tG*-*T*-sheet of only 4 runs as, for each run, we have three additional runs (if the third factor is studied at three levels), and we can then use those to identify peak movements based on peak areas in chromatograms of probably different selectivity. After the 3 *tG-T* sheets are measured, a design cube can be developed. In this cube, the most appropriate WP can easily be identified.

2.3.3 *Column data*

The efficiency (peak capacity) of gradient separations depends on several variables. The variations in column length, inner diameter, particle size and flow-rate might be important to model, as resolution changes when these factors are varied. Column dimension affects the peak variance and thus has a strong impact on the critical resolution.

When maintaining the gradient steepness constant, the peak capacity is related to the square root of the column length. Therefore, to improve kinetic efficiency under a given gradient program, the column length has to be increased in agreement with the linear solvent strength (LSS) theory or the geometrical scaling transfer rules. On the other hand, using serially connected columns for the optimization of LC stationary phase selectivity can also be useful.

Peak width are considered by the model on the basis of particle size and column length, but it can also be defined individually for each compound. Moreover model peak widths can be adjusted by indicating an average plate number. Similar to peak width, the experimentally measured column dead time can also be considered.

The next example shows a comparison of a separation where the column length was shortened from 15 to 7.5 cm (Figs. 2.3(a) and 2.3(b), resulting in a loss of separation and the formation of 2 double peaks (co-elution) and changes in relative peak distances. After the gradient time was reduced by a factor of 2 (Figs. 2.3(a) and 2.3(b)) and the dwell volume was adjusted, the separation selectivity returned to the original one, but was now 2 times faster (Fig. 2.3(c)).

Many HPLC users in the lab try to make the analysis quicker by shortening the analysis time (*tG*) and are surprised by the results. The geometrical method transfer rules should be used for this. An example is shown in Fig. 2.4. The original separation (Fig. 2.4(a)) was accelerated by increasing the flow-rate from 0.8 to 1.6 mL/min. A change in selectivity has been obtained (Fig. 2.4(b)); as the flow-rate is changing, the selectivity

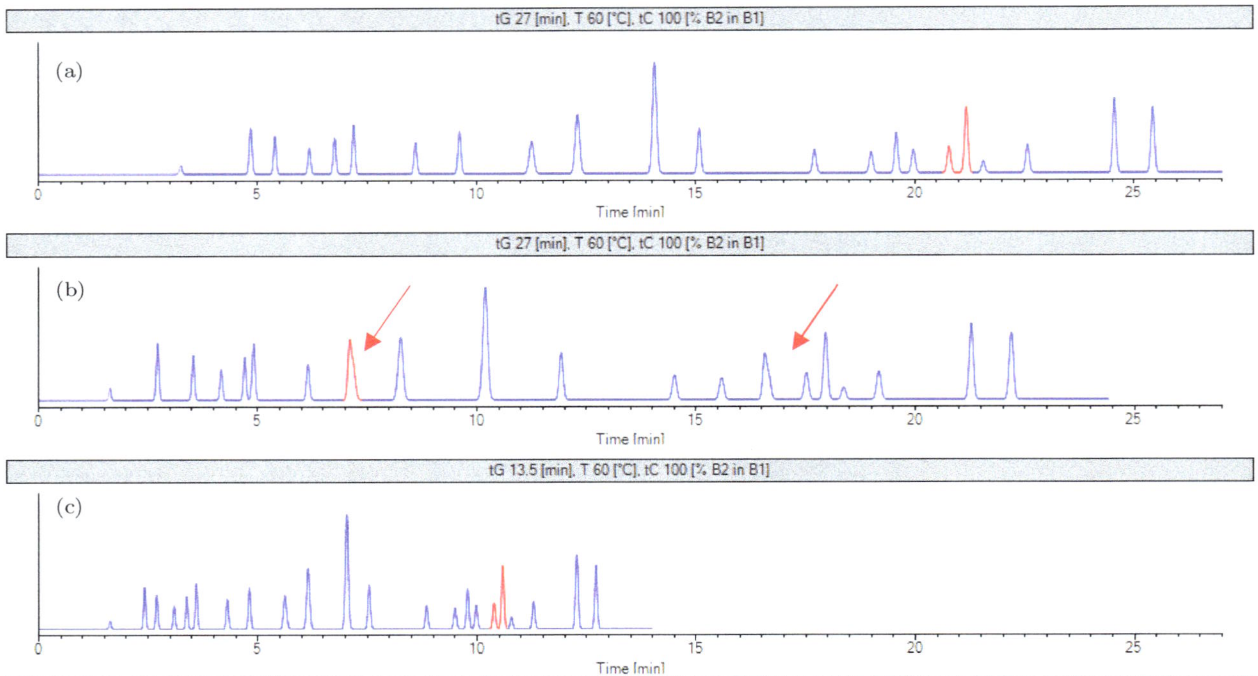

Figure 2.3: Reduction of analysis time by reducing column length from 15 cm (a) to 7.5 cm (b) and (c). The separation selectivity is changing dramatically and two double peaks are formed when maintaining the gradient time (b). Selectivity compensation was however possible by modeling and reducing the gradient time (*tG*) and the dwell volume (V_d) by the same factor (2), which restores the original selectivity (c) and also the original critical peak pair. Note that the analysis time is reduced by a factor 2 only between (a) and (c). The critical peak pair is shown in red.

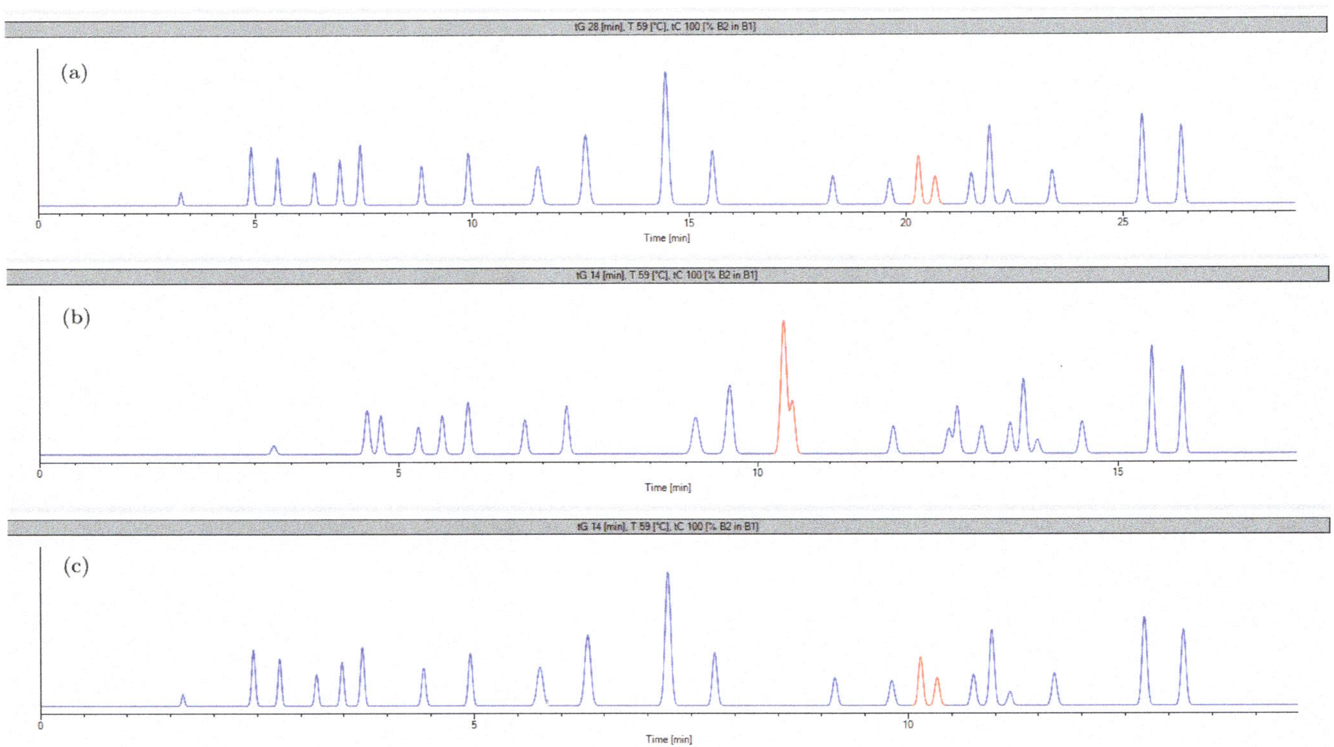

Figure 2.4: Modeling the influence of gradient time (tG) reduction by a factor 2 from $tG = 28$ min (a) to $tG = 14$ min (b) on separation selectivity whilst maintaining the flow-rate. Selectivity compensation was possible by increasing the flow-rate by the same factor 2, from $F = 0.8$ (a) and (b) to 1.6 mL/min (c), which restores the original selectivity and also the original critical peak pair. Note that the analysis time is reduced by a factor 2 only between (a) and (c). The critical peak pair is shown in red.

is altered. But if the gradient time *tG* is reduced by the same factor, the original selectivity can be reset (Fig. 2.4(c)). Finally, the analysis time has also been reduced by a factor of 2.

2.3.4 *Instrument data*

It is important to note the name and type of the instrument and indicate the accurate dwell and extra-column volume of the system. Differences in dwell volume occur quite often through method transfer processes between different laboratories and plant locations and are often a reason for OoS results, due to their influence on separation selectivity.

The dwell volume (V_d) of a system represents the volume from the point where the solvents mix to the head of the analytical column. After the gradient has begun, a delay is observed until the selected proportion of solvent reaches the column inlet. The sample is thus subjected to an undesired additional isocratic migration in the initial mobile phase composition. Two types of pumping systems are available for HPLC operations, i.e. (1) high-pressure mixing systems, where the dwell volume comprises the mixing chamber, the connecting tubing and the auto-sampler loop; and (2) low-pressure mixing systems, combining the solvents upstream from the pump, where additional tubing as well as volume of the pump head are added to the components of the high-pressure mixing system [5]. In the case of conventional HPLC systems, typical dwell volumes are in the range of 0.5–2 mL and 1–5 mL for high-pressure and low-pressure mixing systems, respectively. The dwell volume may differ from one instrument to another, but it can be easily measured by several procedures described in the literature [5].

In comparison with conventional HPLC instruments, which possess V_d between 0.5 and 5 mL, UHPLC systems have a V_d of ca. 300–400 μL, with the best UHPLC systems having V_d of ca. 100 μL, and up to ca. 1000 μL for some UHPLC instruments. Two main concerns related to large system dwell volume when performing fast separations in LC are observed, including (1) unreliable gradient method transfer between columns of different geometries and (2) ultra-fast separations, which require more time than expected.

QC laboratories are often equipped with conventional LC instruments, and so the developed UHPLC methods have to be transferred to HPLC. Whatever the need, i.e. HPLC to UHPLC or UHPLC to HPLC, the methodology for transferring a gradient method from one column geometry to another one remains identical, and some well-established rules have to be applied to scale the injected volume, the mobile phase flow rate, the gradient slope, and the isocratic step duration; pressure considerations also have to be taken into account [6]. However, the system dwell volume also needs to be accounted for during this transfer since it may differ between LC systems. Moreover, the extra-isocratic step created at the beginning of the chromatogram may also be different and could result in retention time variations, affecting the resolution during method transfer. To overcome this issue, the ratio of system dwell time on column dead time (t_d/t_0) must be ideally held constant while changing column dimensions, particle size or mobile phase flow rate [7].

Figure 2.5 shows an example of the impact of dwell volume on the selectivity.

2.3.5 *Eluent data*

The composition and pH of the aqueous buffer, the type of organic modifier, details of the additives used, temperature, etc., should be collected and, together with the gradient range (start and end %B), registered, as changes in these factors alter chromatographic selectivity in gradient methods.

It is also important to note that only peaks which elute in the increasing part of the gradient can be modeled properly with high precision. Just ahead of the gradient start, the so-called "pre-eluted" close to the void volume cannot be modeled. Note, that the gradient composition in the optimization process should be measured in the detector cell and not at the column inlet. This procedural difference is considered as modeling gradients in DryLab®4.

2.3.6 *Creation of experimental data*

Data can be imported in the international export format of the Analytical Instrument Association (AIA-format) (*.cdf) Available in every

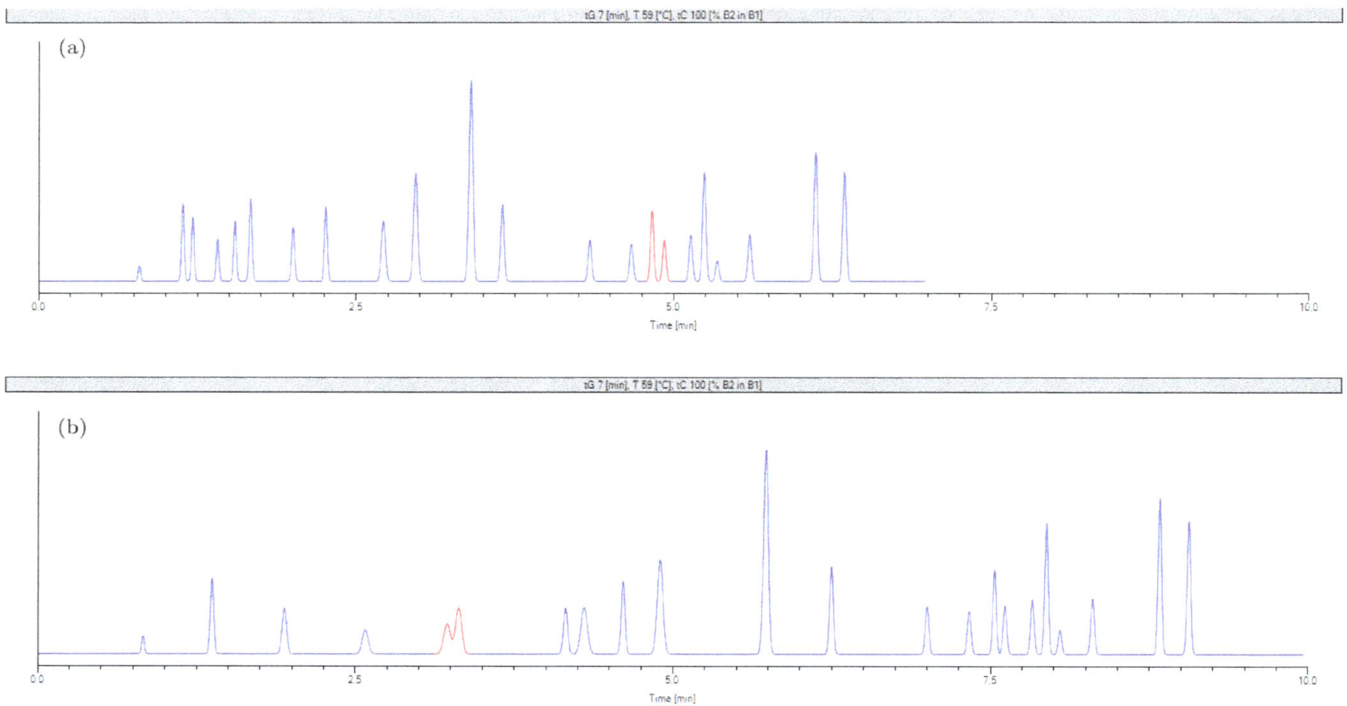

Figure 2.5: Differences in selectivity due to changes in dwell volume. V_d is 0.40 mL in (a) and it is 5.5 mL in (b). Not only are retention times different but also critical resolution values and the critical peak pair vary. As long in (a) the critical peak pair is 15–16 ($R_{s,\text{crit}} = 1.71$) while in (b) the critical peak pair is 5–6 ($R_{s,\text{crit}} = 0.79$).

chromatographic data system (CDS) or entered from an Excel table by copy/paste. Data can also be entered manually.

Extensive chromatographic modeling requires a larger set of experimental test runs. To avoid errors in setting up the runs and in the re-import of the experimental data into the modeling software, it is desirable to provide automated data generation. The analyst then specifies the general method conditions as well as type and ranges of parameters to be optimized. The software then creates the necessary method and batch files and, after the batch is run on the instrument, imports the results into the software. In this way, the input runs will also be done in the most efficient way. A generic order of experiments is to run all experiments at a lower temperature first and, finally, at a higher temperature to avoid extreme changes in method conditions and to keep the equilibration times as short as possible.

Similarly, the optimized separation conditions found from the calculated separation model can be downloaded to the instrument to generate a confirmation run and compare it with the predicted chromatogram.

2.4 Peak Tracking

2.4.1 *Experimental prerequisites*

Before starting the experiments, the column has to be cleaned, by running a gradient of 0–100% acetonitrile or methanol several times without sample injection, until a clean baseline without any ghost peaks is achieved. Furthermore, one should run a "scouting gradient" with the sample to see what the sample composition is. The same sample and identical injection volume are the prerequisites for successful peak tracking process and for subsequent correct model calculation.

2.4.2 *Dealing with the data table*

After running the required experiments (e.g. 12 runs for a Cube) needed to construct a resolution map/cube, one can proceed with the peak tracking procedure. First of all, the elution order of the peaks in the different experiments has to be fixed. It can be done by aligning the flat gradient runs at low temperature (experiments 2, 6 and 10) and fixing the elution order of the peaks in a given order, which will be kept the same in the other 9 experiments also. Then, the 4 data sets in one *tG-T*-sheet can be studied.

Peak areas are used in the peak tracking process to identify peak movements. In this case, they are not meant for quantitation. Peak areas have concentration × volume = mass units as long the flow-rate is constant. In practice, the peak area sums per run are fairly stable, with a standard deviation of typically $<5\%$ RSD — so far, well suited to control peak movements.

Figure 2.6 shows the peak-tracking module. By clicking on the cube at the bottom left-hand side of the display, the planes of the DoE can be selected, and the corresponding peak areas appear in the table on the right-hand side. At the top of the display, the experimentally observed chromatograms can be seen and the integration of the peaks can be modified. Peaks can be added or removed and peak positions can easily be changed. In the peak table, beside peak retention times and areas, the peak symmetry and width can also be included and considered in order to perform more accurate calculations.

2.4.3 *Mass spectrometry-supported peak tracking*

Chromatographic modeling, including the possibility to calculate a real chromatogram, requires peak tracking. This means, we need to know where each peak in a chromatogram elutes in all the basic runs, as this information is necessary to calculate a separation model. This can be done by comparing the peak areas of the sample components, because peak areas are expected to remain constant in a *tG-T*-model as long as the flow-rate is kept constant. This procedure, however, becomes more difficult if the peaks have similar areas or if the integration leads to stronger varying peak areas, as is often the case with very small peaks. In case of similar peak areas, UV spectra might support the correct peak assignments, but in some cases due to pH shifts and analyzing ionizable components, the UV spectra may change, making peak tracking challenging.

The best alternative peak property to find a certain peak in different chromatograms is probably the compound molecular mass, owing to its high specificity [8]. Liquid chromatography-mass spectrometry (LC-MS) is

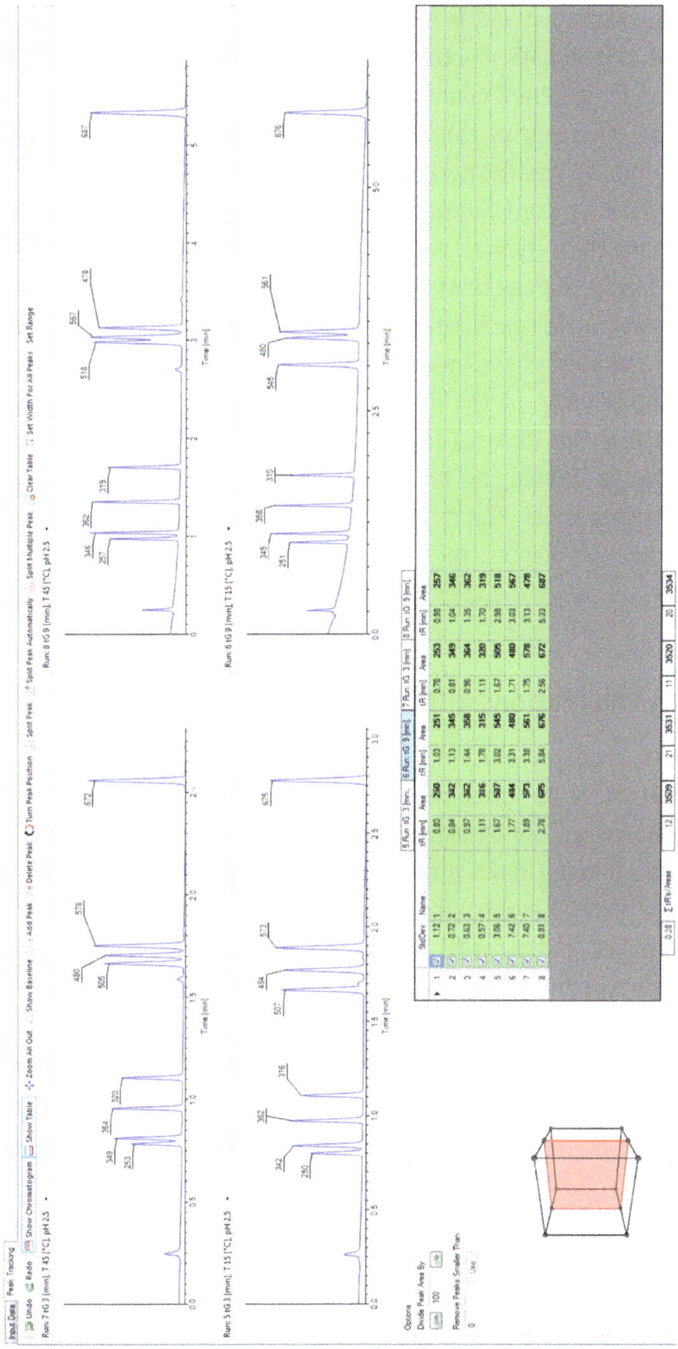

Figure 2.6: Peak tracking window of the DryLab software.

now a proven technique and is routinely available in HPLC labs. The main disadvantage of using MS spectra for peak tracking may be that (i) not all compounds under different conditions will give a suitable MS signal and/or (ii) ionization efficiency of the affected peaks may change as a function of the mobile phase composition and pH of the mobile phase.

The ideal case to use MS data for peak tracking would be if all sample compounds and their masses are known. But even if we have unknown compounds, but have peak retention times from UV detection, we can look for those masses under each peak, which follows the typical peak shape, increasing at the start of the peak, going through a maximum and decreasing to a baseline value at the end. This mass would then belong to that peak, and so one can look for this mass in the different chromatograms. It is advisable to enable the enter user to the molecular mass into the peak tracking data table. The latest DryLab®4 allows using both, UV-peak areas and molecular masses, for tracking peak movements for more robust methods.

2.5 Model Calculations and Validation

The theory of gradient elution makes possible the accurate estimation of band broadening as functions of mobile phase flow rate and column dimension. Moreover, the change in resolution at various retention factors can also be predicted. On the other hand, the development of a model that predicts plate number and bandwidth as a function of conditions for small molecule samples, and a model for predicting large molecule separations in gradient elution mode, particularly for RPLC, was needed for many HPLC users in life sciences [9]. The theoretical background of RPLC, considering solvophobic retention forces, has been studied and explained in detail by Horváth, Melander and Molnár, by investigating the thermodynamics of free energy in a chromatographic process [10, 11].

DryLab allows for the reliable calculation of separations in gradient elution as a function of experimental conditions. Thus, there is a predictable effect on separation of gradient steepness (%B/min), initial and final values of %B in the gradient, gradient shape, flow rate and column dimensions; other conditions such as temperature, mobile phase pH, etc.

are assumed to be held constant while other conditions are changed. The effect of temperature is often taken into account by using the van't Hoff theory of Gibbs free energy.

Computer simulation makes use of the abovementioned calculations as a result of two or more initial gradient separations in order to then predict any isocratic or (step-) gradient separation as a function of different experimental conditions. In this way, the process of developing a gradient RPLC method can be made more efficient, with resulting methods that are better as well as less costly to develop [12].

Here, we do not go into details with the mathematical equations, but interested readers can find more details in the referred works/papers.

2.5.1 *Calculation and visualization of the resolution cube*

DryLab's resolution cube extends the previous 2D resolution map into the third dimension, providing a Method Operable Design Region (MODR) comprised of three variables in which the multifactorial variability for robust HPLC conditions is visualized. In addition, it is possible to model up to eight other variables, including column dimensions, flow rate, gradient points (for segmented or stepwise gradients), and instrument parameters. The 3D resolution cube offers an intuitive display of how simultaneous changes to multiple method parameters affect the critical resolution and selectivity. A special view shows the 3D regions that fulfil the predefined resolution criteria (for example, baseline separation of all peaks). The cube can be scrolled through to see the chromatogram for any set of conditions within the 3D region.

Figure 2.7 shows a resolution cube and the corresponding MODR. The 3 *tG*-*T*-sheets are the basis for the calculation for a number of calculated sheets between them. After the calculation is finished, the cube is "filled out" and the impact of parameters *x*, *y*, *z* on the resolution is demonstrated.

When all WPs (i.e. combinations of measured parameters) with a critical resolution below the threshold of 1.5 ($R_{s,\text{crit}} < 1.5$) are removed from the resolution spaces, robustness regions can be identified and the robustness of the separation can be visualized as irregular geometric bodies.

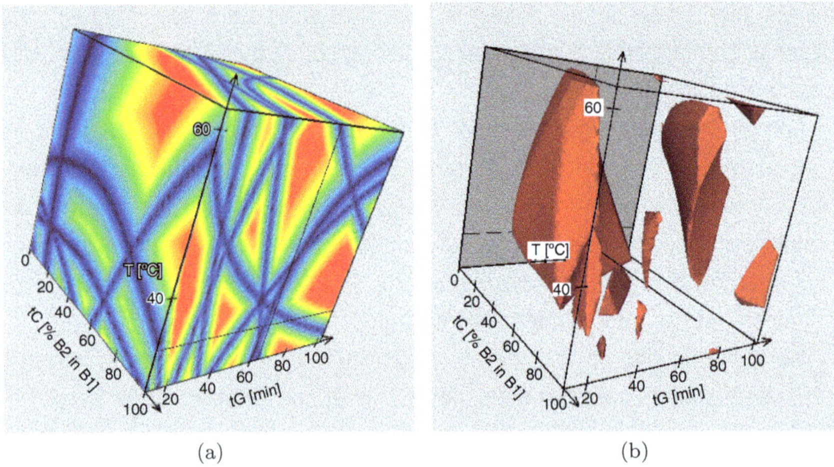

Figure 2.7: The illustration of a 3D resolution cube: Baseline resolution regions are displayed in red; they are the MODR, visualized DS, blue regions indicating the method failure, where the critical resolution is = 0, which corresponds to peak overlaps (a). Baseline resolution regions are shown in red (b). The different geometric bodies form a DS, which allows altering the position of the set point (WP) without the need for a new validation, as the alteration of the WP inside the DS is not considered to be a "change", so far no change management is necessary. The robustness of the individual WP's is different between the different red regions.

2.5.2 *Validation of the model*

After determining the WP, the model can be verified with only one experimental run. In most of the cases, the experimental run — as a confirmation step — is run under the condition that provides the highest possible critical resolution. Then the experimental and calculated chromatograms (retention times) are compared. The relative error of predicted retention times is typically lower than 0.5%.

Further optimization of flow-rate and/or gradient shape can be performed. It is possible to transfer the chromatogram at the best set of conditions to the gradient editor and make further adjustments such as changing the gradient program, the initial and final mobile phase composition, column dimension, flow-rate or dwell-volume.

Figure 2.8 shows an example on the model confirmation. Predicted (calculated) and experimentally observed chromatograms are compared [13].

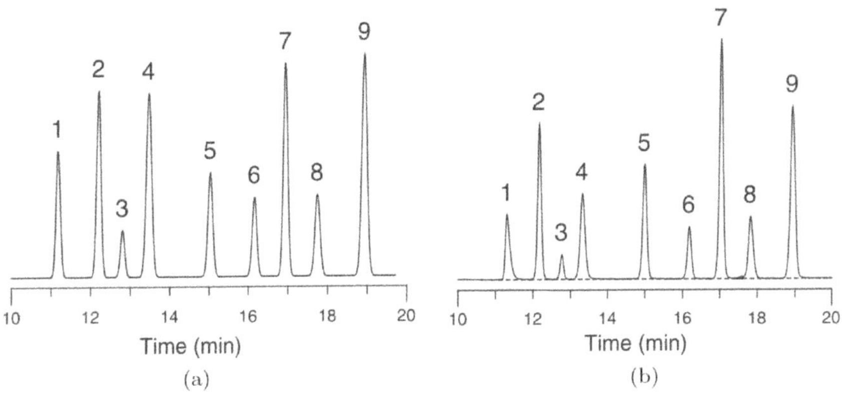

Figure 2.8: Example of model validation, comparison of predicted (a) and experimentally observed (b) chromatograms.

2.5.3 *Robustness calculations: How successful is the method in routine QC work?*

A fundamental criterion of quality in an HPLC separation is robustness. Guidelines define the robustness of an analytical procedure as "a measure of its capacity to remain unaffected by small, but deliberate variations in method parameters..." providing "...an indication of its reliability during normal usage" [14]. Historically, robustness testing was usually carried out as the final step of a method development process, during the validation stage, which often led to unexpected observations [15,16]. However, since a method considered as non-robust should be adapted/redeveloped and revalidated, this could lead to a substantial increase of development time and costs. Therefore, robustness is verified earlier in the lifetime of a method, i.e. at the method development stage or at the beginning of the validation procedure [16–19].

Generally two approaches are used to evaluate robustness according to the ICH definition in pharmaceutical analytical practice. Either a one-factor-at-a-time (OFAT) procedure or an experimental design (DoE) could be applied. When applying a DoE, the effect of a given variable is calculated at several level combinations of the other variables, while with the OFAT approach this is only at one level. Thus, in DoE, a reported variable effect is an average value for the whole domain, and it represents more globally what is happening around the nominal situation. In pharmaceutical industrial practice, the DoE approach is clearly preferred. Plackett–Burman, full factorial, nested factorial, fractionated factorial and asymmetrical factorial experimental designs are often carried out [20–23]. These types of robustness testing typically allow for the investigation of 3–15 factors (variables) based on 8–16 experiments. Besides continuous quantitative variables, e.g. gradient program, mobile phase composition, pH, temperature or flow rate, the effect of non-continuous (discrete) factors such as column or instrument (laboratory) could also be studied and are typically included in ruggedness testing of the method [24].

It is now possible to perform a modeled robustness testing thanks to modeling software. Besides the three main model variables (typically *tG*, *T*, pH or *tC*), the flow rate as well as initial and final compositions of the mobile phase represent the investigated variables in a built up model. The effect of these six variables can be calculated at three levels, corresponding to $3^6 = 729$ variants in selectivity. The modeled deviations from the nominal values can be set arbitrarily. Then, the 729 experiments can be simulated in less than 1 min. A criterion of $R_{s,\text{crit}} > 1.5$ is set and considered as a standard value. The results of the virtual experiments can be expressed in frequency as a function of critical R_s. Figure 2.9 shows an example of the simulated robustness testing. As can be seen, the most frequent resolution was $R_{s,\text{crit}} = 2.76$ (21 conditions provided this R_s value), while the lowest predicted resolution was $R_{s,\text{crit}} = 2.18$. Therefore, the method can be considered as 100% robust in the studied DS. Another feature of the modeling software employed in this study is the calculation of individual and interaction parameter effects. Figure 2.9(b) describes the importance of each variable, related to the selected deviation from the nominal value, for the critical resolution. This figure shows that the "start %B" (initial mobile phase composition) has the most significant influence on the critical resolution (i.e. a negative change in "start %B" would significantly change the critical resolution), followed by *T*, *tG*, final %B and pH as the most important parameters. Some interactions between the factors, i.e. T^* initial %B, have also an impact on critical resolution. The experimental verification of simulated robustness testing was demonstrated recently [24].

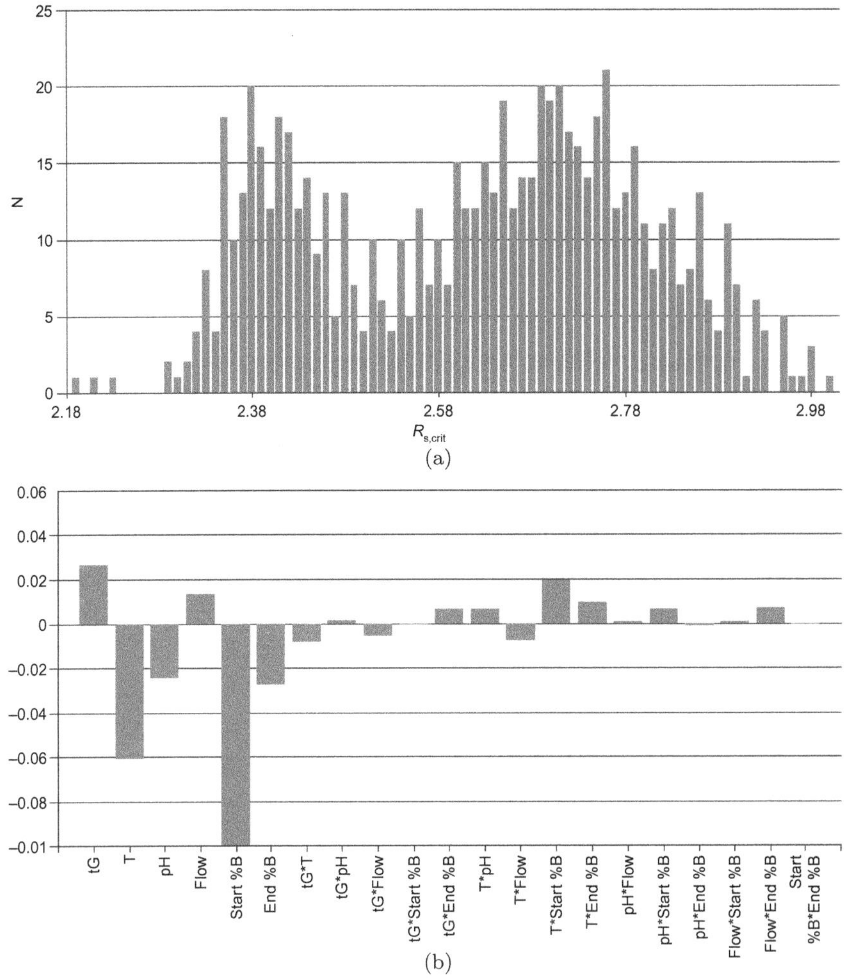

Figure 2.9: Frequency of resolution values of most critical peak pairs (a) and the relative effects of the chromatographic parameters on the critical resolution (b) (adapted from Ref. [24], with permission).

After selecting the WP, the next important question is how robust this WP would be during the lifetime of the method. A rough and mostly qualitative answer is already given by the robust resolution maps, as shown in Fig. 2.9(b). Selecting a WP in the center of a bigger shape means that the method conditions may deviate from the nominal value to some extent without losing baseline separation as long as the WP remains inside the shape.

The result of a simulated robustness evaluation will tell us how often (in percent) a chromatogram will fall outside of the required critical resolution range. Moreover, it provides information on which variable has the highest influence on robustness.

In the latest version of the software, the robustness of multilinear gradients can also be studied by changing the mobile phase composition at the segment points. The changes during the separation can virtually be studied.

2.5.4 *Complete method knowledge management*

If method development is done in a highly regulated environment, which is typically the case in the pharmaceutical industry, a great deal of documentation is required for every step during the process. Regulatory institutions require the complete detailed documentation of the method development procedure, similar to that for the validation (validation protocol and report). Now, thanks to modeling software, a comprehensive method development report can be generated in an automated manner. This "Method Knowledge Management Document" collects all relevant method data directly from all experiments and operations in DryLab and offers a platform for comments and the justifications of method criteria. It produces a GMP-compatible method development documentation that the encourages a QbD approach and ensures that the method conforms to the standards by providing a comprehensive method report, including a step-by-step justification of the method choices.

The report, in pdf format, contains all the experimentally observed data with resolution maps, the peak tracking tables, the proposed WPs, the experimental validations of the models and the conclusions at each step of the method development process. In addition, the robustness of the final method is also tested; the impact of individual variables and their tolerance limits is calculated, the potential failure rate is determined and the method operation design region is provided. Figure 2.10 shows an example of a method robustness calculation included in a knowledge management report.

tG	3 +/-	1	[min]	Flow Rate	0.5 +/-	0.1	[mL/min]
T	60 +/-	1	[°C]	Start %B	30 +/-	1	[%B]
tC	50 +/-	1	[% B2 in B1]	End %B	90 +/-	1	[%B]

Required Resolution	1.5
Successful Experiments	729
Success Rate	100 %
No of Factors	6
No of Levels	3
No of Experiments	729

Figure 2.10: Results of method robustness calculation provided by the Method Knowledge Management module.

2.6 Working with DryLab

In this section, we provide a brief guide on typical RPLC method development from the first experiments till the final optimization and robustness study.

2.6.1 *Running the first experiments*

As the most generic scouting experiments, the best choice is to probably start the method development with two gradient runs at two temperatures (*tG-T model*). This model offers the easiest way of peak tracking, and therefore much information can be obtained on the basis of only four experiments. Peak movements and retention behavior can easily be understood. The set value for *tG* is a function of column volume, length and flow rate. For example, for a conventional 150 × 4.6 mm column $tG_1 = 20$ min and $tG_2 = 60$ min (e.g. from 5% to 95% B) are typical values when operating the column at normal flow rate (e.g. 1 mL/min). For UHPLC applications, when operating a 50 × 2.1 mm column, $tG_1 = 3$ min and $tG_2 = 9$ min are appropriate values. It is suggested to set the temperature at $T_1 = 30$ and $T_2 = 70$°C. If column or sample thermal stability do not to work proceed allow to proceed work at 70°C then, of course, a lower value should be set (e.g. $T_2 = 60$°C). The gradient range (elution strength) must ensure that all the peaks elute in the increasing part of the gradient program. Pre- and post-eluting peaks cannot be processed. Pre-elution conditions can be avoided by decreasing the strength of the initial mobile phase composition, while post-elution can be avoided by increasing the strength of the final mobile phase composition. Except *tG* and *T*, all other parameters which can influence the separation (mobile phase pH, buffer concentration, flow rate, etc.) must be identical in all four runs. After performing the experiments, all the peaks of interest must be integrated correctly and only then must the chromatograms be exported in AIA (AnDI or cdf) format. Finally, the chromatograms can be directly imported to DryLab and the retention model created.

Figure 2.11(a) shows the "input data" page of the software. From left to right, users have to define: (1) the mode (*tG-T* in this example) and the values of the method variables ($tG_1 = 20$ min, $tG_2 = 60$ min, $T_1 = 30°C$ and $T_2 = 60°C$), (2) column data (dimension, flow rate and injected volume), (3) instrument data (dwell volume, extra-column volume, detector time constant and wavelength) and (4) eluent data (mobile phase composition and gradient program). Finally, the experimentally measured chromatograms can be imported in the correct order. By clicking on the "peak tracking" button, the chromatograms and peak tracking table can be displayed (Fig. 2.11(b)).

After defining and providing all the required input data, the DryLab model can be build up by clicking on the button "OK Calculate DryLab Model".

In most cases, based on these four experiments the retention behavior of the peaks can be understood and the separation can be optimized or other models can be created — if adequate separation was not reached — to further optimize the separation.

2.6.2 *Selecting a retention model (experimental design)*

The users can select from several built-in models (Fig. 2.12). 1D, 2D or 3D models (designs) are available. Depending on the sample, the chromatographer has to select the most suitable model. If the sample contains ionizable solutes (acids or basics), then the effect of mobile phase pH on the separation is probably worth studying. If polar-neutral compounds have to be separated — containing functional groups which can evolve H-bonding with the stationary phase or with the solvent — the impact of

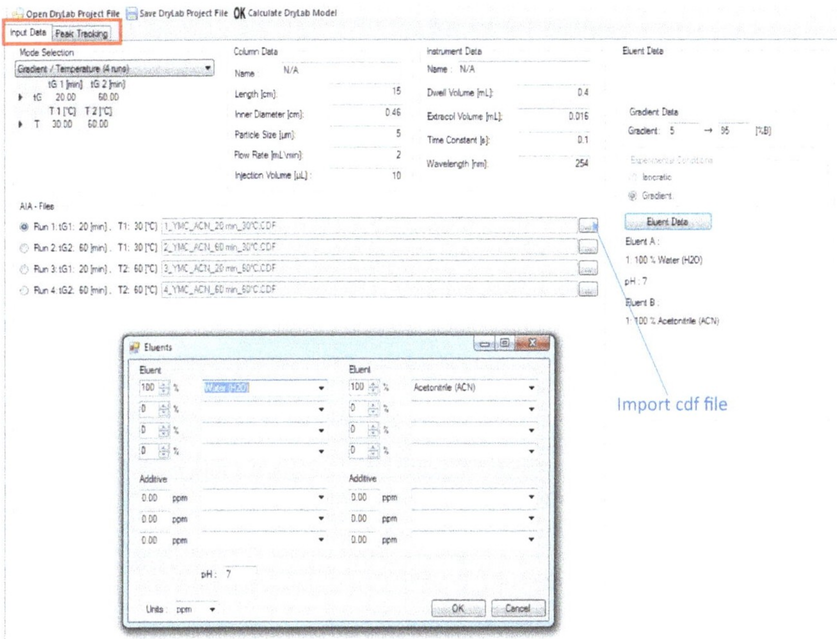

(a)

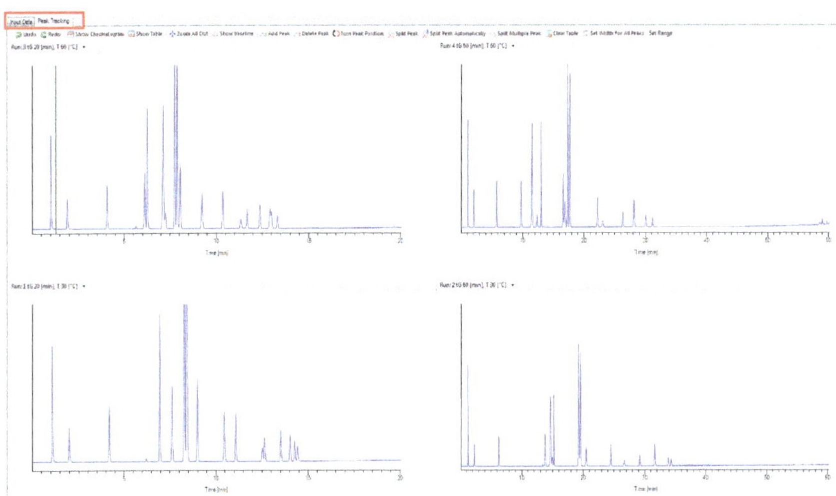

(b)

Figure 2.11: The DryLab "input data" (a) and "peak tracking" (b) pages (*tG-T* model).

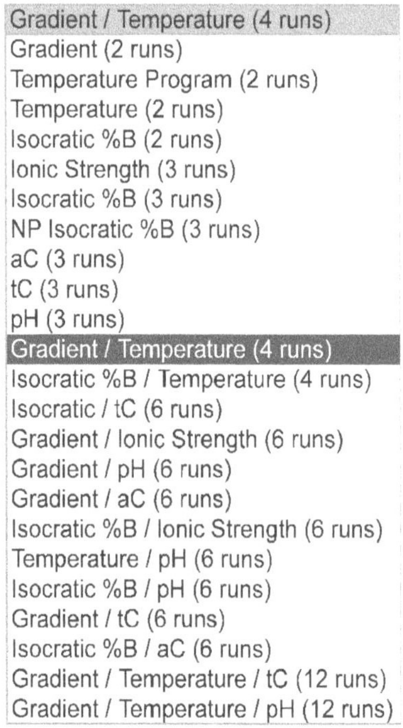

Figure 2.12: Selecting retention model (experimental design) in DryLab.

organic modifier (protic or aprotic) can be studied with a ternary composition model. But many other factors can be selected as model variables including ionic strength, additive concentration, isocratic composition or ternary concentration. The impact of several method variables (e.g. *tG*, *T*, isocratic %B) on retention can be linearized through mathematical transformations, and therefore they necessitate a study of their impact only at two levels. However, other factors may have nonlinear effects on retention (e.g. pH, ternary composition, ionic strength), thus requiring a measurement of their impact at three levels (or more). Accordingly, when combining two "linear" variables in a 2D retention model, $2 \times 2 = 4$ experiments are needed, but if "linear" and a "nonlinear" factors are combined in a design then $2 \times 3 = 6$ experiments are needed. For 3D retention models (typically *tG*-*T*-pH or *tG*-*T*-*tC*), two "linear" and one "nonlinear" variables are combined, thus requiring $2 \times 2 \times 3 = 12$ experiments.

In the previous section, it has been mentioned that the most common 2D model is the *tG-T* one, but obviously more information can be gained by building up a 3D retention model. The increased demand for QbD and DS methodology in recent years has prompted the development of a new concept of HPLC method modeling with three different measured chromatographic variables, *tG-T* and either ternary, pH, ionic strength, etc., at the same time, to generate the so-called Resolution Cube which represents up to 10^6 and more virtual experiments. The advantage of this approach is the reduction of trial and error as only 12 runs are needed to generate 10^6 precise predictions. By performing these experiments, precise knowledge can be obtained about the interaction of critical parameters that most strongly affect selectivity.

As a starting point, for a *tG-T-tC* model, the ternary composition should set as $tC_1 = 100\%$ acetonitrile, $tC_2 = 50\%$ acetonitrile $+50\%$ methanol and $tC_3 = 100\%$ methanol. For a *tG-T-pH* model — for most pharmaceutical applications — the effect of pH is worth studying at $pH_1 = 2.0$, $pH_2 = 2.6$ and $pH_3 = 3.2$. Obviously, other pH ranges can also be selected and studied.

2.6.3 *Performing a 3D optimization* (**tG-T-pH/tC model**)

The factors of the 3D DoE depend on the nature of the sample. In all cases, it is recommended to optimize *tG* and *T*, and additionally when the sample contains neutral compounds, the DoE should include organic modifier composition (*tC*) as the third variable, while for ionizable compounds the pH of the mobile phase should be selected as the third variable. An interesting case is when the sample contains neutral, acidic and basic compounds or unknown solutes; in this case, the pH of the mobile phase "A" and the *tC* of mobile phase "B" can be optimized simultaneously by performing 3×12 experiments as illustrated in Fig. 2.13. This last approach may be more time-consuming, but considering UHPLC conditions (50×2.1 mm column) it requires approximately 5–6 h (18×3 min $+18 \times 9$ min $= 216$ min plus equilibration) of experimental works.

The 12 experiments for one cube should be carried out in a certain order: first the six low-temperature experiments at T_1 should be carried out, then, keeping the same pH (or ternary composition, *tC*), the temperature should

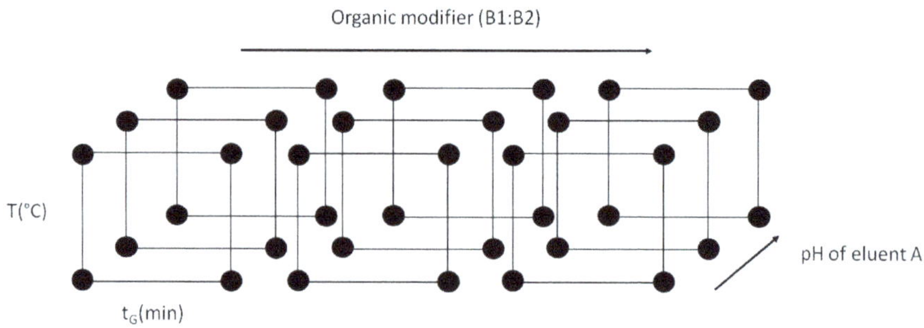

Figure 2.13: Suggested 3D DoE for a sample containing neutral, acidic, basic or unknown compounds.

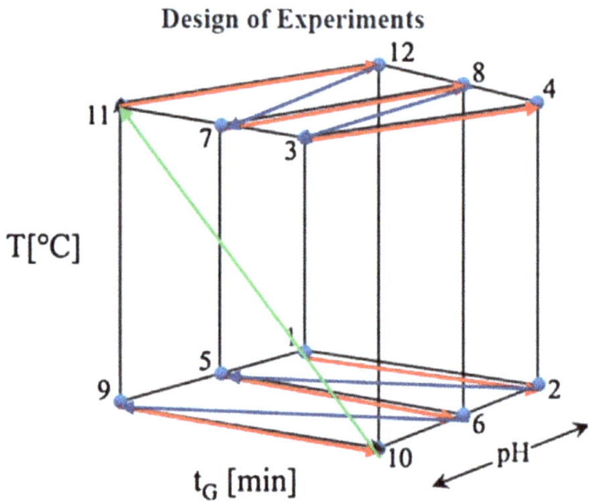

Figure 2.14: Suggested process to measure the DS: First the lower-temperature (T_1) experiments are selected and then the higher-temperature (T_2) ones. The red arrows represent the change of the gradient time, the blue arrows represent the change of the buffers and the green arrow represents the change of the temperature.

be increased to perform the six higher-temperature experiments at T_2, as shown in Fig. 2.14. We have to make sure that enough time is allotted for column equilibration. Changing the organic eluent is the fastest process. Changing the pH takes a bit more time; however, with columns of 2.1 mm ID, this process is also relatively fast. The longest re-equilibration time is required when the user jumps after the 10th run to the 11th run at the higher temperature. When it is doubtful whether the reproducibility is good, the experiments should be repeated several times, until the last two runs are exactly identical.

After performing the experiments, the peak tracking procedure needs to be done. Here, an example is presented on building up a *tG-T-tC* peak tracking table (12 runs). Initially, the order of elution is established at the experimental points at 2, 6 and 10 (Fig. 2.15).

A peak tracking table of a *tG-T-tC* model shows different elution profiles of the same mixture of 18 compounds in fewer than 12 different conditions. The peak areas in those runs have a standard deviation of ca. 2% on

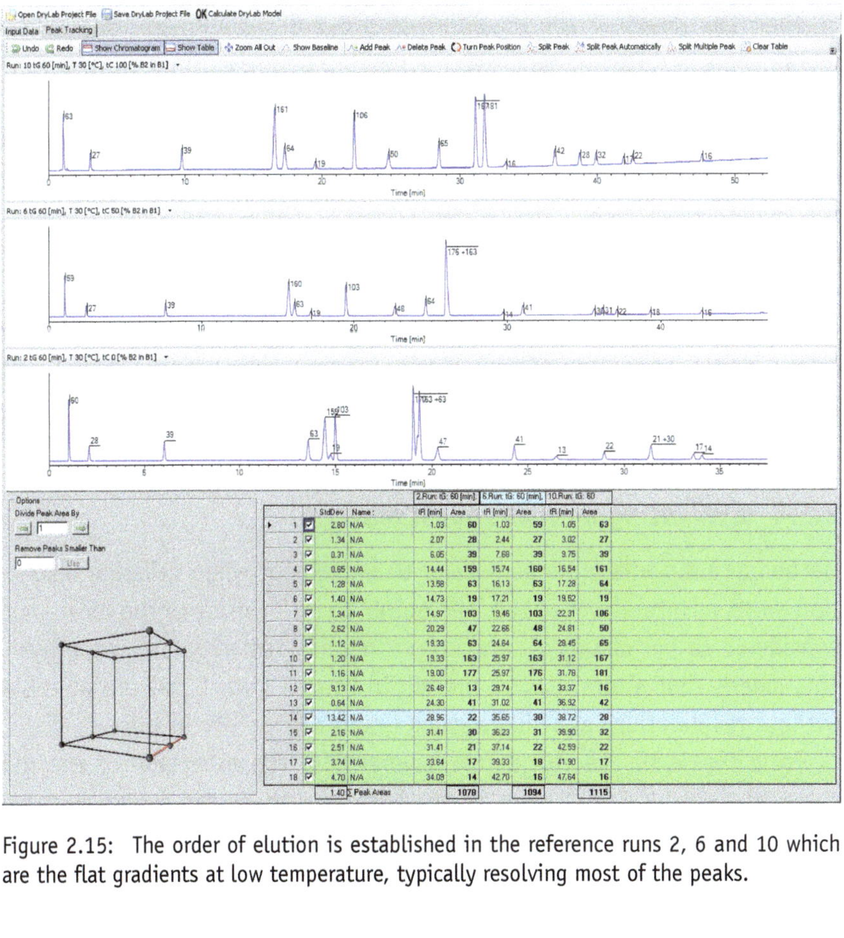

Figure 2.15: The order of elution is established in the reference runs 2, 6 and 10 which are the flat gradients at low temperature, typically resolving most of the peaks.

average (depending on the experience of the user in peak tracking) and can therefore efficiently be used to track moving peaks and establish robust conditions for routine applications. The next step is to align the 12 runs in the 3 *tG-T*-sheets. This is a process of looking at peak movements, peak overlaps and peak turnovers. Peak identification is mostly based on peak areas, which represent the injected amount of sample. Keeping the amount constant, we get constant peak areas for a given compound in the 12 basic experiments. Peak areas are concentration $\times$ volume $=$ mass, and are well suited to identify a moving peak. In peak overlaps, the areas are additive as the masses are too. In Fig. 2.16, the runs 1-2-3-4 are shown, where the organic eluent B1 is acetonitrile. Note the selectivity differences between runs.

Then the peaks of the experiments 5-6-7-8 are aligned (Fig. 2.17). Again, there is different selectivity generated and several co-eluting peak pairs observed.

At the end (the last sheet of runs 9-10-11-12, which is the 100% methanol sheet), all peaks are fully tracked (Fig. 2.18). When peak tracking is complete, we can calculate between the 3 *tG-T* sheets another 97 sheets, filling out the total space so we will be able to model any chromatogram at any point in the whole space with more than 10^6 virtual chromatograms. The results are highly precise, up to 99.8% accurate chromatograms in terms of their retention times, which is comparable to the operational accuracies of most UHPLC instruments.

2.6.4 *Evaluating method robustness*

Method adjustments are much easier to implement when utilizing resolution maps, as alterations of the "set-point" or "WP" inside of the DS are not considered to be "changes", and do not require post-regulatory approval. This means that alterations of the WP in the DS (Fig. 2.19) are possible without revalidation, allowing a much greater flexibility in the lab.

From Fig. 2.19, we can define several DSs. The extension of the red areas (the possible DSs) gives us the first idea about the robustness. We could also find a suitable method parameter in methanol (front sheet of the cube in Fig. 2.19) as well as in acetonitrile (back sheet in Fig. 2.19).

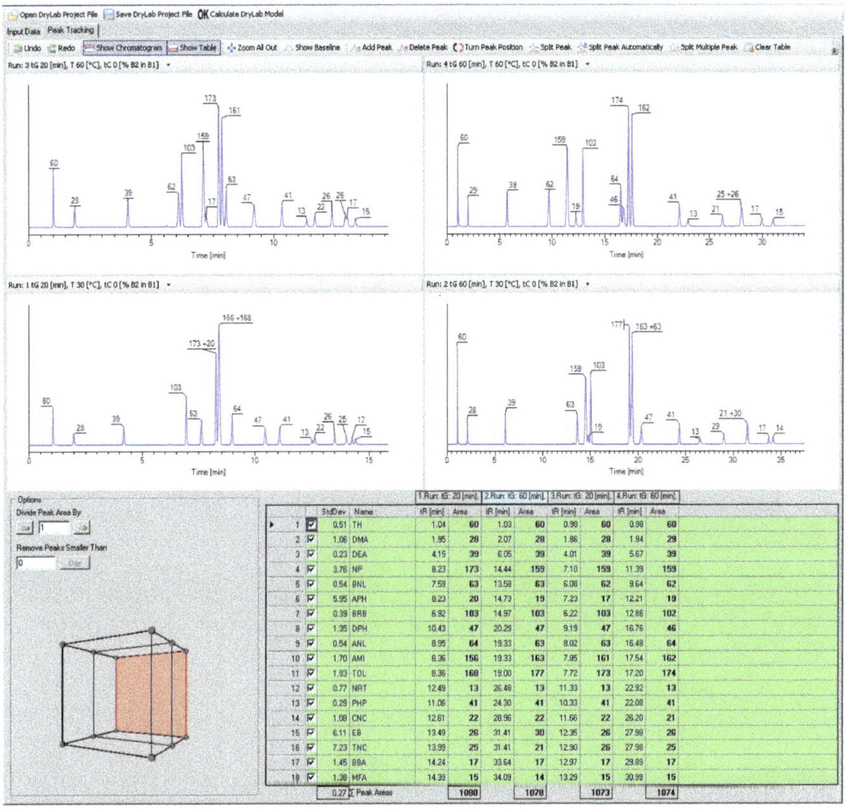

		StdDev	Name	1.Run: tG: 20 [min], tR [min]	Area	2.Run: tG: 60 [min], tR [min]	Area	3.Run: tG: 20 [min], tR [min]	Area	4.Run: tG: 60 [min], tR [min]	Area
1	☑	0.51	TH	1.04	60	1.03	60	0.98	60	0.98	60
2	☑	1.06	DMA	1.95	28	2.07	28	1.86	28	1.94	29
3	☑	0.23	DEA	4.15	39	6.05	39	4.01	39	5.67	39
4	☑	3.76	NP	8.23	173	14.44	159	7.10	159	11.39	159
5	☑	0.54	BNL	7.59	63	13.58	63	6.08	62	9.64	62
6	☑	5.95	APH	8.23	20	14.73	19	7.23	17	12.21	19
7	☑	0.39	BRB	6.92	103	14.97	103	6.22	103	12.86	102
8	☑	1.35	DPH	10.43	47	20.29	47	9.19	47	16.76	46
9	☑	0.54	ANL	8.95	64	19.33	63	8.02	63	16.48	64
10	☑	1.70	AMI	8.36	156	19.33	163	7.85	161	17.54	162
11	☑	1.93	TOL	8.36	168	19.00	177	7.72	173	17.20	174
12	☑	0.77	NRT	12.49	13	26.48	13	11.33	13	22.92	13
13	☑	0.29	PHP	11.06	41	24.30	41	10.33	41	22.08	41
14	☑	1.08	CNC	12.61	22	28.96	22	11.66	22	26.20	21
15	☑	6.11	EB	13.49	26	31.41	30	12.35	26	27.98	26
16	☑	7.23	TNC	13.99	25	31.41	21	12.90	26	27.98	25
17	☑	1.45	BBA	14.24	17	33.64	17	12.97	17	29.89	17
18	☑	1.38	MFA	14.39	15	34.09	14	13.29	15	30.98	15
		0.27	Σ Peak Areas		1080		1078		1073		1074

Figure 2.16: Next, the peaks are aligned in runs 1, 2, 3 and 4 (the first *tG*-*T*-sheet) with reference to the fixed elution order of run 2, shown in Fig. 2.2. The organic eluent was acetonitrile. Note the differences in selectivity in the runs, indicating changes in relative peak positions, which must be understood before the method is validated. Each peak has to be aligned in a horizontal line. The error between peak areas in such a line should be less than 5–10%. The standard deviation of the sum of peak areas per run is also quite stable, in the above case it is excellent, 0.27%. The prerequisite of high accuracy is to inject the same sample solution with all compounds included (names are not needed) and maintain the same injection volume in all runs.

From the DS, we can get robustness information only for the measured parameters: gradient time, pH and *tC* (%B2 in %B1) where B1 is acetonitrile and B2 is methanol. However, as DryLab4 is able to calculate other changes, which might occur at the same time, it is possible to calculate the influence of additional parameters like flow rate or start- and end-%B of the gradient

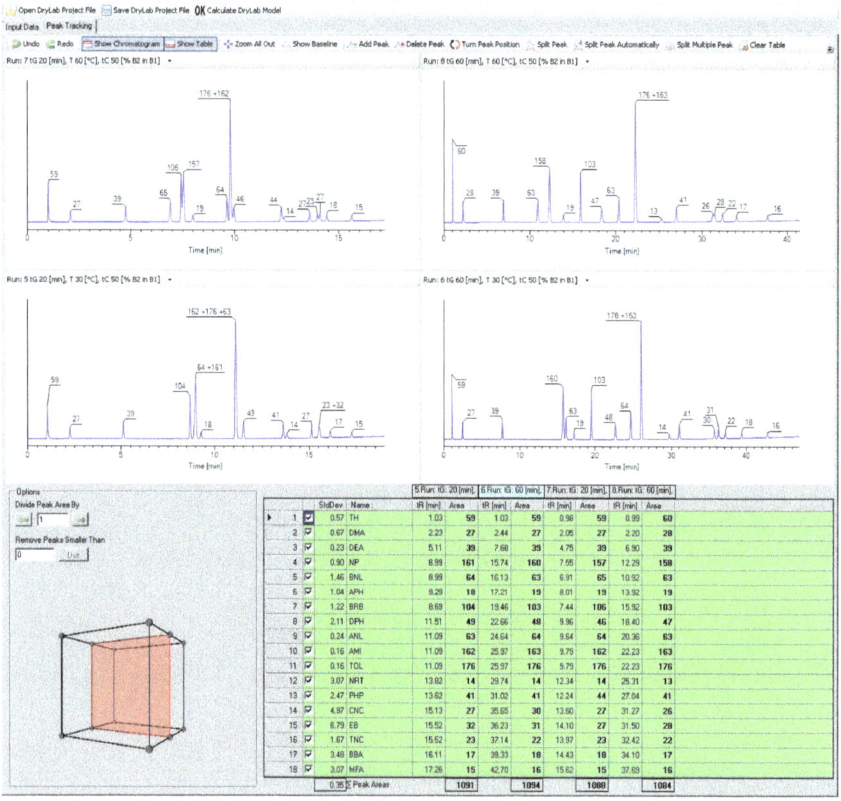

Figure 2.17: Next the peaks of the experiments 5-6-7-8 with the organic eluent (acetonitrile/methanol = 50/50 (V/V)) were aligned. The peak table indicates some double peaks, having the same retention time. These peak pairs are well separated in the other *tG*-*T*-sheets however, indicating the advantages of investigating selectivity changes by varying the eluent B between methanol and acetonitrile (or some other eluent combinations).

(initial and final mobile phase composition). No additional experiments are necessary for this kind of robustness calculation. The result is shown in Fig. 2.20.

On top of the graph, the selected method variables (tG = 46 min, T = 30°C and tC = 100% methanol as organic modifier) with estimated possible deviations from the nominal value are shown. The temperature

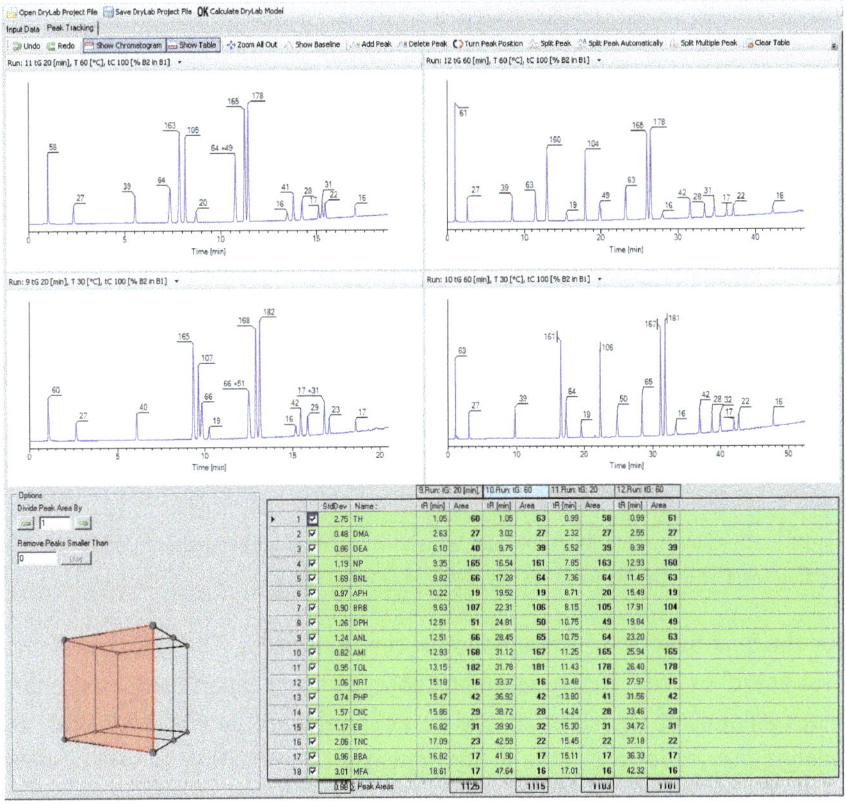

Figure 2.18: The last sheet is the one with 100% methanol as eluent B, delivering the best separations and a decent DS, as we will see in the following figures. As we can see, methanol was better suited for this separation than acetonitrile as there are significantly less double peaks.

is assumed to deviate from the nominal value of 30°C by not more than +/−2°C, (i.e. the true temperature is assumed to be in any experiment between 28°C and 32°C). In the graph on the left, the 'Frequency Distribution' shows how often (N) a certain critical resolution ($R_{s,\mathrm{crit}}$) is found out of the 729 experiments under any combination of possible, true parameter values. As can be seen from the graph, the success rate, i.e. the number of experiments, that are fulfilling the required critical resolution $R_{s,\mathrm{crit}} = 1.5$, is = 100%. This means that practically all experiments are acceptable

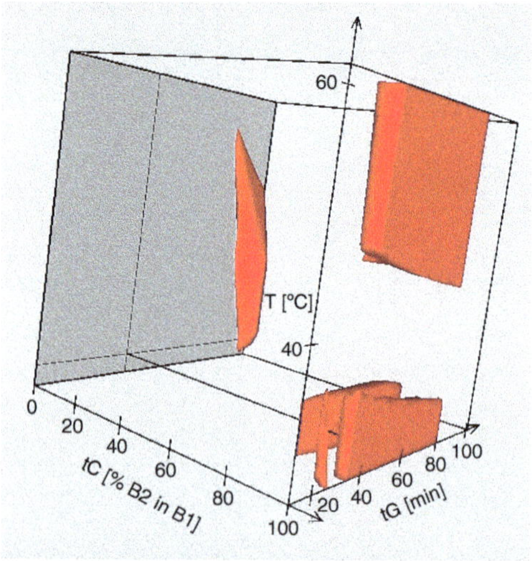

Figure 2.19: Robust regions in the cube are shown as irregular geometric forms of the DS, in which baseline resolution of all components is possible.

for the qualification of the product in the QC process. The position of the "set point" or "WP" is of great importance, as many experiments cost enormous amount of resources. If the point is selected by trial and error, an analyst may have to change it and repeat a large number of experiments to find a new optimum. The so-called PACP also keeps regulatory authorities unnecessarily busy and generates unnecessary costs. Figure 2.20(b) ("regression coefficients") describes the importance of each experimental variable and their combinations, related to the selected deviation from the nominal value for the critical resolution. As can be seen from the graph, temperature has the most important influence; a lower temperature gives a higher critical resolution.

2.7 Method Transfer

To increase sample throughput, a conventional HPLC analysis can be transferred to UHPLC. Alternatively, to decrease method development time, separation can be developed in UHPLC conditions and then transferred

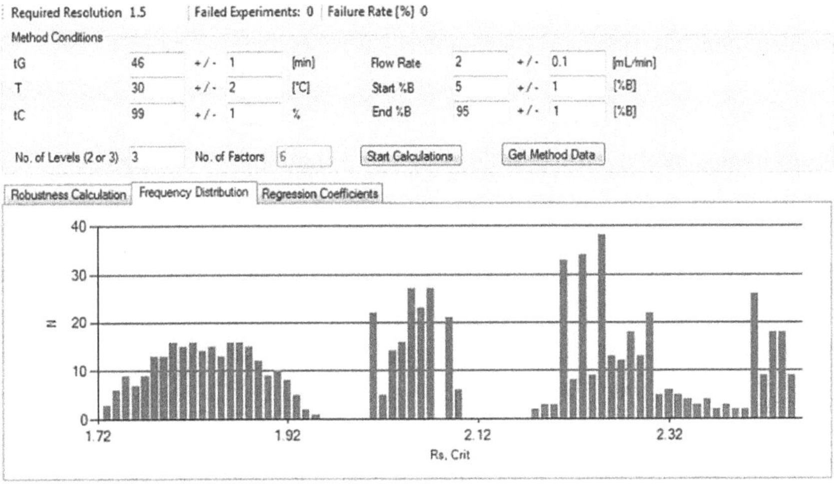

(a)

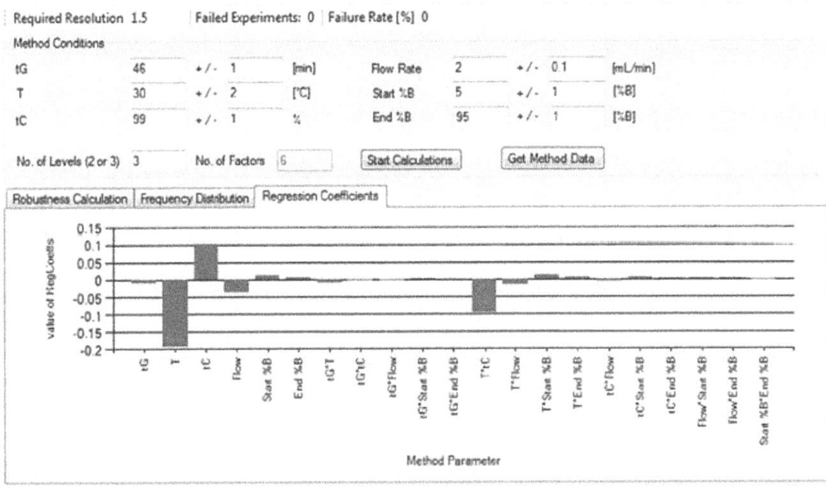

(b)

Figure 2.20: Extended robustness calculation for three measured and three additional parameters. (a) Frequency distribution of critical resolution values and (b) regression coefficients.

to conventional HPLC for routine analysis. In both cases, when transferring an analysis from conventional HPLC to UHPLC — or vice-versa — comparable method parameters must be used to maintain equivalent separations [25].

To have a method compatible with any column dimensions and HPLC or UHPLC instruments, the optimized methods can virtually be transferred to other columns of different lengths, inner diameters and particle sizes and for various system gradient delays and extra-column volumes. By utilizing HPLC modeling software, it is possible to automatically calculate and predict the effect of column parameters (length, diameter, particle size), system dwell volume and extra-column volume on the resolution. Moreover, experimentally observed column porosity can also be considered for making the transfer more accurate (porosity of HPLC and UHPLC columns is often slightly different). For such a simulated method transfer, the initial data acquired on a given column and system have to be changed virtually and the geometrical method transfer rules have to be considered. The accuracy of UHPLC to HPLC method transfer was found to be excellent in previous studies [25].

Here, we illustrate the transfer between different LC systems when using very efficient 50 × 2.1 mm columns. It is often the situation seen when a UHPLC method was developed in a research laboratory and then transferred to the QC laboratory where the UHPLC system might be different (e.g. different provider, different extra-column volume, binary or quaternary pumps, etc.). The original method was developed on an optimized UHPLC system, possessing 100 μL gradient delay (dwell) volume (binary pumping system) and around 10 μL extra-column volume (e.g. 0.065 mm ID connector tubes). The goal was to highlight what happens when using this method on a non-optimized UHPLC system possessing 350 μL gradient delay volume (quaternary pumping system) and 40 μL extra-column volume (e.g. 0.125 mm ID connector tubes). Figure 2.21 shows the simulated separations on the two different UHPLC systems. As expected, there is a systematic shift in the retention times in proportion to the difference in gradient delay volumes. In addition — due to the differences in extra-column volumes — the apparent efficiency of the separation performed

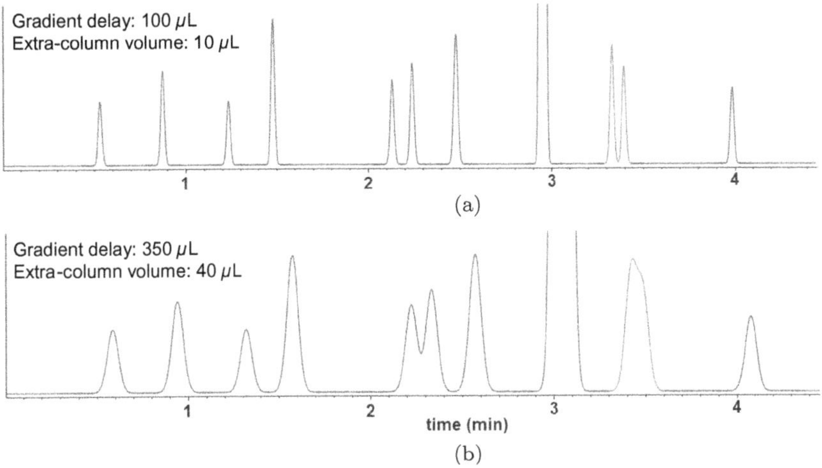

Figure 2.21: Modeled method transfer between optimized UHPLC (a) and non-optimized UHPLC (b) systems (adapted from Ref. [25], with permission).

on the non-optimized system decreased drastically. By performing such a virtual system transfer, surprises during method transfer between different laboratories can be avoided.

References

[1] I. Molnár, H.J. Rieger, K.E. Monks, Aspects of the "Design Space" in high pressure liquid chromatography method development, *J. Chromatogr. A* 1217 (2010) 3193–3200.

[2] ICH Q8 (R2) — Guidance for Industry, Pharmaceutical Development, 2009.

[3] F. Erni, Presentation at the Scientific Workshop, *Computerized Design of Robust Separations in HPLC and CE*, 31 July 2008, Molnár-Institute, Berlin, Germany.

[4] K.E. Monks, H.J. Rieger, I. Molnár, Expanding the term "Design Space" in high performance liquid chromatography (I), *J. Pharm. Biomed. Anal.* 56 (2011) 874–879.

[5] J.W. Dolan, Dwell volume revisited, *LCGC North Am.* 24 (2006) 458–466.

[6] S. Fekete. I. Kohler, S. Rudaz, D. Guillarme, Importance of instrumentation for fast liquid chromatography in pharmaceutical analysis, *J. Pharm. Biomed. Anal.* 87 (2014) 105–119.

[7] J.W. Dolan, L.R. Snyder, Maintaining fixed band spacing when changing column dimensions in gradient elution, *J. Chromatogr. A* 799 (1998) 21–34.

[8] D. Spaggiari, F. Mehl, V. Desfontaine, A.G.G. Perrenoud, S. Fekete, S. Rudaz, D. Guillarme, Comparison of liquid chromatography and supercritical fluidchromatography coupled to compact single quadrupole massspectrometer for targeted *in vitro* metabolism assay, *J. Chromatogr. A* 1371 (2014) 244–256.

[9] I. Molnár, Computerized design of separation strategies by reversed-phase liquid chromatography: Development of DryLab software, *J. Chromatogr. A* 965 (2002) 175–194.
[10] Cs. Horváth, W. Melander, I. Molnár, Solvophobic Interactions in liquid chromatography with nonpolar stationary phases (solvophobic theory of reversed phase chromatography, Part I.), *J. Chromatogr.* 125 (1976) 129–156.
[11] Cs. Horváth, W. Melander, I. Molnár, Liquid chromatography of ionogenic substances with nonpolar stationary phases (Part II.), *Anal. Chem.* 49 (1977) 142–154.
[12] L.R. Snyder, J.W. Dolan, *High Performance Gradient Elution, The Practical Application of the Linear-Solvent-Strength Model*, John Wiley & Sons, Inc., Hoboken, New Jersey, 2007.
[13] I. Molnár. K.E. Monks, From Csaba Horváth to Quality by Design: Visualizing design space in selectivity exploration of HPLC separations, *Chromatographia*, 73 (2011) S5–S14.
[14] International Conference on Harmonisation of Technical Requirements for Registration of Pharmaceuticals for Human Use, ICH Harmonised Tripartite Guideline, Validation of Analytical Procedures: Text and Methodology Q2(R1), Current Step 4 version, Parent Guideline dated 27 October 1994 (Complementary Guideline on Methodology dated 6 November 1996 incorporated in November 2005).
[15] K. Monks, I. Molnár, H.J. Rieger, B. Bogáti, E. Szabó, Quality by Design: Multidimensional exploration of the design space in high performance liquid chromatography method development for better robustness before validation, *J. Chromatogr. A* 1232 (2012) 218–230.
[16] D.M. Bliesner, *Validating Chromatographic Methods*, Wiley-Interscience, New Jersey, 2006.
[17] B. Dejaegher, Y. Vander Heyden, Ruggedness and robustness testing, *J. Chromatogr. A* 1158 (2007) 138–157.
[18] Y. Vander Heyden, A. Nijhuis, J. Smeyers-Verbeke, B.G.M. Vandeginste, D.L. Massart, Guidance for robustness/ruggedness tests in method validation, *J. Pharm. Biomed. Anal.* 24 (2001) 723–753.
[19] M.W. Dong, *Modern HPLC for Practicing Scientists*, Wiley-Interscience, New Jersey, 2006.
[20] J.J. Hou, W.Y. Wu, J. Da, S. Yao, H.L. Long, Z. Yang, L.Y. Cai, M. Yang, X. Liu, B.H. J., D.A. Guo, Ruggedness and robustness of conversion factors in method of simultaneous determination of multi-components with single reference standard, *J. Chromatogr. A* 1218 (2011) 5618–5627.
[21] M. Novokmet, M. Pučić, I. Redžić, A. Mužinić, O. Gornik, Robustness testing of the high throughput HPLC-based analysis of plasma N-glycans, *Biochim. Biophys. Acta* 1820 (2012) 1399–1404.
[22] R. Ragonese, M. Mulholland, J. Kalman, Full and fractionated experimental designs for robustness testing in the high-performance liquid chromatographic analysis of codeine phosphate, pseudoephedrine hydrochloride and chlorpheniramine maleate in a pharmaceutical preparation, *J. Chromatogr. A* 870 (2000) 45–51.
[23] E. Hund, Y. Vander Heyden, M. Haustein, D.L. Massart, J. Smeyers-Verbeke, Robustness testing of a reversed-phase high-performance liquid chromatographic assay:

comparison of fractional and asymmetrical factorial designs, *J. Chromatogr. A* 874 (2000) 167–185.

[24] R. Kormány, J. Fekete, D. Guillarme, S. Fekete, Reliability of simulated robustness testing in fast liquid chromatography, using state-of-the-art column technology, instrumentation and modelling software, *J. Pharm. Biomed. Anal.* 89 (2014) 67–75.

[25] S. Fekete, R. Kormány, D.Guillarme, Computer assisted method development for small and large molecules, *LC-GC, HPLC* 2017 30(supplement) (2017) 14–21.

Chapter 3

ChromSword®: Software for Method Development in Liquid Chromatography

Sergey V. Galushko*,‡, Irina Shishkina*, Evalds Urtans† and Oksana Rotkaja†

*ChromSword, Dr. Galushko Software Entwicklung GmbH Im Wiesengrund 49B, 64367, Muehltal, Germany

†ChromSword Baltic, Antonijas 22-1, Riga, LV-5041, Latvia

‡galushko@chromsword.de

3.1 Introduction

Method development in chromatography can be considered as a process studying the empirical relationships between the quality of a chromatogram and the chromatographic conditions. A chromatographer changes conditions to find an acceptable method to achieve separation in a reasonable time. The time required to find optimal conditions or to make any conclusion can be substantially reduced by using computer programs for method development. HPLC method development programs can be utilized interactively (off-line) and for automatic optimization (online). ChromSword® for off-line computer-assisted method development was launched in 1994 as an extension of ChromDream® software [1]. During 1998–2000, the first version for unattended method development was started [2]. The latest version of ChromSword® combines different technologies of method development in one software platform:

- Computer-assisted
- Automated optimization

- Automated robustness studies
- Scouting to screen different column, solvents, buffers and methods

It is possible for a chromatographer to use only the computer-assisted (off-line) or automated method development approach or to use both interactive and unattended optimization.

ChromSword® off-line can be used for optimizing separations in reversed-phase (RPLC), normal-phase (NPLC) and ion-exchange (IEX) liquid chromatography (LC). In the off-line mode, chromatogram simulations and optimizations as a function of one or more variables are possible. The off-line mode includes two possibilities for optimization in RPLC.

The approach which takes into account the characteristics of compounds and column/solvent properties is the solvatic or solvophobic model of RPLC.

The traditional method for optimizing separation using only retention data of analytes is the linear solvent strength (LSS) model and other polynomial models.

In the automated mode, the software operates as a chromatography data system controlling HPLC instruments and executes a sequence of runs. The user can predefine such a sequence of runs — this is a scouting approach to screen different stationary phases (SPs) or mobile phases (MPs) or statistical design of experiments (DoE) according to some statistical rules to study the effect of method variables on the separation. This method is defined as robotic process automation. Another approach is intelligent automation. Intelligent automation automates non-routine tasks like optimizations involving complex data processing and reasoning. ChromSword® supports both types of automation to assist chromatographers for routine and intelligent method development workflow.

To support various method development workflows ChromSwordAuto® package contains modules dedicated to different scenarios and tasks:

ChromSword®	for computer-assisted method development
ChromDraw®	chemical editor for drawing and processing structural formulae

ColumnViewer®	reversed-phase column properties data base
ChromSword® Scout	for automated method screening
ChromSword® Developer	for automated method optimization
AutoRobust®	for automated robustness study and method transfer
ReportViewer®	for data browsing, chromatogram and spectra processing, project management and report generation

3.2 Automated Method Development

Most automated HPLC method development approaches can be divided into three classes:

- Mechanistic or model-based optimization
- Statistic or direct process optimization
- Screening or running a large number of column/solvent/method combinations to identify those with a reasonable separation

In the model-based optimization, mathematical models are utilized to reduce the number of experiments. The development of mechanistic models requires good chromatography understanding, reliable tests for parameter estimations and peak tracking. Limiting factors are computational time and reliability of the models that are applied for simulation and optimum search. The determination of mechanistic model parameters can be complicated for computer-assisted (off-line) method development and requires time and operator qualification for optimization of multi-component mixtures. Automatic optimization with mechanistic DoE incorporates engineering knowledge in the form of constrains, expert-rules and known fundamental relationships of LC; therefore, this technology can find optimal conditions faster than the off-line approach. One of the main advantages of the automatic optimization is that a chromatographer can avoid complex tasks of the off-line computer-assisted optimization — peak tracking, data input, method and sequence specifications and other routine and non-routine operations. It should be noted that in the recent final guidance for industries with regard to the analytical method development, the U.S. Food and Drug Administration (FDA) recommends submission of data to indicate a mechanistic understanding of the basic methodology [3].

An alternative to the mechanistic model-based approach is to directly identify process optima based on the results of experiments that are planned by statistical software such as repeated DoE. In contrast to model-based strategies, no mathematical process model is required, which is a significant advantage for many operators, and it is also better to use when the theory of LC and separation process interactions are not yet fully understood. Unfortunately for complex mixtures, when retention models cross each other in different regions of method variables, the direct approach can find the optimum only accidently. Usually, this type of DoE is used in a case where no, or little, prior process knowledge is available. However, for separation processes where a high degree of knowledge is available, statistical DoE is often not the most efficient strategy. Nevertheless, experimental results from the direct approach can be successfully used to identify a local optimal separation region for simple mixtures and to estimate the sensitivity of method quality to specific parameter changes within the design space (DS). Special software that include both features to create DoE and control of LC instruments to execute the DoE have substantial advantages against statistical software which have only options to plan DoE.

An alternative to the mechanistic and statistic approaches is to run the high-throughput screening to test combinations of method variables and factors — columns, solvents, buffers, gradients, etc. In contrast to the model-based and the statistical strategies, neither mathematical process model nor statistical DoE is required for the scouting approach. A chromatographer needs to only create a large sequence and then run it for new samples, thus relying on these few combinations of method variables and factors that will provide practically reasonable separations. The scouting approach is used frequently for chiral separations and samples when specific optimization is not necessary. Specialized software for automated method scouting are practically useful to create and edit long sequences rapidly and run them automatically.

For analytical method development, all three approaches proved to be practically useful, and any combination of them increase the probability of finding more suitable methods. To support various automated method development workflows, ChromSwordAuto® can operate in three modes: scouting, model-oriented optimization and statistic (direct optimization). Each mode can be applied separately or in various combinations depending on the preferred strategy of method development at a particular laboratory and project stage. Each mode is operated with a dedicated module.

3.2.1 *Instrument control and software configurations*

ChromSwordAuto® can operate as a chromatography method development data system (CDS) or as a third-party software. Functioning as the CDS ChromSwordAuto® controls Agilent, Waters and Hitachi HPLC and UHPLC systems. To control these instruments, no other CDS is necessary, and a stand-alone or a client–server configuration of ChromSwordAuto® can be chosen during installation. For the client–server configuration, data are collected on the local network or the internet file server (Fig. 3.1). The client–server configuration satisfies the requirements for data integrity with regard to applicable regulations like FDA 21 CFR Part 11.

Operating as a third-party software, ChromSwordAuto® controls Agilent, Waters and Dionex instruments thorough OpenLab/ChemStation, Empower or Chromeleon CDS. These CDS can work in the stand-alone, network or client–server environments.

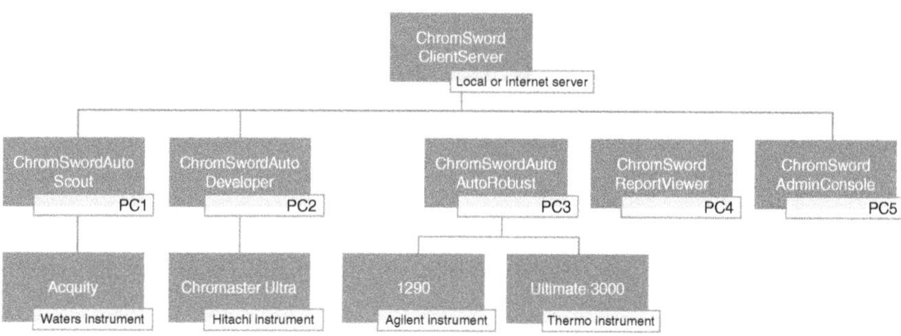

Figure 3.1: ChromSwordAuto® client–server configuration.

Different configurations of HPLC and UHPLC instruments can be used for automated method development. The most simple method development system consists of a binary pump, UV detector and autosampler; however, typically, method development systems contain 4–8 columns and 2–6 solvent channels to test different stationary and MPs.

ChromSwordAuto® incorporates automation of routine operations: column equilibration, column wash-out methods, system purging and column and solvent switching sequences.

3.2.2 *Strategies of automated method development*

Different strategies can be applied for automated method development. Strategies can combine screening, optimization and robustness study steps. One of the successful strategies for development of RPLC methods with ChromSwordAuto® has been used for dug candidates. It includes an automated screening step to identify the best column and solvent followed by an optimization step to fine-tune the separation [4, 5]. A similar strategy was used to apply ChromSwordAuto® for optimization of chiral separations in NPLC [6] and RPLC [7]. In another approach, the rapid optimization mode can be used for several predefined SP and MP combinations which are accepted at a lab as a standard method development column set, and then the fine optimization mode is applied for the most promising combination. Robustness studies can be included optionally for late-stages projects or methods to be transferred to other laboratories. The steps of such a strategy are shown in Fig. 3.2.

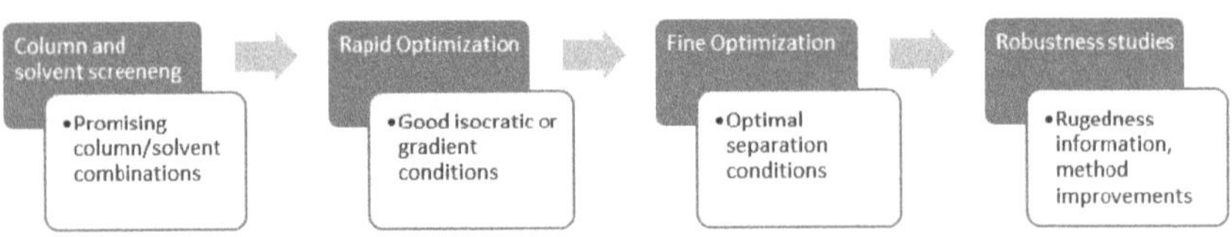

Figure 3.2: The strategy of method development for the latest stages of product developments.

3.2.3 *Automated method screening with ChromSwordAuto® Scout*

Automated screening of SP and MP are used to find practically a acceptable separation and run time when full optimization is not necessary. The screening can also be the first step in a multi-step method development strategy to identify promising combinations of columns and MPs.

ChromSwordAuto® Scout screening module generates sequences automatically and runs them to scout different gradients, columns, solvents, buffers, temperatures and other method variables for one or several samples. For multi-column and multi-solvent instruments, ChromSwordAuto® Scout controls several column compartments with 4–8 columns in each compartment and several (4–12 position) solvent switching valves connected to a binary or a quaternary pump. ChromSwordAuto® Scout analyzes 2D and 3D data acquired from two detectors simultaneously.

ChromSwordAuto® Scout application incorporates automation of column equilibration, column wash-out methods, system purging and column and solvent switching sequences for changing solvents, buffers, columns and other chromatographic process variables and factors.

3.2.4 *Automated model-based method optimization with ChromSwordAuto® Developer*

ChromSwordAuto® Developer module can be used for automated method optimization in RPLC, NPLC, IEX, HIC, HILIC, size exclusion chromatography (SEC) and supercritical fluid chromatography (SFC). For SEC, ChromSwordAuto® optimizes isocratic conditions, and for other types of chromatography, both isocratic and gradient separations can be optimized. Retention models that are used for different type of LC are described in Sec. 3.3.

ChromSword® is used for automated optimization of various mixtures; however, most frequently, it is applied for method development in the pharmaceutical industry. Typical applications are the development of stability-indicating and quality control methods (e.g. impurity profiling,

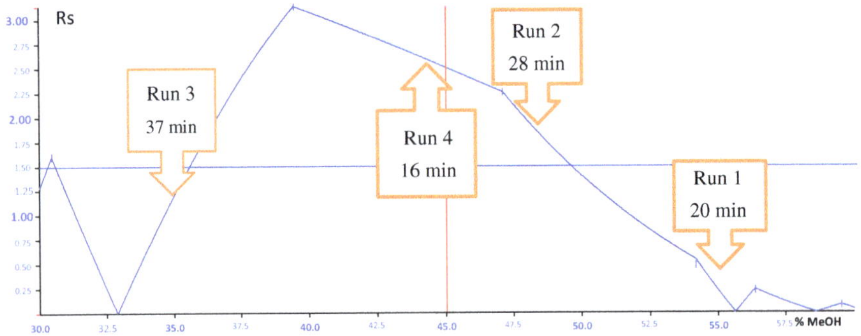

Figure 3.3: Runs shown on the resolution map that the software performs searching for optimal conditions in the unattended mode. Method development for a mixture of nine beta-blockers. Column: Purospher RP 18e, 5 μm, 150 $\times$ 4 mm. Mobile phase: 0.05 M phosphate buffer, pH $=$ 3.0 — methanol. The goals: $R_s \geq 2.0$ and run time ≤ 20 min.

assay, cleaning control, etc.). For automatic optimization, a user should specify the starting conditions: the column, solvent, flow rate, injection volume and the task type — rapid or the fine optimization. A chromatographer can also specify the development of either isocratic or isocratic and gradient methods. For both procedures, the optimization process includes the study of a sample to build retention models followed by application of the optimization procedure to find the optimal conditions. For planning new runs, the software processes the results of the previous runs and takes them into account. In Fig. 3.3, the method by which the software searches for optimal conditions developing the isocratic methods is shown.

For optimizations of gradient methods, both the studying and optimization runs can be linear and multi-step gradients. For optimization of separation, the Monte Carlo, genetic algorithms and the neural network methods are used. For the rapid optimization algorithm, the software performs 3–4 runs (Figs. 3.4–3.6), and for the fine optimization algorithm more runs are executed to study a sample and optimize the separation.

3.2.4.1 *Method development for large molecules*

Large molecules like proteins exhibit substantially different retention behavior than small analytes [8]. For these samples a small shift in chromatographic conditions can lead to high changes in retention and

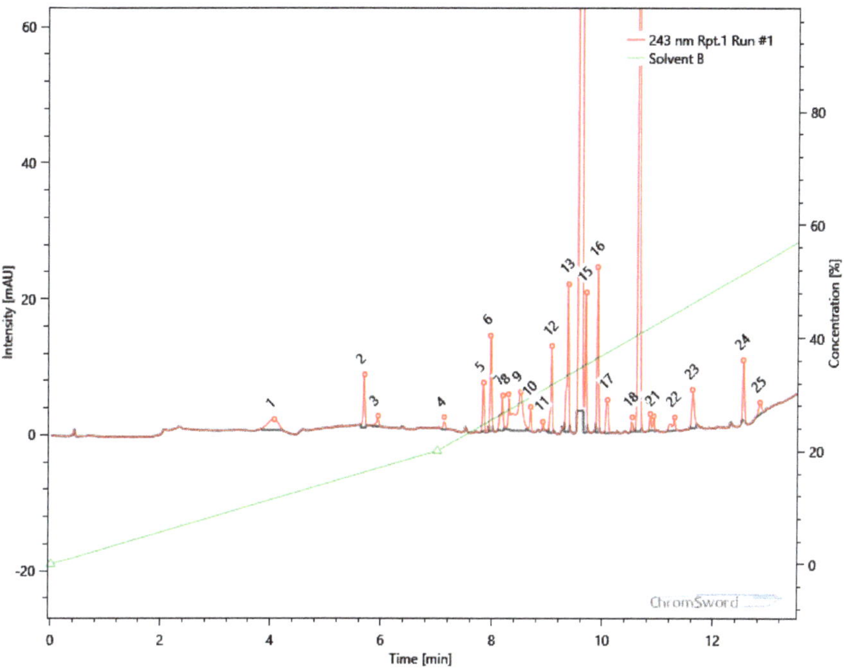

Figure 3.4: The first run of the automatic rapid optimization of the force degradation test mixture. Column: Zorbax Eclipse C18, 1.8 μm, 50 $\times$ 2.1 mm, flow rate 0.6 mL/min.

efficiency. The other point is that these compounds have practically identical UV spectra and cannot be used for peak tracking. Recently computer-assisted (off-line) method optimizations were reported for monoclonal antibodies (mAbs) and their domains in RPLC and IEX using 2D model as the gradient time–temperature model [9, 10]. It should be noted however, that the computer-assisted method optimization can be a time consuming process when many samples, columns and effects of different method variables require evaluation. An effective approach to circumvent and increase productivity is automated method development. In this instance, an analyst defines a strategy and an "*intelligent*" chromatography method development data system plans and performs many routine and optimization experiments autonomously. Various strategies of automated method development for mixtures of large molecules can be realized with ChromSwordAuto®. These can combine automated screening experiments

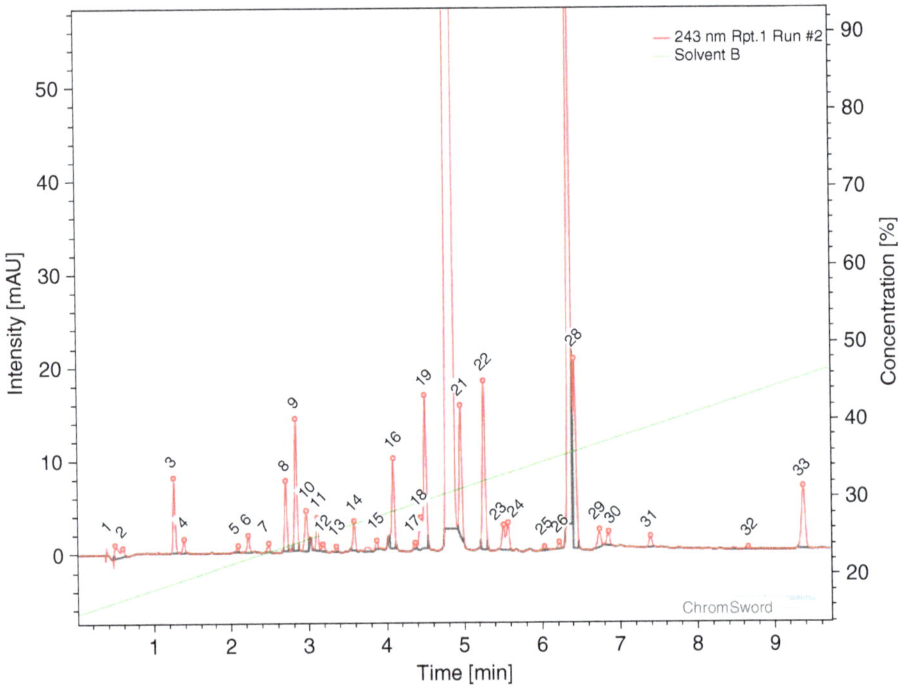

Figure 3.5: The second run of the automatic rapid optimization. Conditions are same as described for Fig. 3.4.

with unattended optimization, which is then followed by robustness studies using different DoEs. Results can also be used for off-line simulation and optimization. Such a strategy is used in different laboratories for automated RPLC method development using ChromSwordAuto® for the separation of variants and degradation products of the recombinant mAbs. The aim of method development for such projects is to study the domain-specific oxidation and develop stability-indicating methods that separate degradation products. For complex mixtures the optimization program can run multi-step gradients to separate more components (Fig. 3.7).

An important point to be considered is the column length for optimization of small and large molecules. It is known that the column efficiency for small compounds like peptides, after the digestion of proteins, is improved by increasing the column length. In contrast, the retention behavior of large proteins is different, and their bandwidth can be almost

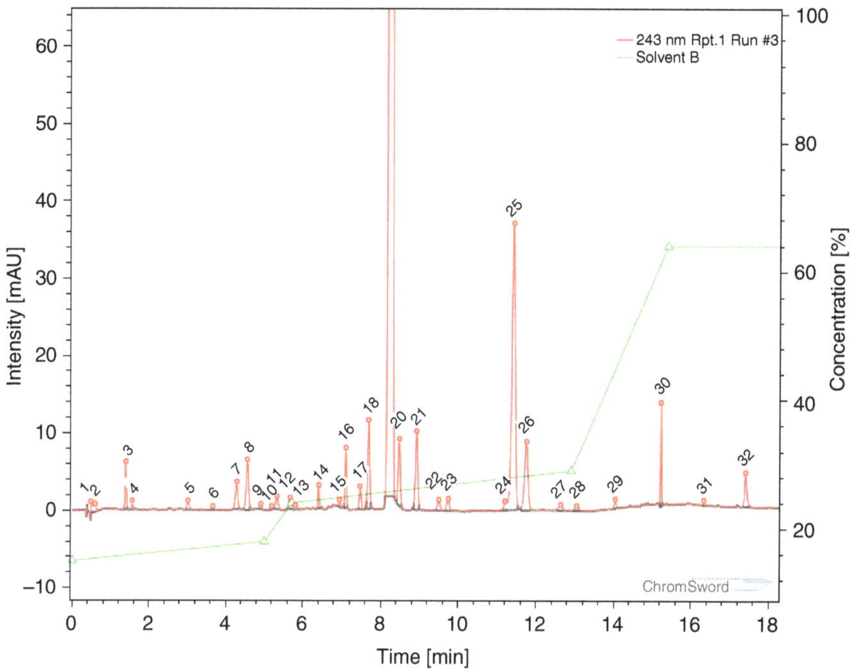

Figure 3.6: The third run of the automatic rapid optimization. Conditions are same as described for Fig. 3.4.

constant for all practical column lengths in the range 50–250 mm [11]. For such samples, longer columns do not provide higher separation efficiency [11], and therefore a short column can be a good alternative. Results in Figs. 3.7 and 3.8 show that the automated procedure can successfully find conditions to separate proteins on small columns. It should be noted that the optimization procedure is not related strictly to the column length. It is related to the target resolution and practical run time; therefore, shorter run times can be obtained on a long column and longer run time on a short column. In Fig. 3.8(a) the initial three study runs and in Fig. 3.8(b) the final gradient run are shown to separate monoclonal antibodies, under RPLC conditions. It should be noted that no optimal linear gradient for this mixture could be found in the temperature range of 70–80°C where reasonable peak width is observed and the column can be operated.

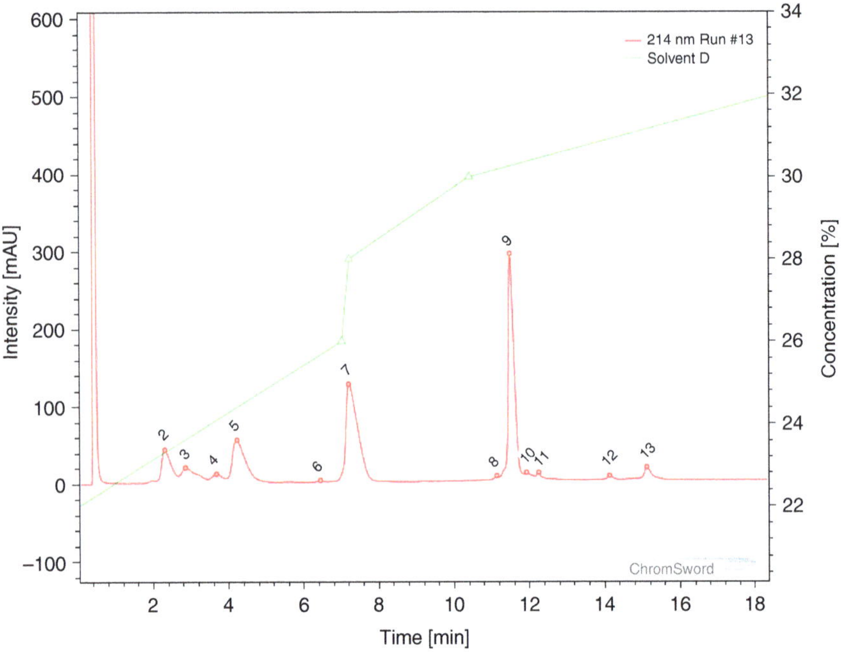

Figure 3.7: Partially digested (using IdeS) and reduced (using dithiotreitol, DTT) mAb sample. Peaks 2–4 — oxidation products of the crystallizable fragment (Fc/2); peak 5 — (Fc/2); peak 7 — the light chain (LC); peak 9 — the N-terminal half of one heavy chain (Fd). Column: 50 mm × 2.1 mm AdvanceBio RP mAb C8. Mobile phase A: Water + 0.1% TFA, B: ACN + 0.1% TFA. Temperature was set to 70°C, flow rate = 0.3 mL/min.

3.2.5 *Automated robustness studies and statistical DoE with ChromSword® AutoRobust*

ChromSword® AutoRobust is a specialized application for automatic evaluation of robustness of HPLC methods. According to the ICH guidelines [12] "Validation of Analytical Procedures: Methodology (Q2B)", the robustness of an analytical procedure is defined as a measure of its capacity to remain unaffected by small, but deliberate variations in method parameters and provides an indication of its reliability during normal usage. The robustness should be considered at an appropriate stage in the development of the analytical procedure [12]. AutoRobust is a software tool for automation of robustness experiments to study the influence of variations in method parameters on chromatographic results.

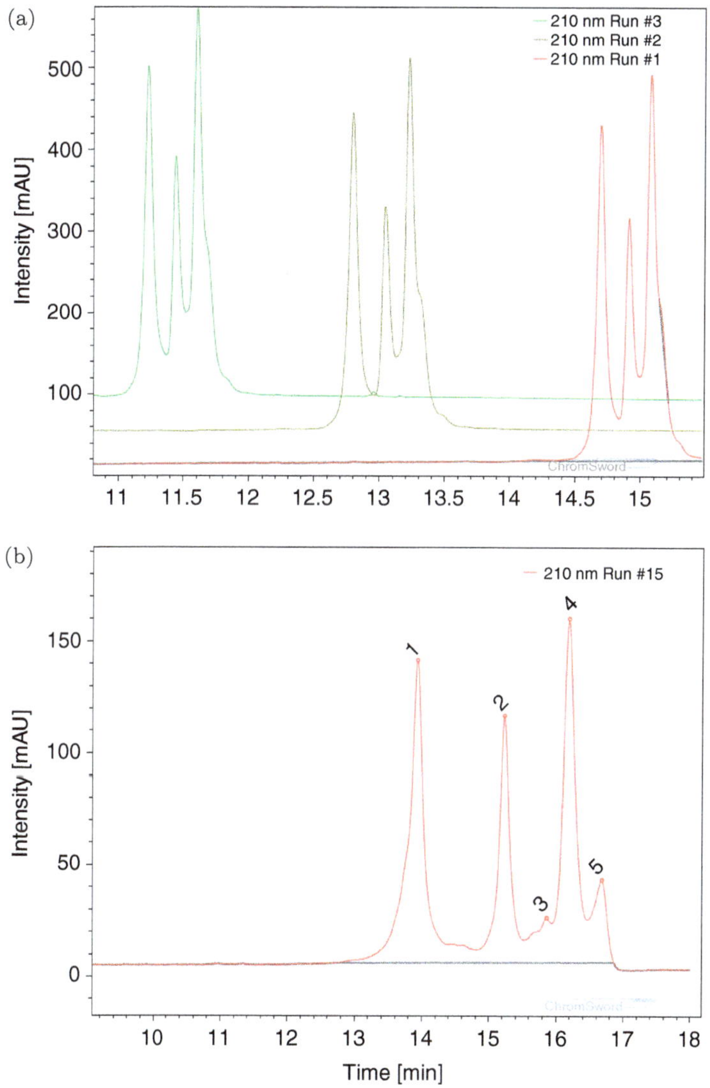

Figure 3.8: Column: 50 × 2.1 mm Zorbax 300 SB-Diphenyl. Mobile phase A: water + 0.1% TFA, B: ACN + 0.1% TFA. Flow rate: 0.25 mL/min; Temperature: 80°C. Sample: test mixture of mAbs (mAb1, mAb2 (confidential), Erbitux and Avastin). (a) Initial study runs of unattended optimization for separation; gradients: 1. 30–70% B in 25 min; 2. 36–66% in 22 min; 3. 36–66% in 19 min. (b) The final run of the unattended optimization; gradient: 0 min — 50% B in 2.2 min — 51% B; 16.6 min — 54% B; 18 min — 55% B.

Robustness of a method is extremely important for providing method transfer to other laboratories and instruments. Typically, robustness tests are performed at late stages of drug development projects; however, performing robustness tests at later stages involves the risk that when a method is found to not be robust, it should be redeveloped and optimized. Therefore, it is better to perform robustness tests at an earlier stage of method development. Different critical quality attributes (CQAs) of a method can be tested — including area, area%, retention time, resolution and other CQAs. One of the most important CQAs for HPLC methods is the resolution between peaks of target compounds. The resolution characteristic of a method should be within appropriate limits to ensure the drug product quality.

The following steps can be identified for robustness tests projects:

(1) selection of the factors to be tested,
(2) selection of the experimental design,
(3) definition of the different levels of the factors,
(4) creation of the experimental set-up,
(5) execution of the experiments,
(6) calculation of effects,
(7) statistical and graphical analysis of the effects,
(8) drawing conclusions from the statistical analysis and
(9) if necessary, improving the performance of the method.

These different steps are considered in more detail below.

3.2.5.1 *Selection of the factors*

For robustness tests, different operation factors can be considered. The selected factors can be quantitative (continuous) like the temperature or the concentration or qualitative (discrete) like the column batch. These factors should represent those that can be changed when a method is transferred between laboratories, analysts or instruments and that potentially could affect the response of the method. Typically, the following factors can be included in the robustness tests:

- gradient time and slope of linear gradients,
- initial and final concertation of linear gradients,

- time and concentration of each gradient node (step) for multi-step gradients,
- flow rate,
- column compartment temperature,
- pH of the MP,
- wavelength,
- column batch,
- method equilibration time,
- injection volume.

All these parameters and factors are supported by automated DoE with ChromSword® AutoRobust module. A chromatographer can optionally specify all or several factors to be included in the DoE.

The difference in flow rate, concentration and gradient time affect the resolution when different type of pumps (low- or high-pressure mixing systems), different solvent mixers and pumps from different manufacturers are used. The effective temperature inside a column can be different due to the difference in construction of compartments (forced air or still air oven). The small difference in glass electrodes and standard buffers can lead to differences in pH of a MP and selectivity of separation of basic and acidic compounds. If concentration of a sample is too low or too high, then increasing the injection volumes can lead to peak distortion.

3.2.5.2 *Selection of the experimental design*

The one-factor-at-a-time (OFAT), full factorial design (FFD) and the Plackett–Burman partial factorial design (PFD) can be used for robustness tests. The OFAT is the fastest design; however, it cannot estimate interactions of different variables without preliminary studies. The FFD is the most comprehensive design to determine interactions of factors and describe the response surface for finding optimum factor-values; however, it requires substantially more experiments. The PBD can be used as an alternative to FFD, but arrays of data points after the PBD cannot typically be used to solve the system of equations to determine chromatographic retention model parameters. In this case, a less reliable, simplified model is usually used to calculate response; however, deviations between the predicted and experimental value of a critical quality parameter can be too high. Another problem is a possible confounding of effects due to reducing the number of runs in PBD. In this case, the effects of different factors or interaction factors cannot be evaluated individually and the interpretation of the results becomes difficult and even incorrect.

We consider that the robustness projects should include two designs:

(1) The OFAT design which can rapidly identify which of tested variables has a significant effect on the response.
(2) The FFD of the critical variables which were identified in (1).

Both steps can be executed in a completely automatic manner with a reasonable number of experiments. The PBD can be planned when the number of runs is too high and it is not practically reasonable to run the FFD designs.

3.2.5.3 *Definition of the levels for the factors*

The factor levels of variables to be tested should be set around the nominal values specified in the operating (basic) method. The interval chosen between the extreme values represents the limits between which the factors are expected to vary when a method is transferred. It should be noted that the levels should be defined by the analyst according to the results of a preliminary study of chromatographic retention behavior of compounds and instrument specifications taking into account the precision and the uncertainty with which a factor can be set and reset. To define the factor levels for the temperature, concentration and time of gradient steps, it is recommended to study the effect of these variables in more detail.

3.2.5.4 *Creation of the experimental set-up*

Each variable is studied in the experimental design, which is selected as a function of the number of factors and of levels to investigate. Two-level screening designs are a simple approach that can screen a relatively large number of factors in a relatively small number of experiments. More informative are the two-level designs with center points for effects of concentration and gradient time or the four-level designs with center points for effects of flow rate and temperature. Such designs are optional in AutoRobust and allow the analyst to establish a linear or nonlinear retention model. Creation of the experimental design manually takes substantial time, even for OFAT. For planning FFD and PBD, normally special statistical software are used and then the design plan should be transferred into a sequence of runs of a chromatography data system. This is also a time-consuming process, and is practically very important that robustness test software can create DoE and transfer it into a sequence of runs automatically. The AutoRobust software module in ChromSword® provides a simple and rapid automated set-up of up to eight variables with 2–7 levels for OFAT, FFD and PBD. An unlimited number of qualitative factors (column, solvent batches, etc.) can also be included in the DoE.

3.2.5.5 *Execution of experiments*

It is important for reproducible robustness experiments to provide constant parameters both for injection and conditioning runs. Column and instrument wash-out, and purging and conditioning runs should be set up according to the instrument and column specifications. Sufficient time for column equilibration, not less than 10 column volume have a paramount importance especially for large proteins to obtain reproducible results. For more confidence, it is recommended to include the column equilibration time as a variable in the robustness tests DoE.

The planned DoE is executed automatically with AutoRobust. The method development system performs these runs while interacting with a chromatography data system or directly with the modules. For estimation of time effects and stability of the instrument and the column, a number of additional experiments at nominal levels can be added to the planned DoE. These replicate experiments are performed before, at regular time intervals between, and after the robustness test experiments. These experiments allow checking whether the method performs well at the beginning and at the end of the experiments and to estimate for drift and column stability.

The results of runs are used to calculate effects of variables and determine the response.

3.2.5.6 *Calculation of effects and response determined*

From the performed experiments, a number of responses can be determined. For chromatographic methods, responses describing a quantity such as the content of main substance and by-products and effects of variables on peak area% and areas should be evaluated. The responses determined during the robustness test can be one of the following: the resolution between each pair of neighboring peaks, the retention time, the area and the area% of compound peaks. These parameters allow for evaluating the quality of a method and the effects of variables and factors.

The automated data processing procedure additionally calculates the relative retention, the peak asymmetry, the peak height and number of theoretical plates, which can also be included in the robustness study results.

3.2.5.7 *Numerical and graphical analysis of the effects*

One of the most important CQAs for HPLC methods is the resolution between peaks of target compounds. The resolution characteristic of a method should be within appropriate limits to ensure the drug product quality. As mentioned earlier, two approaches can be used to evaluate the effect of method variables on resolution — descriptive and mechanistic. Traditional statistically based software uses the descriptive approach and models the response surfaces with quadratic polynomials [12]. The main advantage of this approach is the simple and easy data processing procedure. This approach does not use physical models of the separation process and peak tracking from run to run. However, from the theory and practice of computer-assisted HPLC method development, it is well known that the quadratic dependence between resolution and method variables (concentration of organic modifier, gradient profile, temperature, pH) is more an exception rather than a rule for complex mixtures with irregular retention models [8]. Retention models of compounds can cross each other, and dependences $R_s = f$ (temperature, concentration, gradient time, pH) can have one or several maxima and minima. Figure 3.9 shows the resolution plots for limited pairs of a mixture of nine beta-blockers as a function of the concentration of methanol in the mobile phase. It is obvious that

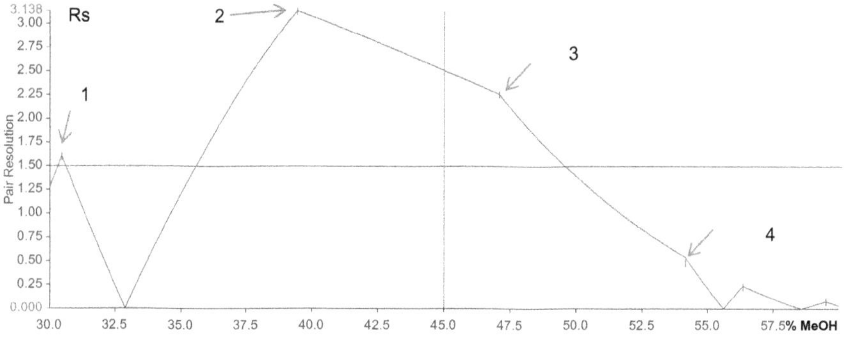

Figure 3.9: Resolution map: Effect of methanol concentration in MP on resolution of a mixture of nine beta-blockers. The arrows show the change of the limited pair in different regions of methanol concentration. 1 metipranolol/alprenolol; 1–2 propranolol/metipranolol.; 2–3 carazolol/celiprolol; 3–4 metoprolol/celiprolol; alprenolol — carvedilol.

modeling of the resolution response without peak tracking in this case will lead to wrong conclusions regarding optimal conditions and robustness of the method. The mechanistic approach uses parameters of the chromatographic process responsible for the response; however, retention behavior of the compounds must be studied to describe the effect of variables on the resolution. These include peak tracking from run to run, evaluation of parameters of retention modes in gradient elution and under different temperatures, and building a system of equations and solving them.

The mechanistic approach that applies relations from the theory of LC is supported in the AutoRobust software. After the design of experiments is created and performed in automated mode, data are processed for statistical and graphical analysis of responses. Method variables can have a substantial effect on resolution, and knowledge of the effect of the combination of these variables is necessary to study the robustness and to build up a DS of the method. The example of the effect of two variables with a fixed nominal value for two other variables is shown in Fig. 3.10.

3.2.5.8 *Improving the performance of the method*

Analysis of the resolution maps for a combination of three different variables enables visualization of areas where resolution can be increased or decreased. For example, the resolution map shows that temperature

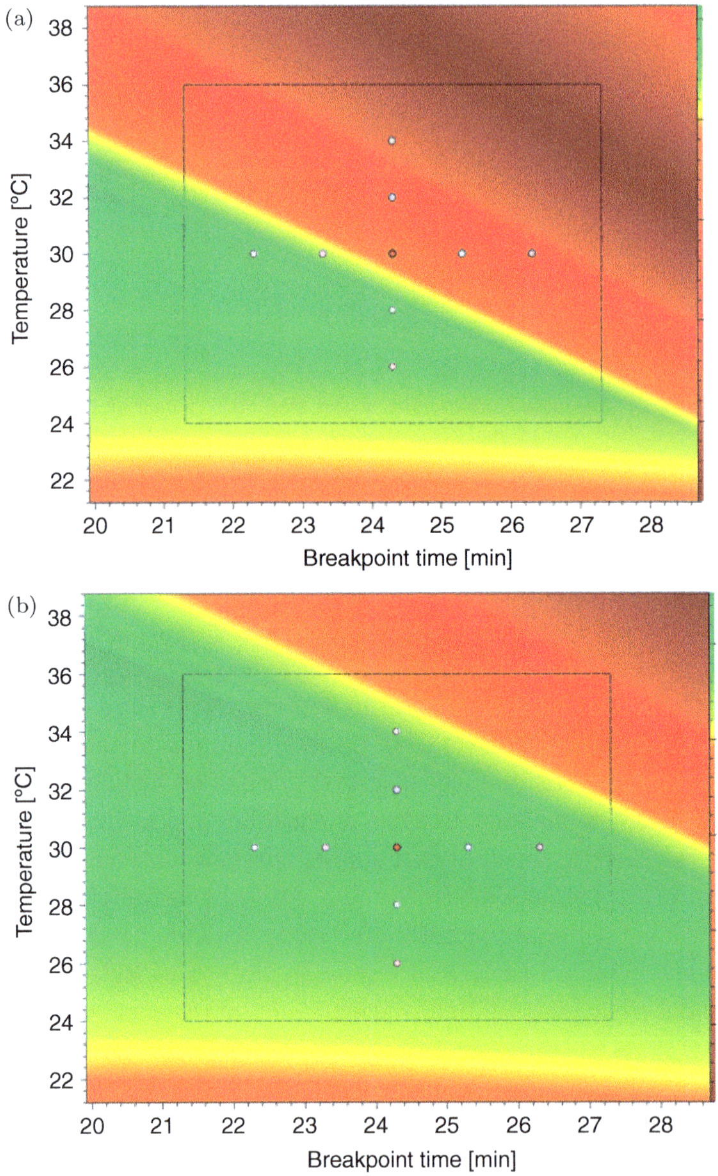

Figure 3.10: Resolution maps: Effect of the temperature and the gradient breakpoint time on resolution of a limited pair at the flow rate of 1.0 mL/min (a) and 0.8 mL/min (b). Mixture: 10 hair dyes. Column: ACE Excel C18-Amide 100 × 4.6 mm, 3 μm.

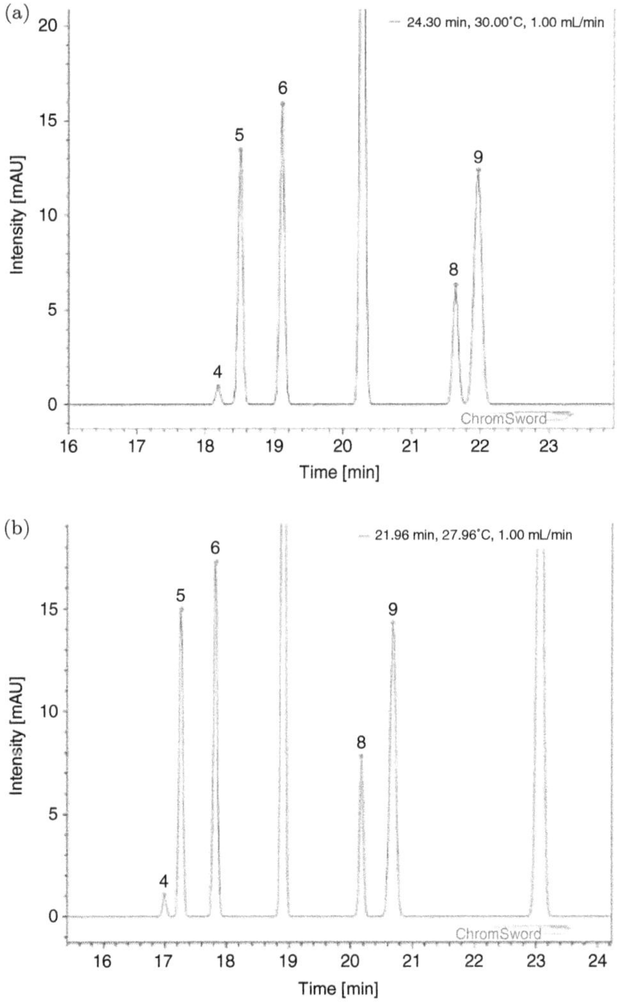

Figure 3.11: Chomatograms at a temperature 30°C, flow rate of 1.0 mL/min and gradient time of 24 min (a) and at 28°C, 0.80 mL/min and 22 min, respectively (b).

at 28°C, the flow rate of 0.80 mL/min and the gradient time of 22 min will provide a more robust method with higher resolution than one that was used after optimization (30°C, 1.0 mL/min and 24 min, respectively) (Figs. 3.10(b) and 3.11). Thus, robustness studies can also be considered as an additional tool to improve the performance of the method.

3.3 Computer-assisted Method Development

ChromSword® in the off-line mode can be used for optimizing separations in RPLC, NPLC and IEX.

If the structural formulae of compounds are known then, ChromSword® can predict the conditions of isocratic or gradient elution for acceptable retention to be obtained. No preliminary experiments need to be performed for the virtual chromatography. If the structural formulae of compounds being separated are known, then it is possible to start optimization of resolution after the first run. In this case, after inputting the experimental retention data for the first run, parameters of solutes will be refined to predict the best conditions for the separation. Entering experimental retention data for the second and the following runs makes possible a more precise prediction.

For solutes with unknown structures, ChromSword® can determine, from chromatographic experiments, their characteristics (molecular volume, the energy of interaction with water, nature (acid, base, neutral, p*Ka* value) and then predict their retention times on different reversed-phase columns and with different MPs.

Prediction is the first step in method development. The subsequent steps are optimization of retention and separation. ChromSword® enables a user to optimize the concentration of a modifier in a MP, pH value, temperature, gradient profile and column coupling. To optimize the separation of a mixture in gradient elution mode, stochastic methods like Monte Carlo and genetic algorithms are used.

For NPLC, it is possible to optimize the concentration of a stronger solvent in a weaker one when the retention data for two or more runs are entered. For IEX, the buffer or salt concentration in a MP can be optimized. Optimization of temperature is possible both for NPLC and IEX.

Optimization of method variables are organized in different modules of the software. The results depend on the information that a user enters into the software (Table 3.1).

ChromSword® can work with massive amounts of data. One sample file can contain up to 100 compounds including structural formulae and the data for up to 20 runs in the each module.

Table 3.1: Input/output of ChromSword® in the off-line mode.

Minimal input	Expected output
Structural formulae are considered	
Structural formulae (up to 100 in a file)	Starting conditions for RPLC: column type, eluent.
Structural formulae and data of one run	Optimal eluent for separation of a mixture in isocratic RPLC on a column being used. Optimal gradient profile. Starting conditions of RPLC for other column types and an eluent.
Structural formulae are not considered	
Data of two runs with different concentrations of an organic solvent in a MP (RPLC)	Optimal eluent for separation of a mixture in isocratic RPLC on a column being used. Starting conditions of RPLC for other column types and an eluent. Evaluation of the analyte parameters (molecular volume, polarity).
Data of two runs with different concentrations of an organic solvent or a buffer in a MP	Optimal eluent for separation of a mixture in isocratic RPLC, NPLC and IEX. Optimal gradient profile.
Data of two runs with different gradient profiles	Optimal gradient profile for separation of a mixture in gradient HPLC. Optimal eluent for separation of a mixture in isocratic HPLC.
Data of two runs with different temperatures of a column	Optimal temperature for separation of a mixture in isocratic HPLC. Enthalpy sorption of analytes.
Data of two runs with different pH of a MP	Optimal pH for separation of a mixture in isocratic RPLC. Optimal pH for separation of a mixture in isocratic RPLC.
Data of three runs with different pH of a MP	Nature of analytes (base, acid, neutral). pK value of analytes.
Two variable optimizations	
Data of three and four runs with different concentrations, pH, temperatures, columns, solvents, gradient profiles	Optimal gradient profile and temperature; concentration and pH; concentration and temperature; pH and temperature; concentration of two different organic solvents; optimal connection of two columns with different selectivity and concentration, gradient profile, pH or temperature.

3.3.1 *Concepts and procedures for developing HPLC methods*

The central idea of the computer-assisted method development is to input information about the mixture to be separated and then to apply a computer simulation to predict results for different chromatographic conditions, thus finding the acceptable conditions for separating the mixture.

One of the options is to use structural formulae as input for a computer program and to predict acceptable chromatographic conditions by analyzing information concerning their structures. It is an easy way for the user, but it is one of the most complicated problems in chromatographic science to predict acceptable conditions from a chemical structure. A much less complicated problem is to predict the results of chromatographic experiments by analyzing the results of several experiments previously performed.

It is understandable that the less information a computer program receives, the less precise the prediction that is obtained. If the input is only the structural formulae of compounds, the level of predictability is much less than that we would have after entering the results of several chromatographic experiments and their conditions. On the other hand, the fewer experimental results the computer program requires to produce acceptable prediction, the less time we have to spend developing the method.

It is hard to obtain an exact prediction of the retention time values from the structural formulae. The task of working with structural formulae is not to enable the precise prediction of retention in the first-guess experiment but to predict the concentration of an organic solvent in a MP (or a gradient profile) for acceptable retention to be obtained. Successful prediction of the concentration or the gradient profile will save time and the amount of solvent used in the experimental work. From a practical point of view, it is not important at this stage to predict the retention factor values precisely. The most important issue is to obtain these values within the acceptable practical limits of 1–20.

A practically reasonable approach is to start method development with only the information about structure, to receive the first prediction of chromatographic conditions (the first-guess method), to inject the sample and then to use experimental retention results for correcting the first-guess prediction. In this case, a good chance exists to find acceptable conditions within a minimal amount of time. However, in many cases, a chromatographer has no information about compounds in a mixture or the structure parameters are not known. This situation is typical for developing stability indicating methods, reaction monitoring, separation of bio-mixtures and large molecules. In this case, it is necessary to obtain retention times for two or more experiments and then start computer experiments.

3.3.2 *Retention models*

The retention model in ChromSword® is defined as a type of a mathematical equation which describes the relationship between the retention of a compound and its properties as well as the conditions appertaining to the chromatographic experiments.

It is the focal point in method development software to determine retention models that adequately describe the effect of chromatographic conditions on the retention of compounds in a sample. In this case, based on only a few experiments, the software can predict the results of many other experiments under different conditions, thus allowing a chromatographer to simulate experiments with a computer and find the conditions for acceptable or best separation.

ChromSword® supports two approaches for the determination of retention models in RPLC. These are as follows:

(A) A traditional formal approach which applies linear, quadratic, cubic or other polynomial models for describing the relationship between the retention of solutes and the concentration of an organic solvent in a MP:

$$\ln k = a + b(C) \tag{1}$$

$$\ln k = a + b(C) + d(C)^2 \tag{2}$$

$$\ln k = a + b(C) + d(C)^2 + e(C)^3 \tag{3}$$

where k is the retention factor of a compound, C is the concentration of an organic solvent in a MP and a, b, d, and e are parameters of equations that must be determined by the software for each compound from the retention data obtained by using different concentrations of an organic solvent in a MP.

The simplest is the first linear model, which is known as the LSS model. It requires two initial experiments to start the optimization, but sometimes it does not completely predict correctly the effect of concentration of an organic solvent in a MP. This can be observed for basic and acidic compounds that contain highly polar and charged structural fragments. Such fragments are typically observed in natural and pharmaceutical compounds, and retention models for such compounds are nonlinear in many cases. Additional experiments as a rule do not lead to improvement in the accuracy of the linear model when it is applied for nonlinear functions.

The quadratic model describes retention more adequately. Additional experiments improve the accuracy, but three initial experiments are required to start computer optimization. The higher the power of a model, the more complex retention behavior can be described and the more initial experiments must be performed to start optimization of separation.

ChromSword® supports optimizing separation for polynomial models up to power 6. A chromatographer optionally can choose from powers 1 to 6. Typically, the powers 1–3 are most commonly used; however, the most complex retention can be described and separation optimized with the higher polynomial powers.

All polynomial models predict the retention of solutes rather precisely in the interpolation region of those concentrations studied. These models are less reliable in the extrapolation region. For example, if experiments were performed with 40% and 50% of the organic solvent in a MP, one can expect rather a good prediction of retention and separation in the region between of these concentrations and less accuracy in the regions of 30–35% and 50–55%. Extrapolation within wider limits very often leads to substantial deviations between predicted and experimental data.

(B) An approach that takes into account both the features of solutes being separated and the characteristics of the stationary and MPs being used:

In this method, the two-layer continuum solvatic retention model was proposed [14, 15] as an extension of the solvophobic model of RPLC [16]:

- The surface of a modifier sorbent in RPLC has a surface layer that involves hydrocarbon radicals and some of the components of a MP.

- The surface layers are assumed as being quasi-liquid having their own physical characteristics i.e. surface tension and dielectric permittivity.
- The surface characteristics vary with varying the MP composition and SP properties.
- Molecules of retained substances penetrate into the surface layer.
- The retention is determined by the difference in molecule solvation energies in the mobile and SPs.

In this model, the retention of a solute is derived as

$$\ln k = a(V)^{2/3} + b(\Delta G) + c \tag{4}$$

where V is the molecular volume of a solute, ΔG is the energy of interaction of a solute with water, and a, b and c are the parameters which are determined by the characteristics of a reversed-phase column in the eluent being used, i.e. surface tension, dielectric permittivity and others. This approach works more precisely and rapidly than that based on formal linear and quadratic polynomial models, but it requires that both the parameters of the solutes (volume and energy of interaction with water) and the characteristics of the reversed-phase column under experimental conditions be known.

The characteristics of different commercially available RPLC columns were experimentally determined initially in a wide range of concentrations of methanol and acetonitrile in water. ChromSword® contains a database of characteristics for more than 150 commercially available reversed-phase columns in these eluents; they load automatically when a column and an eluent are chosen from the software menu.

ChromSword® calculates the parameters of compounds from the structural formulae. If structural formulae of the compounds being studied is not known or a user decides not to draw them, these parameters can be determined by ChromSword® from the two chromatographic experiments with different concentrations of an organic solvent in a MP.

This approach enables ChromSword® to predict regular or irregular retention behavior of solutes separated and enables a chromatographer to move rapidly to achieve maximal separation in minimal time. Each additional experiment leads to an improvement in the predictability.

Thus, this approach enables a chromatographer to start optimizing retention without any preliminary tests if the structural formulae of the compounds are known and also enables one to start optimization of separation on entering the retention data for only one run.

For solutes with unknown or undefined structures, this approach can also be used after entering the retention data and chromatographic conditions for two runs.

The main advantage of the structure and column properties related approach is that it "fills" both a column and compound features. It works precisely in the interpolation region and reliably in the extrapolation region. Figure 3.12 and Table 3.2 show that the solvatic model provides a good enough prediction of retention behavior for highly polar compounds that contain both uncharged and charged highly polar fragments.

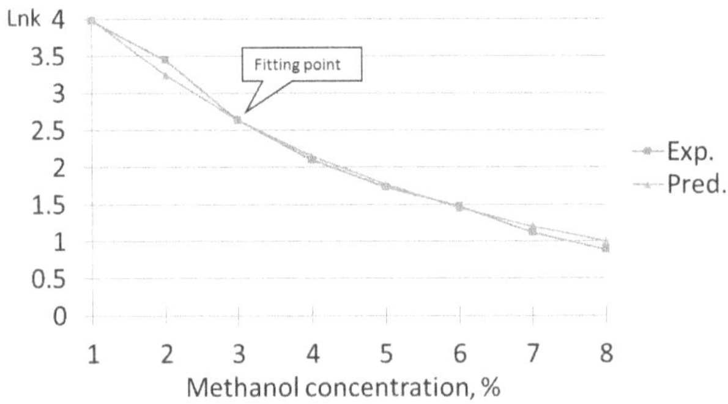

Figure 3.12: Adenosine monophosphate: predicted and experimental retention. Input: structure and data of one run at 3% MeOH. Column: Purospher RP-18e, 5 μm. MP: MeOH – phosphate buffer, pH = 2.5.

Table 3.2: Predicted and experimental retention of the beta-blocker carazolole in the extrapolated region of concentration of MeOH in a MP.

MeOH (%)	k_{exp}	K_{linear}	Dev (%)	$K_{quadratic}$	Dev (%)	$k_{Solvatic}$	Dev (%)
60	4.62	4.62		4.62			4.62
50	6.33	6.33		6.33			6.33
45	8.83	7.71	−12.7	8.83		8.26	−0.33
30	33.57	19.70	−41.3	38.74	15.4	31.90	−4.97

Notes: Retention values at 60% and 50% were used as input for the linear and solvatic models and at 60, 50 and 45% for the quadratic model. Column: Purospher RP 18e, 5 μm, 150 × 4 mm. MP: MeOH – 50 mM phosphate buffer, pH = 3.5.

3.3.3 *Procedure for optimizing pH in RPLC*

When a sample contains basic or acidic compounds with ionizable atoms or groups, pH is a very effective tool for optimizing the separation. ChromSword® supports two mathematical procedures for optimizing pH in RPLC. The first procedure is based on applying polynomials with powers up to 6 and the second procedure determines, using the retention data obtained with different pH values of a MP, the nature of solutes (neutral, acidic, basic), their p*Ka* value and then builds their retention models.

3.3.3.1 *Polynomial models*

The first three members are:

$$\ln k = a + b(pH) \tag{5}$$

$$\ln k = a + b(pH) + d(pH)^2 \tag{6}$$

$$\ln k = a + b(pH) + d(pH)^2 + e(pH)^3 \tag{7}$$

The powers 4–6 optionally can be employed for describing the most complex dependencies between retention and pH value of a mobile phase.

In order to optimize pH, a user must enter experimental retention data for two or more isocratic or gradient runs with different pH value of a MP. By analyzing retention data, ChromSword® determines and then refines the parameters of the retention model for the column being used and predicts the conditions for the best separation.

Tasks of a user are the same as that for optimizing separation in RPLC using a polynomial model and is described in Chapter 2 "procedure" for method development in HPLC using polynomial models.

3.3.3.2 *Fit pKa optimizing procedure*

This procedure determines, using the retention data obtained with different pH values of a MP, the nature of solutes (neutral, acidic, basic), their p*Ka* values and then builds their retention models:

$$k = k(0) + k(i)/(1 + F) \tag{8}$$

where $k(i)$ is the retention factor of an ionic form of a solute, $k(0)$ is the retention factor of a molecular form of a solute, and F is $Ka/[H^+]$ for acids and $[H^+]/Ka$ for bases, where Ka is the dissociation constant of a solute.

In order to optimize the pH value using the fit p*Ka* procedure, a user must enter experimental retention and efficiency data for three or more isocratic or gradient runs with different pH values of a mobile phase. By analyzing retention data, ChromSword® determines the nature of the compounds (base, acid, neutral) studied at pH intervals, calculates the p*Ka* values and then refines the parameters of the retention models for the column being used (Table 3.3, Fig. 3.13).

Substantial differences can be seen for retention time of basic and acidic compounds predicted by the p*Ka* and quadratic retention models. The p*Ka*-related model typically predicts retention for acidic and basic compounds better (Table 3.4).

Deviations in predicted retention can lead to a substantial difference in predicted optimal pH value for separation of a mixture with basic and acidic compounds. In Figs. 3.14 and 3.15, the resolution maps as functions of the quadratic and the fit p*Ka* models are shown for optimization of separation of a mixture of sweeteners and preservatives.

The Fit p*Ka* procedure enables a user to not only optimize the separation but also determine the nature of the compounds and evaluate their p*Ka*

Table 3.3: The p*Ka*-related model parameters determined for mixtures of nucleobases and nucleosides.

	Compound	Nature	k_0	k_i	*pKa*
1	Uracil	Neutral	1.12		
2	Cytosine	Base	0.78	0.51	5.63
3	Thymine	Neutral	3.77		
4	Uridine (U)	Neutral	3.10		
5	Cytidine (C)	Base	2.06	1.34	4.45
6	Ara-U	Neutral	4.43		
7	Ara-C	Base	2.68	1.68	4.17
8	6-azauridine	Acid	1.54	1.20	5.62
9	6-azacytidine	Neutral	0.98		
10	5-azacytidine	Base	2.21	1.47	4.04

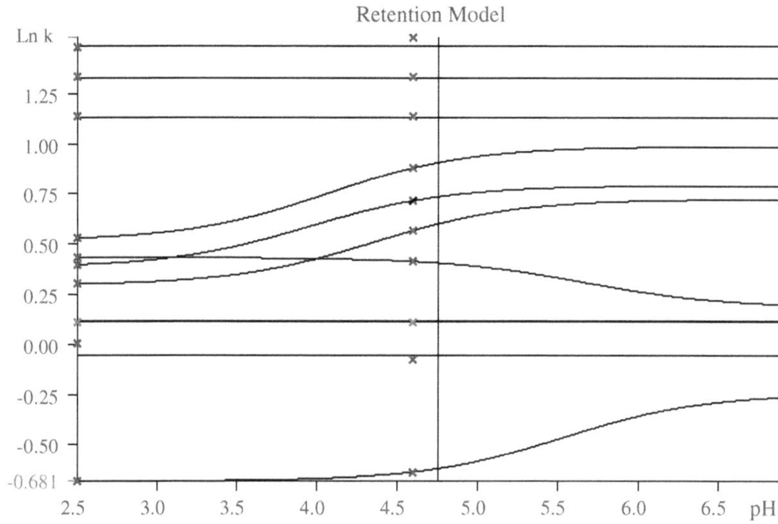

Figure 3.13: Retention models ($\ln k = f(pH)$) built with the Fit p*Ka* procedure for the compounds listed in Table 3.3 Column: Purospher RP18e, 5 μm, 125 × 4 mm. MP: 20 mM phosphate buffer pH = 2.5; 4.6, 7.0. Flow rate 0.8 mL/min, $T = 35$°C.

Table 3.4: Predicted retention time with the quadratic (RT_q) and p*Ka*-related (RT_{pK}) model. RT_e — experimental values.

	Compound	RT_q	RT_{pK}	RT_e	p*Ka*
1	Sorbic acid	7.54	10.00	10.00	4.67
2	Benzoic acid	4.76	5.41	5.37	4.19
3	Acesulfame	2.63	2.61	2.64	
4	Saccharine	3.41	3.45	3.43	
5	Aspartame	14.18	14.41	14.35	
6	Caffeine	7.07	7.06	7.08	

values under the conditions of a chromatographic experiment. In Tables 3.3 and 3.4, the p*Ka* values calculated from the experimental data are listed. It should be noted that the chromatographic method for the determination of p*Ka* values has advantages over other methods because it can be applied for mixtures and requires only a small amount of compounds.

It is necessary to take into account that the fit p*Ka* procedure assumes solutes to be monoprotic; therefore, for diprotic (and more) solutes as well as for zwitterions, p*Ka* values can be considered as conditional.

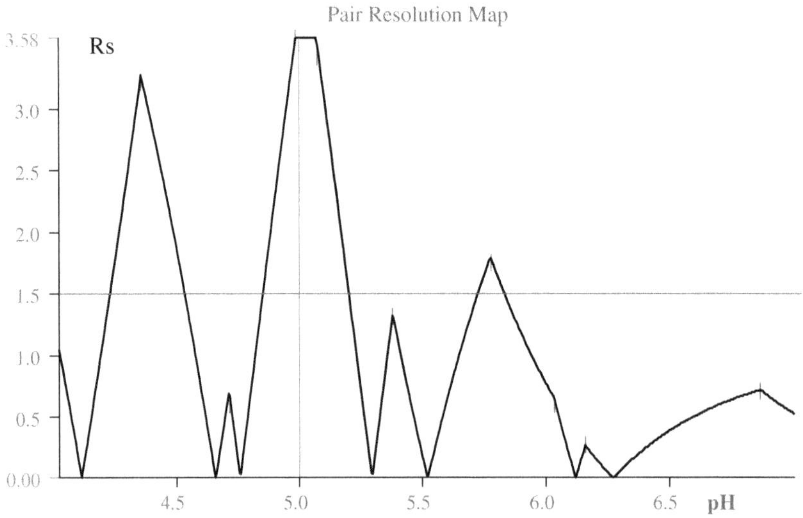

Figure 3.14: Resolution map built with the Fit p*Ka* procedure. Separation of the caffeine, acesulfame, saccharine and benzoic and sorbic acids. Column: Purospher RP18e, 5 μm, 125 × 4 mm. MP: 10% ACN/90% 20 mM phosphate buffer, pH = 7.01; 4.02, 5.75. Flow rate = 0.8 mL/min, T = 30°C.

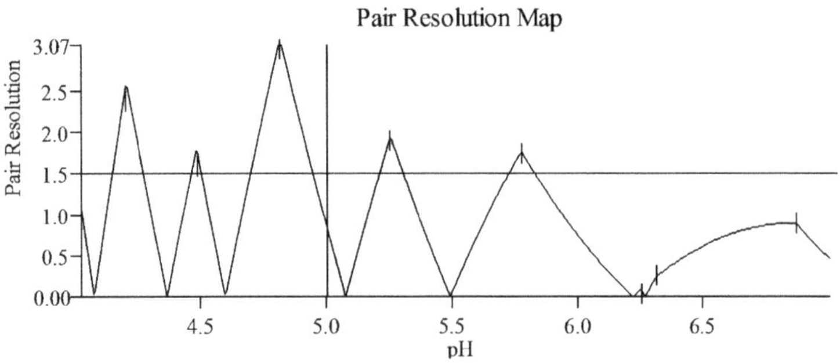

Figure 3.15: Resolution map built with the quadratic model. Conditions and mixture as described for Fig. 3.14.

Nevertheless, this procedure can give valuable information about unknown compounds.

3.3.4 *Optimization of NPLC methods*

For optimization of the separation in NPLC, ChromSword® now supports only polynomial retention models. Retention in the NPLC can be described rather adequately by bilogarithmic models. ChromSword® supports polynomials up to a power of 6. The first three are the following:

$$\ln k = a + b(\ln C) \tag{9}$$

$$\ln k = a + b(\ln C) + d(\ln C)^2 \tag{10}$$

$$\ln k = a + b(\ln C) + d(\ln C)^2 + e(\ln C)^3 \tag{11}$$

where C is the concentration of the stronger solvent in the mobile phase.

The powers 4–6 can be employed for describing the most complex dependencies between retention and concentration of a modifier in a MP.

In order to optimize a separation in NPLC, it is necessary to enter experimental retention and efficiency data for two or more runs with different concentrations of a strong solvent in the MP. By analyzing the retention data, ChromSword® determines and then refines the parameters of the retention model for a column being used and predicts the conditions for the best separation.

User tasks are the same as for optimizing separation in RPLC by using polynomial model and described in Chapter 2 "*procedure*" for method development in HPLC using polynomial models.

3.3.5 *Optimization of IEX methods*

The effect of the buffer concentration in the MP on retention in IEX can be described adequately by the same functions as for NPLC. Thus, a user can utilize the same procedure both for normal-phase and for IEXLC.

In order to optimize a separation in IEXLC, the user must enter experimental retention and efficiency data for two or more isocratic or gradient runs with different concentrations of a counter-ion in the MP. By analyzing the retention data, ChromSword® determines and then refines the parameters of the retention model elution for the column being used and predicts the conditions for the best separation.

3.3.6 *Optimization of the temperature*

Optimizing the temperature can be an effective tool if the conformation of solutes changes with temperature. This phenomenon can be observed rather often in the case of large molecules such as peptides, proteins or for molecules with bulky substituents. In general, the effect of temperature on the logarithmic retention factor can be described by the simple equation $\ln k = a + b(1/T)$ for any mode of chromatography including gas chromatography. But if a solute changes its conformation, the function $\ln k = f(1/T)$ can be much more complex.

To optimize the temperature of a chromatographic separation, ChromSword® uses up to six power polynomials. The first three are the following:

$$\ln k = a + b(1/T) \tag{12}$$

$$\ln k = a + b(1/T) + d(1/T)^2 \tag{13}$$

$$\ln k = a + b(1/T) + d(1/T)^2 + e(1/T)^3 \tag{14}$$

where T is the temperature of the MP.

For optimizing the temperature, the same procedure as for optimizing the concentration of a modifier in RPLC, NPLC and IEX can be used.

In order to optimize a separation, the user must enter experimental retention and efficiency data for two or more runs with different temperatures of the MP. By analyzing the retention data, ChromSword® determines and then refines the parameters of the retention model elution for the column being used and predicts the conditions for the best separation.

If the model with the power one is applied, then ChromSword® also determines the enthalpy of sorption from the retention model:

$$\ln k = \ln k_0 + \Delta H/(RT) \tag{15}$$

where ΔH is the enthalpy of sorption of a solute in kJ/mol and R is the universal gas constant.

Thus, ChromSword® can be applied not only for optimizing a separation but for physico-chemical studies of compounds. For unknown compounds, ΔH values can be useful for elucidation of their structure.

3.3.7 *Optimization of the gradient*

There are different approaches to optimize gradient profiles after the determination of the retention models. The most frequently used approach is the optimization of linear gradient profiles when two runs with linear gradients and different gradient times are used as input. These runs are used to build retention models. The initial and final concentrations of a modifier are fixed both for input and optimization. In this case, only the gradient time is optimized. This is simple approach that can easily be combined with the optimization of other variable like the temperature of a MP. However, complex mixtures in many cases can be separated only with multi-step gradient profiles. These include natural samples or samples after force degradation tests in pharmaceutical research and development laboratories. Every gradient node can be characterized by two parameters — time and concentration — and the position of every node in the time and the concentration dimensions should be optimized. Such multi-step gradients can be optimized by simulating chromatograms for different multi-step gradient profiles; however, this is not a fast method.

To build retention models, ChromSword® can process two or more runs with linear or (and) multi-step gradients. In this case, every new run can be used to refine retention models. For the optimization of both linear and multi-segment gradient profiles, the Monte Carlo and genetic algorithms are used. A user needs to enter the parameters of optimization, desired run time, separation and target peaks to be separated, and the stochastic procedure will find the best gradient profile automatically, assuming the separation is possible. The more segments on the gradient profile and compounds in a sample, the more time for optimizing is necessary. Typically, ChromSword® spends only a few minutes with conventional PCs finding the best multi-segmented gradient profile.

3.3.8 *Optimizing two variables simultaneously*

Optimization of two variables is an effective tool for improving and developing HPLC methods. ChromSword® provides all necessary interface and mathematical procedures for optimization of two chromatographic variables simultaneously. The following two variables can be optimized with ChromSword®:

Using one column:

- gradient profile and temperature
- concentration of a modifier in a MP and temperature

- pH and temperature
- concentration of an organic solvent and pH
- concentration of two different organic solvents

Using up to four connected columns with different selectivity (*column coupling, column combination*):

- gradient profile and ratio of columns
- concentration of an organic modifier and column ratio for RPLC
- concentration of an organic modifier and column ratio for NPLC
- pH and column ratio for RPLC
- temperature and column ratio for RPLC and NPLC

3.3.9 *Simultaneous optimization of a gradient profile and temperature*

Gradient and temperature optimization procedure allows the user to predict retention and to optimize the separation in gradient elution by entering retention data and experimental conditions for three or more gradient runs with different slopes and temperature. It is practically useful that the gradient profiles can be both linear and multi-step. One of the possible plans of the experiments can be for a user to perform two linear gradients with different slopes and same temperature and the third linear gradient with a different temperature. The slopes should be substantially different.

- Run 1: 20 min linear gradient with concentration of an organic solvent ranging from 5% to 95% at temperature 30°C.
- Run 2: 40 min linear gradient with concentration of an organic solvent ranging from 5% to 95% at temperature 30°C.
- Run 3: 40 min linear gradient with concentration of an organic solvent ranging from 5% to 95% at temperature 40°C.

The difference in temperature should be, in the majority cases, not less than 10°C between gradients.

When the user inputs data of experimental runs (retention, efficiency, area) and conditions (gradient profiles, temperature, column dead time, the dwell time of the HPLC system), ChromSword® builds retention models and the user can compute simulate experiments with different profiles and temperatures. It is also possible to search for optimal gradient profile and temperature using the automatic procedure. The simplest approach that is used in different method development software is to build resolution maps where the resolution is a function of the gradient time and temperature. In this case, the initial and final gradient time values are fixed and cannot be optimized automatically. The user should change the initial and final MP compositions and observe their impact on the resolution map. This manual procedure takes substantial time, even for simple linear gradients. For example, to study the effect of initial and final concentrations for +/−5% it is necessary to simulate 100 resolution maps for all combinations of the initial and final concentration. For multi-step gradients, the number of computer experiments to simulate the position of every gradient point and their combinations is enormous. Automated optimization procedures that are implemented in ChromSword® have no such limitations and enable a user to optimize simultaneously the initial and final concentrations, gradient time and temperature for linear gradient profiles or the temperature and position of all nodes in multi-step gradient profiles.

When a user finds a promising gradient profile with ChromSword® and performs the run, it is also possible to input the obtained retention data to refine retention models and then repeat the computer simulation and optimization.

3.3.10 *Optimization of separation using supervised machine learning*

In recent years, machine learning-based models have been able to solve problems that previously could be resolved only by experts [17, 18]. Deep machine learning models on limited datasets were applied for the prediction of retention time of peptides in RPLC [11]. In earlier publications, outdated artificial neural network methods were utilized to predict retention time of simple samples and a few linear gradients [19, 20]. None of these contributions attempted at finding multi-step solvent gradient for separation of compounds. We applied machine learning as one of the optimization methods in ChromSword®. The deep machine learning technology was not utilized widely in chromatography. We consider that some information on its possibility in gradient optimization can be interesting for both computer scientists and specialists in computer-assisted method development.

For the deep learning model, we used the recurrent neural network (RNN). An efficient algorithm for RNN is the long short-term memory (LSTM) cells [21]. The LSTM cell in an RNN-based model is a recursive function that uses a set of sub-functions. This function receives input data from the training set for every time step. In our case, the time steps are training runs in a sequence of runs. Then, this function tries to forecast the desired result as an optimal solvent gradient to achieve good separation of compounds. Parameters of sub-functions inside the LSTM cell are trained using modern variations of stochastic gradient descent (SGD) algorithms. It should be noted that the LSTM cannot be applied directly to produce usable method conditions because the resulting value will be in a range from -1 to 1 (*tanh* function). To use LSTM layers, we need to normalize input and output data vector to appropriate scale or to use as a last layer linear regression of deep learning model. The linear regression layer would then produce usable values for concentration in a range between 0 and 100. As for the input part of the LSTM, we use the convolutional neural network (ConvNet) [22] to embed features of scouting runs like data points of the chromatogram, spectra, retention time of compounds, solvent concentration gradient, temperature, etc. A very promising development in machine learning research in recent years has been made in the field of deep reinforcement learning [23]. These algorithms use a model that learns regression task when it tries to forecast the cumulative reward of the whole trajectory of actions to perform a predefined task. It means that we can train a model to generate method conditions for a sequence of runs that will gradually lead to the best separation of compounds. For each run, the quality of the result (reward) can be estimated using the sum of pair resolution values for each peak in a run. The model calculates cumulative reward value for each run in a sequence. Using these rewards, the model learns to construct a gradient, extract knowledge from the acquired chromatogram and then construct the next gradient that will have a higher reward value

$$R = \sum_{t=0}^{n} \gamma^t r_t \tag{16}$$

Cumulative reward R is calculated by summing all rewards r_t of runs multiplied by discount constant γ^t that reduces the importance of future rewards at the present state. We cannot use R value directly to train our model, because it takes into account only executed actions. For example, it calculates rewards from runs with method conditions in a specific sequence, but we would like to construct utility function to train our model to include more possible method conditions. To realize this, we can construct the quality method value (Q)-based model using Bellman's equation Q_π that takes advantage of partial Markov decision process property:

$$Q_\pi(s_t, a_t) \leftarrow Q_\pi(s_t, a_t) + \alpha\Big(r_t + \max_a Q_\pi(s_{t+1}, s_{a+1}) - Q_\pi(s_t, a_t)\Big) \tag{17}$$

State s_t contains retention time, width of peaks, pair resolutions and other important method quality characteristics. State a_t contains proposed a concentration gradient and other method conditions. We try to maximize Q-value that is approximated cumulative reward by changing method conditions.

To train the deep reinforcement learning model, we used physical retention models generated by ChromSword® as a training environment. The retention models were determined from retention behavior of different families of compounds, like small molecules and proteins. Then, a special procedure generated a large dataset of runs and simulated chromatograms for the training. In fact, the pattern of chromatograms as a function of solvent gradients and other conditions like temperature or pH can be used for the training. When beginning the training set, the Q-value model produces random method conditions; however, after training — using distributed computing — it can be applied to new samples. Our results showed that after training with simulated samples, the procedure can process the results of scouting runs of real samples and predict gradient profiles to provide a reasonable separation.

3.3.11 *Column coupling*

ChromSword® provides support in the case of the most complex mixtures when no acceptable conditions were found with several types of columns. In this case, the chromatographer can try to separate a mixture by coupling columns with different selectivities. To optimize separation on coupled columns, it is possible to use data that were obtained separately for different single columns. Typically, columns with 2, 5, 7.5, 10, 15 and 25 cm lengths are commercially available and can be easily combined by using dead volume connectors or column cartridges. In this case, the generic procedure can be applied. This is done as follows:

- Make several runs with different concentrations of a modifier or gradient profiles in column 1.
- Input data of the runs for the column 1 page.
- Build retention models for compounds being separated.
- Build the pair resolution map, search for promising regions and simulate chromatograms.
- If no acceptable conditions are found, a user has choice for the next step:
  - Try an other type of column (columns 2, 3, 4).
  - Try an other solvent and pH or/and temperature with column 1.

If the chromatographer chooses the first option (change a column), it is possible to repeat the same steps 1–4 to try to optimize the concentration of a modifier in the MP or the gradient profile with that of the column 2. The other conditions must be the same as used for column 1 (solvent type, temperature, pH). If no good separation was found with column 2, the user can perform a computer simulation on:

- Coupling of columns 1 and 2 (a maximum of four columns can be virtually coupled) and optimizing the ratio of column lengths or column segments.
- Effect of the concentration of organic modifiers or the gradient profile on the separation for coupled columns.

The same procedure can be used for optimizations of pH or temperature and column coupling simultaneously.

3.4 Conclusions

ChromSwordAuto® is a software package which includes a chromatography method development data system and ChromSword® module for off-line computer-assisted method development.

ChromSwordAuto® is used for automatic method development of small and large molecules and supports mechanistic and statistic approaches for the optimization of method variables. ChromSwordAuto® also contains a module for high-throughput screening of many SP and MP combinations.

ChromSword® and ChromSwordAuto® are used for method development and optimization in practically all types of LC.

References

[1] S.V. Galushko, A.A. Kamenchuk, G.L. Pit, The calculation of retention and selectivity in RP LC. IV. Software for selection of initial conditions and for simulating chromatographic behaviour, *J. Chromatogr.* 660 (1994) 47–59.

[2] S.V. Galushko, V. Tanchuk, I. Shishkina, O. Pylypchenko, W.D. Beinert, ChromSword® software for automated and computer-assisted development of HPLC methods, In *HPLC Made to Measure: A Practical Handbook for Optimization*, ed. Stavros Kromidas, WILEY-VCH Verlag GmbH & Co. KgaA, 2006, pp. 557–570.

[3] Industry Analytical Procedures and Methods Validation for Drugs and Biologics. https://www.fda.gov/downloads/drugs/guidances/ucm386366.pdf.

[4] E. Hewitt, P. Lukulay, Implementation of a rapid and automated high performance liquid chromatography method development strategy for pharmaceutical drug candidates, *J. Chromatogr. A* 1107 (2006) 79–87.

[5] K.P. Xiao, Y. Xiong, F.Z. Liu, A.M. Rustum, Efficient method development strategy for challenging separation of pharmaceutical molecules using advanced chromatographic technologies, *J. Chromatogr. A* 1163 (2007) 145–156.

[6] S. Larson, G. Gunawardana, M. Preigh, Automated method development in HPLC. Evaluation of the ChromSword® software package. HPLC 2007 Abstract book, P23.06. http://www.chromatographyonline.com/efficient-chiral-hplc-method-development-using-chromsword-software.

[7] F. Vogel, S.V. Galushko, Automated development of reversed-phase HPLC methods for separation of chiral compounds, *Chromatogr. Today* 8 (2015) 54–55.

[8] L.W. Snyder J.W. Dolan, *High Performance Gradient Elution*, John Wiley & Sons, Inc., Hoboken, New Jersey, 2007, p. 228.

[9] S. Fekete, S. Rudaz, J. Fekete, D. Guillarme, Analysis of recombinant monoclonal antibodies by RPLC: Toward a generic method development approach, *J. Pharm. Biomed. Anal.* 70 (2012) 158–168.

[10] S. Fekete, A. Beck, J. Fekete, D. Guillarme, Method development for the separation of monoclonal antibodycharge variants in cation exchange chromatography, Part I: Saltgradient approach, *J. Pharm. Biomed. Anal.* 102 (2015) 33–44.

[11] J. Koyama, J. Nomura, Y. Shiojima, Y. Ohtsu, I. Horii, Effect of column length and elution mechanism on the separation of proteins by reversed-phase high performance liquid chromatography, *J. Chromatogr.* 625 (1992) 217–222.

[12] ICH Topic Q 2 (R1) "Validation of Analytical Procedures".

[13] http://www.smatrix.com/products.html.

[14] S.V. Galushko, The calculation of retention and selectivity in RPLC, *J. Chromatogr.* 552 (1991) 91–102.

[15] S.V. Galushko, The calculation of retention and selectivity in RPLC. II. Methanol-water eluents, *Chromatographia* 36 (1993) 39–41.

[16] Cs. Horvath, W. Melander I. Molnár, Solvophobic interaction in liquid chromatography with nonpolar stationary phases, *J. Chromatogr.* 125 (1976) 129–140.

[17] M. Ren, R. Kiros, R.S. Zemel, Exploring models and data for image question answering, arXiv:1505.02074.

[18] J. Donahue, L.A. Hendricks, M. Rohrbach, S. Venugopalan, S. Guadarrama, K. Saenko, T. Darrell, Long-term recurrent convolutional networks for visual recognition and description, arXiv:1411.4389.

[19] N.H. Tran, X. Zhang, L. Xin, B. Shan, M. Li, De novo peptide sequencing by deep learning, *PNAS* 114 (2017) 8247–8252.

[20] T. Bolanča, Š. Cerjan-Stefanović, M. Novč, Application of artificial neural network and multiple linear regression retention models for optimization of separation in ion chromatography by using several criteria functions, *Chromatographia* 61 (2005) 181–187.

[21] H. Wang, W. Liu, Optimization of a high-performance liquid chromatography system by artificial neural networks for separation and determination of antioxidants, *J. Sep. Sci.* 27 (2004) 1189–1194.

[22] Y. Li, Deep reinforcement learning: An Overview, http://arxiv.org/abs/1701.07274.

[23] K. Arulkumaran, M.P. Deisenroth, M. Brundage, A.A. Bharath, A brief survey of deep reinforcement learning, arXiv:1708.05866.

Chapter 4

Intelligent Systems to Predict Retention from Molecular Properties for Reversed-phase HPLC Separations

György Morovján

Egis Pharmaceuticals PLC, Budapest, Hungary

morovjan.gyorgy@egis.hu

4.1 Introduction

Reversed-phase liquid chromatography (RPLC) has become the most widely used liquid chromatographic (LC) method and, as such, a workhorse for pharmaceutical and biomedical as well as environmental analysis. A practical chromatographer is faced with analytical problems involving solutes/analytes of broad structural variety and a limited time frame for method development (and validation). Therefore, the need arises for computational support of chromatographic separation planning in the broad sense, and expert systems are required that assist the chromatographer in analytical method development.

Several possible theories regarding the separation mechanisms in RPLC have been proposed for interpreting chromatographic retention and selectivity [1,2]. In RPLC, the choice of the mobile phase components is usually limited to water and some solvents, and the surface properties of the hydrophobic stationary phase dominate the interactions resulting in chromatographic resolution. One possible way of explaining the interaction between the stationary phase and the solute is considering a partitioning process of the solute between the stationary and the mobile phases [1].

A model system for predicting the properties of chemical compounds in biological systems, especially in an interaction with biological membranes with respect of lipophilicity, has been developed, based on the determination of the aliphatic alcohol–water partition coefficient of the solute [3]. After selection of 1-octanol as the most useful lipophil solvent in partition studies, a large body of experimental data has become available from physicochemical studies as well as pharmacological and environmental research for the 1-octanol–water partitioning system (K_{ow} and log K_{ow}) for compounds varying in chemical structure [3, 4].

Collander has postulated that the logarithm of the partition coefficient (P, for the system of 1-octanol and water, P_{ow}) is related approximately linearly to the logarithm of partition coefficient in a similar system, assuming that the mechanisms of the solute–solvent interaction are similar in the two systems [3].

Featuring the retention mechanism in an RPLC system according to the partitioning theory may be seen as remarkably similar to the partitioning between the components of a solute in a system of water-immiscible, hydrophobic solvent and water (or water-miscible solvent), despite of the fact that the partitioning theory does not address the factors outside of the scope of the partitioning process, e.g. ionization equilibria, ion-pair formation, silanophil interaction (in case of some silica-based stationary phases), complex formation, hindered diffusion or inadequate wetting of the stationary phase, etc.

Considering the analogy in the partitioning processes and assuming that a Collander-type relationship exists, a linear relationship could be presumed between the log P_{ow} value and the logarithm of the retention factor k [5]. Assuming a partitioning system, e.g. the most-used 1-octanol–water system, and a specific RPLC system, the knowledge of the retention factor would allow for the prediction of the log P_{ow} and vice versa.

One application of this relationship was the method established by the Organization for Economic Co-operation and Development (OECD) Guideline 117 for the determination of log P_{ow} values by RPLC [6]. This method allowed for the determination of log P_{ow} values in the range of 0–6, complementary to other OECD testing methods with different log P_{ow} ranges.

In RPLC method development, however, the problem that the chromatographer is faced with resides in defining an RPLC system in terms of the stationary phase, mobile phase composition, detection and, optionally, operational temperature, which provides the required chromatographic resolution of the analytes in the shortest possible time, with predefined system suitability features. One important issue to be resolved during method development is the development of the mobile phase composition with respect to a given stationary phase, providing resolution in real time, which could further be optimized [2].

Working on the definition of an initial, first-guess RPLC system, especially in terms of mobile phase composition, in most cases, the chromatographer has access to the chemical structure of the analytes. Chemical structure bears information which is *prima facie* important for the estimation for the mobile phase composition, e.g. the presence of grossly hydrophobic groups or ionizable groups. Most chromatographers proceed by almost instinctly recognizing these molecular features and thus formulate a first-guess mobile phase based on experience, and then addressing secondary issues, among which ionization of the analyte is the most important. Therefore, the two principal molecular descriptors relevant for RPLC method development are related to the ionization equilibria and hydrophobicity. Alternatively, a less desirable trial-and-error approach may be followed.

Several studies demonstrated that besides the possibility of measuring the $\log P_{ow}$ parameter, it is possible to estimate them as a linear combination of molecular fragments and their interaction [7–9]. Furthermore, it has been shown that ionization constants (pKa) of acids and bases can also be estimated by linear combination of the pKa value of a basic structure and sum of contribution of the substituents and related reaction constants [10]. pKa value for bases is understood as the dissociation constant of the conjugated acid (protonated base).

The recognition of the relationship between $\log k - \log P_{ow}$, methods for predicting the volume fraction of the organic modifier on such basis [11–14], combined with the availability of methods for estimating $\log P_{ow}$ and pKa values allowed for the construction of an expert system embodying the expert knowledge and skill and provides molecular descriptor-based advice for RPLC method development [15].

4.2 EluEx Software

The EluEx software (CompuDrug Chemistry Ltd., Budapest, Hungary) has been developed as an expert system to assist chromatographers in RPLC mobile phase optimization for achieving resolution of the solutes defined by their chemical structure. The software, also available alone, forms part of the suite of physical–chemical and metabolic property estimation suite Pallas developed by CompuDrug, consisting of (Pro)Log P, pKalc, MetabolExpert and AgroMetabolExpert (the last two software are intended for drawing up metabolic pathways and so are not discussed here). Although the modules of PrologP and pKalc are available separately for property estimation, the functions of these modules are used by the EluEx system, supplemented with specific modules for optimization of separation. EluEx has been written in C and C++ languages.

Besides the modules responsible for prediction of log P_{ow} and pKa values on the basis of structural formulae, EluEx further comprises a module for program control, the initial step module for assessing first-guess mobile phase composition, an optimization module for governing the optimization phase using experimental data and a simulation module for generating a simulated chromatogram on the basis of retention data obtained with different mobile phase compositions and preset system parameters (plate number).

4.2.1 *Setting of basic operational parameters*

This function allows the operator to define the operational parameters, such as plate number, preferred retention factor range, choice of organic modifier, choice of ion-pair reagent, operational pH range (depending on column chemistry and stability of the solutes), and expected minimal resolution.

4.2.2 *Estimating log P_{ow} and pKa based on chemical structure*

The input of the software is a graphical interface that allows for drawing the chemical structure. Thereafter, the structure is analyzed and log P_{ow} and pKa values are calculated. Alternatively, the compound can either be selected from the database or imported as a text file in Molfile representation. Several compounds to be separated can be selected to comprise a set of analytes.

The $\log P$ prediction is based on the approach developed by Rekker and De Kort [7], further developed by CompuDrug. During structure analysis, the formulae of the analytes are fragmented into groups and interactions occurring between these groups. The contribution of each fragment and each interaction are calculated as a linear combination of the contributions of the fragments and interactions, respectively, multiplied by their incidence in the formula. This approach assumes that molecular fragments and interactions result in additive changes in free energy [15]. Contributions of fragments and interactions between fragments are stored in a database.

p*Ka* values are predicted using the Hammett equation for aromatic acids and bases and the Taft equation for aliphatic and alicyclic acids and bases [10], taking into account both electronic and steric effects, i.e. in case of aromatic compounds, substitution is considered. In the case of condensed aromatic systems, the Dewar–Grisdale method is applied [10].

4.2.3 *Selection rules for determining the mobile phase pH*

EluEx utilizes the concept of suppression of ionization of ionizable solutes where feasible within the operating pH range. Since $\log P_{\text{ow}}$ prediction in this software is possible for non-ionized (neutral) forms of ionizable compounds only, in order to make use of suppression of ionization, the p*Ka* calculations must actually precede the calculation of $\log P_{\text{ow}}$ values, and then a pH suitable for suppression is derived from the p*Ka* value. If $\text{pH} > \text{p}Ka - 2$ for an acid or $\text{pH} > \text{p}Ka + 2$ of a base, then ionization is almost completely suppressed. Although the p*Ka* variation caused by the addition of the organic modifier and the ionic strength of the buffer may also be theoretically accounted for [2, 15], the change in p*Ka* values in mixed aqueous–organic medium compared to the same in purely aqueous medium is empirically corrected by increasing the p*Ka* value of acids by a value depending on the value of the volume fraction of the organic modifier and decreasing the p*Ka* value of bases similarly [15]. All pH values should lie within the operational pH range preset by the user (pHmax, pHmin). The selection of mobile phase pH is based on the lowest corrected acidic p*Ka* and the highest corrected basic p*Ka* values and is predicted for the compounds present in the sample according to the following rules:

(a) In the case when neither acidic nor basic ionizable compounds are detected in the sample, no buffer (or pH adjustment) is proposed;
(b) In the case when neutral and weakly acidic compounds are detected in the sample, the mobile phase pH is set to 2 units less than the lowest corrected acidic p*Ka* value, thus effectively suppressing ionization;
(c) In the case when neutral and weakly basic compounds are detected in the sample, the operating pH is set to 2 units higher than the highest corrected basic p*Ka* value, thus suppressing ionization of weak bases. In cases when the highest corrected p*Ka* value is higher than 4, the addition of a silanol masking agent (e.g. triethylamine) is suggested;
(d) In case of the presence of neutral and strongly acidic analytes for which full suppression of ionization can not be achieved in the operational pH range and in the absence of basic compounds, the pH is suggested to be set to the highest from all p*Ka* values of the strong acids. A basic ion-pair reagent is proposed. Dissociation of other weakly acidic compounds is suppressed;
(e) In case the sample comprises neutral and strongly basic compounds for which the full suppression of ionization can not be achieved in the operational pH range, pH is proposed to be set to the lowest p*Ka* value among the bases. An acidic ion-pair reagent is proposed. Ionization of weak bases will be suppressed;
(f) In case the sample comprises weak acids and weak bases besides neutral compounds, the pH is set to the average of highest acidic p*Ka* and lowest basic p*Ka* to suppress ionization. Suppression of ionization, however, may not be complete if the difference between the highest acidic p*Ka* and lowest basic p*Ka* is less than 4 pH units. In case the lowest basic p*Ka* is greater than 4, addition of a masking reagent (e.g. triethylamine) is suggested to prevent silanol effect;
(g) In case of samples comprising strong acids and weak bases, the pH is set to the highest of all p*Ka* values of acids and 2 pH units higher than

the lowest basic p*Ka*, otherwise to the higher end of the operating pH range. A basic ion-pair reagent is proposed, thereby achieving the separation of strong acids by ion-pair chromatography and suppressing the ionization of weak bases. A masking agent may also used;

(h) For samples containing weak acids and strong bases, the pH is set to the lower of the smallest of p*Ka* values of the bases less 2 pH units and the highest of p*Ka* values of acids less 2 units, and the use of acidic ion-pair reagent is proposed. Thereby, the strong bases are separated by ion-pair chromatography, while the ionization of weak acids is suppressed;

(i) The software cannot handle those cases when strong acids and strong bases are present in the sample at the same time. However, such cases are not usually encountered in RPLC and are usually better handled by using different chromatographic methods such as ion chromatography.

4.2.4 *Calculation of initial mobile phase composition*

Calculation of the initial, i.e. first-guess mobile phase composition is carried out by predicting the volume fraction of the organic modifier based on $\log P_{ow}$ value [11, 15] and formulating the mobile phase by optionally adding a buffer of predefined strength and composition (usually a phosphate buffer) having the pH derived from the p*Ka* determination of solutes and applying rules (a)–(i) discussed in Sec. 4.2.3, optionally proposing the use of ion-pair reagent and/or masking reagent.

The first-guess volume fraction of the organic modifier is predicted by averaging the predicted volume fraction to achieve a retention factor of 1 for the least hydrophobic compound in the sample (lowest $\log P_{ow}$) and predicted volume fraction to achieve a retention factor of 5 for the most hydrophobic compound (highest $\log P_{ow}$). In case the difference between the lowest and highest $\log P_{ow}$ values is greater than 5, gradient elution is suggested by the software. In addition, gradient elution is proposed in cases when the isocratic resolution is poor or the elution time is excessive.

Following the initial mobile phase prediction, experimentally determined retention data of the analytes as well as those of the matrix compounds, if any, obtained in a chromatographic run carried out with initial mobile phase composition, are entered into the system. Together with the entry of the retention data, peak symmetry/asymmetry is indicated.

In case one or more peaks are asymmetrical, EluEx proposes pH change and increasing ion-pair reagent concentration or masking agent concentration, depending on the sample type (a)–(i) (see earlier). A further chromatographic run is required to test the effect of the proposed change.

In case the peaks are symmetrical, the volume fraction of the organic modifier is changed to map the function of log *k* vs. the volume fraction of the organic modifier. By this second run, a linear fit can be established, which is later used for calculating the organic volume fraction for optimum resolution and chromatogram simulation.

Figure 4.1 shows the scheme of method development by using EluEx software.

4.2.5 *Isocratic optimization and calculation of resolution*

Calculation of the resolution for closely eluting pairs is performed according to well-established relationships, assuming constant preset plate number. In case the resolution is to be increased or the retention factor is to be decreased, an additional experimental run is proposed and a quadratic model for log *k* vs. organic modifier volume fraction is calculated. On the basis of this model, the resolution map is recalculated. In case the required resolution is achieved within the preset retention factor range, the optimization is ended successfully. Otherwise, gradient optimization is proposed. Simulated chromatograms may be displayed after the second experimental run, allowing the user to visually check the chromatogram to be expected at optimized conditions, especially the resolution of the critical pairs.

4.2.6 *Gradient optimization*

Gradient optimization is based on the methods described in Ref. [16]. Gradient slope can be calculated from the logarithm of the retention factor measured in isocratic runs having organic volume fractions of the initial and final gradient compositions, respectively, and ratio of the void time to the gradient time. Using the isocratic retention factor, gradient slope and

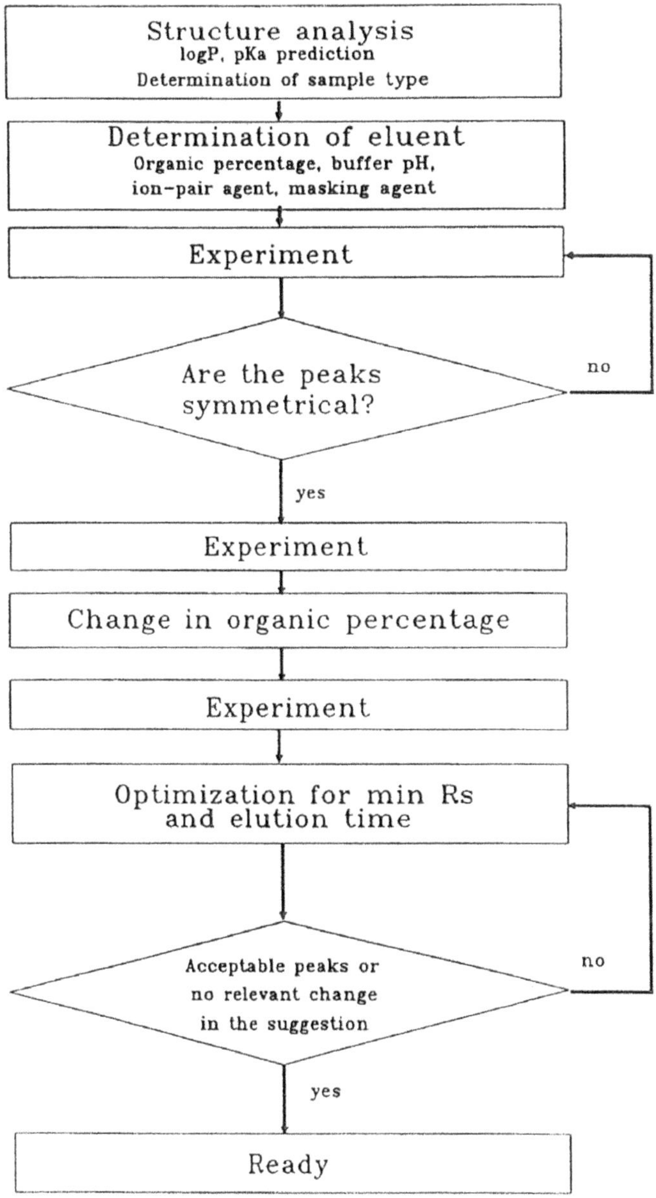

Figure 4.1: Flow scheme of the EluEx program (adapted from Ref. [15], with permission).

void time, the gradient retention time can be calculated. The resolution is calculated similarly to isocratic elution, additionally taking into account that the band dispersion is dependent on band compression factor and gradient slope.

4.2.7 *Applications and advantages*

EluEx has been applied to a vast range of RPLC separation problems and was also tested by simulating RPLC separations documented in the literature.

It has been found that the most important application area of EluEx is initial mobile phase estimation of neutral and weakly acidic or weakly basic analytes. Most biologically active compounds belong to this group. Cases of stronger acids or bases are dealt with ion-pair chromatography in EluEx, which has been less extensively tested. Minimal or no testing could have been done with saccharides, polymers, peptides, proteins and nucleic acids.

As typical examples of application, the analysis of chlorinated phenols in environmental samples and the antibiotic fumagillin in biological matrix is presented [15].

The test compounds 2,4-dichlorophenol, 2,4,6-trichlorophenol and pentachlorophenol have estimated p*Ka* values of 7.9, 6.3 and 4.7, respectively. Their $\log P_{ow}$ values are 2.98, 3.72 and 5.20, respectively. The proposed composition of the initial mobile phase was 79% (v/v) acetonitrile, 50 mM KH_2PO_4 buffer, pH 3.5. Completing the experimental run using a LiChrosorb RP-18, 10 μm column (250 $\times$ 4.6 mm i.d.), at flow rate of 1 mL/min, the retention factors of the solutes were 0.24, 0.40 and 0.80, respectively (using 80% (v/v) acetonitrile). The peaks were symmetrical, but the resolution of 2,4-dichlorophenol and 2,4,6-trichlorophenol was incomplete. The second suggested eluent differed from the first in the volume fraction of the organic modifier, which was 50% (v/v) in the second run. Completing the experiment, the retention factors of the solutes were 1.49, 2.59 and 6.21, respectively. Peak symmetry was appropriate and all analytes were resolved on the baseline. Therefore, the method was found to be acceptable. Figure 4.2 shows the chromatograms obtained with the initial mobile phase and with the second suggested eluent.

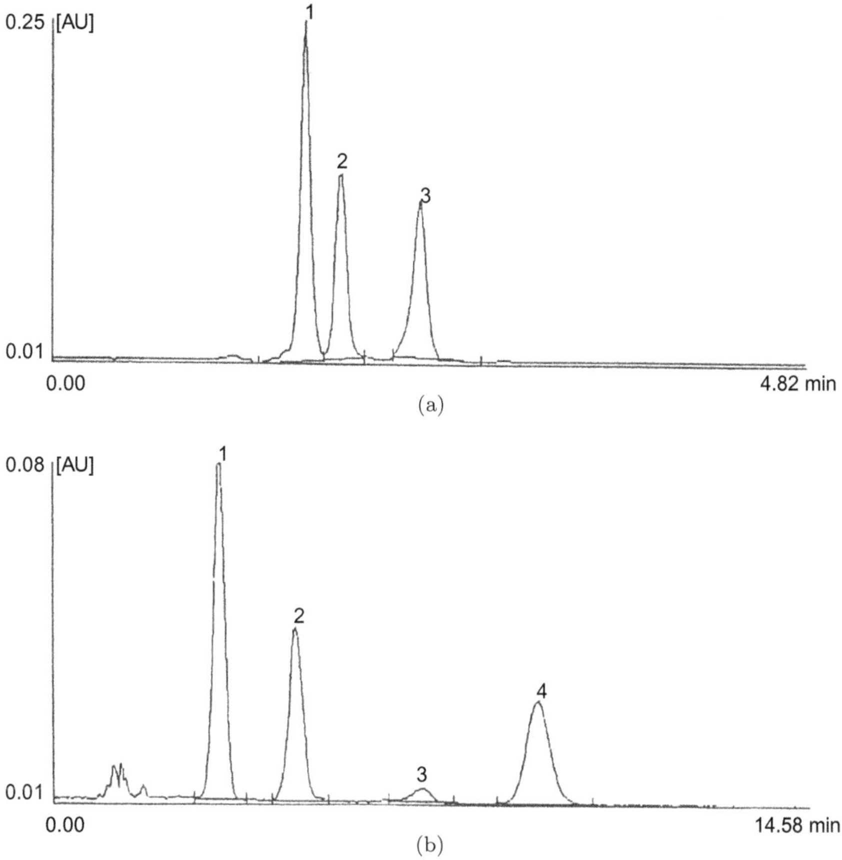

Figure 4.2: Chromatographic separation of chlorophenols (for chromatographic conditions, see text). Upper trace, (a) chromatogram obtained with the initial mobile phase; Lower trace, (b) chromatogram obtained with the second suggested eluent (adapted from Ref. [15], with permission).

The antibiotic fumagillin is a relatively hydrophobic weak acid due to its carboxylic acid moiety, having a p*Ka* of 3.2 and $\log P_{ow}$ of 4.76. The first-guess mobile phase composition comprised 90% (v/v) acetonitrile, 50 mM KH_2PO_4 buffer, pH = 2.1. Using this mobile phase with the same column and flow rate as in the previous example, the retention factor was 0.66. Matrix components interfered with the analysis of fumagillin. In the second chromatographic run, 75% (v/v) acetonitrile was proposed, resulting in a retention factor of 1.29 and still there was no complete resolution of fumagillin. In the third step, using the suggested proportion of the organic modifier, 60% (v/v) acetonitrile, interference from matrix components was eliminated and complete resolution was achieved with a retention factor of 3.45.

4.2.8 *Perspectives for further development and applications*

Despite of the fact that the development of the EluEx software has been terminated, it is worth mentioning some avenues for further development of systems based on molecular descriptors of hydrophobicity and dissociation constant. It appears that these descriptors remain to be applied further in research, and therefore continuing their use in separation science is justified.

The concept of the EluEx system is believed to be a unique approach that is close to chemist's mindset since it treats the analytes in terms of chemical structure and directly conceivable physicochemical descriptors, such as $\log P_{ow}$ and p*Ka*, the effect of which can be directly evaluated both in pharmacological and environmental studies. These properties can be further developed and exploited.

The EluEx software relies on its proprietary databases and algorithms for calculating $\log P_{ow}$ and p*Ka* values. On one hand, there is a vast and ever-increasing body of experimentally determined values that could be directly used for RPLC method development if it were input into the system. On the other hand, new calculation methods have become available since the establishment of the software, which may be valuable alternatives for $\log P_{ow}$ and/or p*Ka* estimations in the current approach [8, 9].

It has been contemplated that there may be no single method that can be applied to solutes of different chemical structure to produce equally good results; there may be a choice of calculation method best fitting the chemical structure. Therefore, comparison of eluent compositions based on different $\log P_{ow}$ and p*Ka* prediction approaches may be a subject of further study for different classes of chemical structures. Calculating molecular descriptors by several different methods may allow for switching between them or assigning a probability range to a descriptor, and initial mobile phase composition based thereon.

Besides applying different approaches for estimating molecular descriptors, the possibility of using chromatographic systems for which the the $\log k - \log P_{ow}$ relationship has been established by analyzing compounds with predetermined molecular descriptors could be devised, thereby reducing the effect of the differences between chromatographic systems. Such a chromatographic system, in turn, would allow the prediction of $\log P_{ow}$ from chromatographic data, similar to the method given in the OECD Guideline 117 [6].

Knowledge of hydrophobic and ionization properties of a solute during the development and optimalization of sample preparation/purification/enrichment by solid-phase extraction using sorbents similar to RPLC stationary phases (e.g. octadecyl-modified silica) is of great importance. The approach of the EluEx software is well applicable for designing such methods, allowing for maximization of the recovery of the solutes of interest and minimizing matrix interference at the same time.

Ion-pair chromatography has been proven to be a valuable separation method for strongly acidic/basic compounds. Recent versions of the EluEx software recognize the possibility of applying the ion-pair reagent. However, there is a need felt in terms of possibilities for optimization of ion-pair reagent concentration (optionally, ion-pair reagent choice) and separation pH by a robust experimental design.

4.3 Conclusions

EluEx software has proved to be a viable approach for determination of initial RPLC mobile phase composition based on molecular descriptors ($\log P_{ow}$, p*Ka*) predicted directly from chemical structure of the analyte, assuming linear relationship between $\log k - \log P_{ow}$. Mobile phase pH is assigned according to the p*Ka* values of the analytes. The software may suggest the use of an ion-pair reagent based on the p*Ka* of the analyte; however, the function for optimizing ion-pair separations needs further development. Tests with this software demonstrated that in most instances, the initial mobile phase composition already provided a good starting point for RPLC method development for neutral or weakly acidic/basic analytes. The software furthermore allows for further optimization and refining of the method. Additionally, the RPLC separation can be optimized and simulated with selected mobile phase compositions.

References

[1] J.G. Dorsey, W.T. Cooper, Retention mechanisms of bonded-phase liquid chromatography, *Anal. Chem.* 66 (1994) 857A–867A.
[2] K. Valkó, L.R. Snyder, J.L. Glajch, Retention in reversed-phase liquid chromatography as a function of mobile phase composition, *J. Chromatogr.* 656 (1993) 501–520.
[3] R. Collander, The partition of organic compounds between higher alcohols and water, *Acta Chemica Scandinavica* 5 (1951) 774–780.
[4] J. Sangster, Octanol-water partition coefficients of simple organic compounds, *J. Phys. Chem. Ref. Data* 18 (1989) 1111–1229.
[5] K. Valko, General approach for the estimation of octanol/water partition coefficient by reversed-phase high-performance liquid chromatography, *J. Liq. Chrom. Rel. Tech.* 7 (1984) 1405–1424.
[6] http://www.oecd-ilibrary.org/environment/test-no-117-partition-coefficient-n-octanol-water-hplc-method_9789264069824-en, accessed: 07 September, 2017.
[7] R.F. Rekker, H.M. de Kort, The hydrophobic fragmental constant; an extension to a 1000 data point set, *Eur. J. Med. Chem.* 14 (1979) 479–488.
[8] R. Mannhold, G.I. Poda, C. Ostermann, I.V. Tetko, Calculation of molecular lipophilicity: State of the art and comparison of log P method on more than 96,000 compounds, *J. Chrom. Sci.* 98 (2009) 861–893.
[9] A. Pyka, M. Babuska, M. Zachariasz, A comparison of theoretical methods of calculation of partition coefficients for selected drugs, *Acta Pol. Pharm.* 63 (2006) 159–167.
[10] D.D. Perrin, B. Dempsey, E.P. Serjeant, *pKa Prediction for Organic Acids and Bases*, Chapman and Hall, London, 1981.
[11] P. Csokán, K. Valkó, F. Darvas, F. Csizmadia, HPLC method development through retention prediction using structural data, *LC-GC* 12 (1994) 40–45.
[12] G. Szepesi, K. Valkó, Prediction of initial high-performance liquid chromatographic conditions for selectivity optimization in pharmaceutical analysis by an expert system approach, *J. Chromatogr.* 550 (1991) 87–100.
[13] K. Valkó, P. Slégel, New chromatographic hydrophobicity index (ϕ_0) based on the slope and the intercept of the log k′ versus organic phase concentration plot, *J. Chromatogr.* 631 (1993) 49–61.
[14] K. Valkó, RP-HPLC retention data for measuring structural similarity of compounds for qsar studies, *J. Liq. Chromatogr.* 10 (1987) 1663–1686.
[15] J. Fekete, Gy. Morovján, F. Csizmadia, F. Darvas, Method development by an expert system advantages and limitations, *J. Chromatogr.* 660 (1994) 33–46.
[16] L.R. Snyder, *High-performance Liquid Chromatography, Advances and Perspectives*, Vol. 1, ed. Cs. Horváth, Academic Press, New York, 1980, p. 207.

Chapter 5

Statistical Methods in Quality by Design Approach to Liquid Chromatography Methods Development

Hermane T. Avohou*, Cédric Hubert*,§, Benjamin Debrus*, Pierre Lebrun†, Serge Rudaz‡, Bruno Boulanger*,† and Philippe Hubert*

**Laboratory of Pharmaceutical Analytical Chemistry, CIRM, Department of Pharmacy, University of Liège, Belgium*

†Arlenda SA, Louvain-la-Neuve, Belgium

‡School of Pharmaceutical Sciences, University of Geneva, University of Lausanne, Switzerland

§chubert@uliege.be

5.1 Introduction

Nowadays, the development of liquid chromatography (LC) methods is still largely performed by the quality by testing (QbT) approach [1] (Fig. 5.1). QbT consists in evaluating the quality (e.g. accuracy and robustness) of an analytical method after its development. Afterward, quality improvement is sought considering additional steps aiming to tune the method parameters (i.e. column chemistry pH of the mobile phase and gradient time) by trial-and-error, mostly based on the prior knowledge of the chromatographer. This generally results in an unstructured search for optimal working conditions and rarely enables an in-depth understanding of the underlying separation processes. The separation process can be defined as the chromatographic behavior of analytes as a function of method parameters [2]. Consequently, efficient optimization of the method, robustness building and quality risks management as required by regulations [3–6] are hardly achieved [1]. Such a level of method knowledge can be efficiently achieved

Figure 5.1: Comparison between the Quality by Testing (a) and the Quality by Design approaches (b) [2].

through an analytical quality by design (AQbD) approach, which is an adaptation of the quality by design (QbD) approach of process development to analytical methods development (Fig. 5.1) [1, 7]. Briefly, AQbD is a systematic and risk-based approach to method development and optimization that begins with predefined objectives, seeks method understanding and defines a method control strategy based on scientific knowledge and quality risk management. A key output of the AQbD strategy is the design space (DS), which defines an envelope of operable region of method parameters that guarantees acceptable method performances [7].

Statistical methods play a prominent role in the QbD development of LC methods. They are the foundation of the learning process, risk management and validation steps [8]. Therefore, any statistical method that is meant to support a LC-QbD method development should not only be statistically correct but also QbD-compliant. To be concrete, this means the statistical method should: (1) enable a deep understanding of the LC method, that is how important method parameters and uncertainty factors combine to affect the method performances; (2) help to build robustness and provide assurance that the method is fit for use in routine [8, 9]. It must be emphasized that mathematical tools are not meant to replace the skills and expertise of analysts but, rather, to support and enhance their understanding of the processes affecting the LC method [10, 11].

Since the seminal paper of Borman *et al.* [7] who applied the concept of QbD and DS to analytical methods, several statistical methods have been proposed to implement the approach in LC, each claiming to be innovative, accurate and QbD-compliant. Unfortunately, very few of these methods are both statistically correct and QbD-compliant. A major part of them is misleading and often falls into pitfalls of poorly statistically defined robustness [12]. These statistical methods do not truly reflect the goals of AQbD strategy and the related concepts of quality assurance, DS and robustness. These poor understandings of the DS and robustness concepts partially result from some regulation ambiguities [12, 13].

The present chapter aims to present a critical review of current and emerging statistical methods supporting the development of LC methods by a QbD approach. In Sec. 5.2, we summarize the concept and components of a QbD approach to method development, with an emphasis on the meaning of the key concepts of robustness and analytical DS. Graphical and mathematical formalisms are provided to help to grasp their significance and to discuss the required statistical properties for any statistical method intended to estimate them. In Secs. 5.3 and 5.4, we present the current and most common statistical methods supporting QbD methods development. We distinguish between two categories of statistical methods. First, the design of experiments (DoE) and semi-empirical retention models-based methods are discussed (Sec. 5.3). These methods combine in a fully automated approach, the DoE and retention models such as linear solvent strength (LSS) or the quantitative structure retention relationships (QSRR) models derived from the solvophobic theory. Second, the DoE and fully empirical (i.e. data-driven) models-based methods which are based on empirical models such as the multivariate multiple linear regression and similar techniques are presented (Sec. 5.4). We perform a critical review of each category with respect to the goals of AQbD strategy. We argue that the empirical models-based methods are totally risk-oriented and interestingly more flexible and open to innovations. Hence, they can be adapted to the diversity of LC techniques and concrete problems faced by LC analysts. In Sec. 5.5, two case studies illustrating the DoE and Bayesian method for DS — what the authors believe is the most appropriate risk-oriented empirical method — in LC methods development are presented.

5.2 Overview of the AQbD Approach to LC Methods Development

As discussed in Sec. 5.1, AQbD is a systematic approach to method development that uses scientific knowledge to enhance understanding of the method, manage risks, and provide a guarantee that the method is fit for its intended use [7, 14]. An important point to be emphasized in this definition is that, a QbD approach to a LC method development is more than a simple optimization of chromatographic separation. The ultimate goal of the QbD development of LC methods is to define a set of working conditions, namely the DS, that guarantee the quality of separation with a sufficient probability. This is achieved by gaining in-depth scientific knowledge of the method.

Typical components of AQbD are described below. The concept of DS and robustness are clarified. Emphasis is placed on the role of the tandem "design of experiments — statistical modeling" in defining a DS compliant with the quality assurance objectives of the guidelines of the International Council on Harmonization (ICH) Q8.

5.2.1 *Analytical target profile and critical quality attributes*

An important step of the AQbD strategy is setting the analytical target profile (ATP) which defines the expectations of a method in terms of chromatographic separation, quantitative performances and robustness.

Consequently, a set of method adequacy criteria to be evaluated are defined along with their specifications. These criteria are called critical quality attributes (CQAs) and may include, for instance, the resolution of critical pairs of peaks or any relevant function of the responses (e.g. retention time, time at the beginning, apex and end of peaks, peak widths, etc.) to be measured during experiments [12, 14].

5.2.2 *Prior knowledge of the analyst*

LC methods development strategy should always start with an assessment of the analyst's prior knowledge of the sample and equipment whenever possible. Indeed, the scientist's know-how and available data about intrinsic physico-chemical properties of targeted molecules (e.g. molecular mass, $\log P$ and p*Ka*) may be a strong basis for setting specifications, assessing risks, pre-selecting several chromatographic parameters or establishing their ranges of investigation.

5.2.3 *Risk assessment and choice of critical method parameters*

After the definition of the ATP and CQAs, a risk assessment of the chromatographic method is performed. This step involves an evaluation of potential sources of variability in method results over the method lifecycle, from sample preparation to data analysis. Key risk factors that can alter the method are identified and prioritized. This enables the selection of several priority or critical factors to be considered for subsequent investigations using DoE. These factors are conventionally called critical method parameters (CMPs). The remaining method parameters are conventionally called nuisance parameters, and they may be further categorized into controllable and unavoidable nuisance parameters [7, 12].

To achieve risk assessment, structured tools such as flowcharts, Ishikawa (fishbone) diagrams, and failure mode and effects analysis (FMEA) are commonly used. Flowcharts partition the method into important steps with associated potential risks. Ishikawa diagrams enable the classification of the identified risks into groups such as instrumentation, material, method, human factor and environmental risks (Fig. 5.2). FMEA is typically

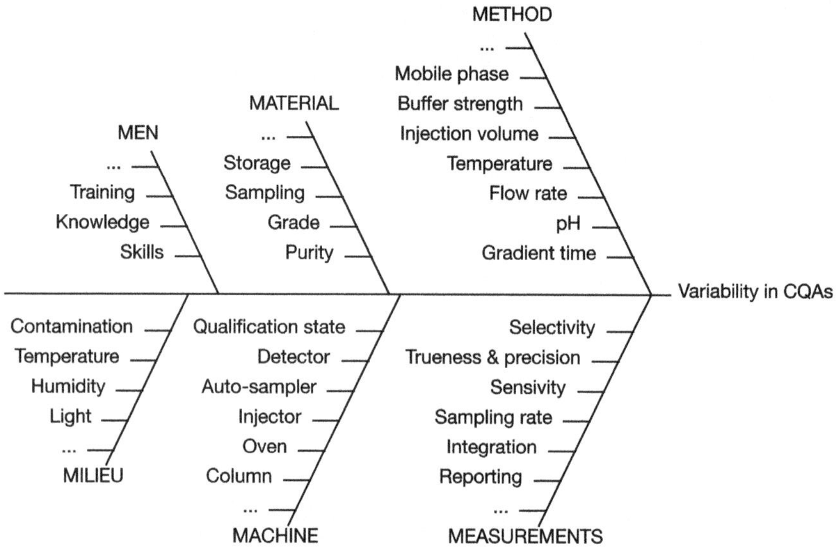

Figure 5.2: Typical fishbone diagram for risk factor categorization.

used to perform risks prioritization [7,12,14]. This tool consists in assigning to each factor three scores measuring, respectively the severity, the likelihood of a failure and the ability to detect it, if it were to occur. Then, a risk priority number equal to the product of these three scores is computed and used to prioritize the factors. Eventually, the priority method and instrumental risks factors are selected for further investigations. The CMPs may include, for example, the composition and pH of the mobile phase, the gradient time, the column temperature, the flow rate and so on. The template and instructions to perform a FMEA analysis are available in the paper of Borman *et al.* [7].

These risk assessment tools may be extremely useful in LC to screen out not only quantitative parameters but also qualitative factors. The inclusion of the quantitative factors in experimental designs for a systematic study has the effect of exploding the number of experiments and thus making the study excessively expensive.

5.2.4 *Design of experiments*

After risk assessment, AQbD proceeds with the in-depth investigation of the CMPs using DoE and statistical modeling tools. DoE is the foundation of the knowledge generation and learning processes in AQbD. It is a structured and organized method for conducting chromatographic experiments with the aim of establishing mathematical relationships between the CMPs and the CQAs. It allows to scientifically and efficiently (economically) define the experimental conditions to be tested so that as much information as possible can be obtained with a minimum number of experiments, to model the separation behaviors of the analytes. In contrast to the one-factor-at-a-time (OFAT) approach, the DoE approach varies simultaneously the investigated CMPs so that their mutual interactions can be assessed. Generally, screening and optimization designs could be used in chromatography depending the number of identified CMPs and the complexity of the estimated mathematical relationships between the CQAs and the CMPs [15, 16]. These designs are briefly described below. The reader is referred to specialized books or papers for details on candidate DoE in chromatography [15, 17, 18]. Especially, the review by Hibbert [15] is an excellent summary to introduce the reader to DoEs in chromatography.

5.2.4.1 *Screening designs*

If a large set of CMPs are identified as critical factors, screening designs such as the Plackett–Burman or other Partial Factorial Design (PFD) might first be used to select the smallest possible subset of CMPs that have the most significant effects on the CQAs. Depending on the selected resolution, these designs enable to fit models including either main effects only (Plackett–Burman) or main effects with a restricted number of low-order interactions models (PFD) [15, 16]. D-optimal designs for the optimal estimation of the main effects only would also be a sensible and more flexible strategy for screening experiments.

It is important to note that there is no need to include all factors, especially the qualitative ones, in screening experiments. As discussed earlier (Sec. 5.2.3), the outcomes of the FMEA analysis may serve as a basis to reduce the number of CMPs. Moreover, the scientist's know-how of the properties of the targeted molecules (e.g. molecular mass, log *P* and p*Ka*) may be a strong basis for selecting some chromatographic parameters. For instance, qualitative parameters such as the elution mode, the stationary phase of the analytical column and the organic modifier may be selected using the knowledge of the analyst. Based on the p*Ka* of molecules, the pH of the aqueous part of the mobile phase could be investigated on a reasonably restricted range or even set at a fixed value. As a result, the number of qualitative parameters studied during subsequent optimization could be minimized as much as possible, and therefore the costs of the method development will be reduced.

5.2.4.2 *Optimization designs*

Screening designs do not generally enable sufficient understanding, optimization and improvement of the method because they assume only simple linear and additive effects of the CMPs [16]. Therefore, optimization designs that support models with curvatures and interactions may then be used to fit functions of CMPs that predict the CQAs with a higher precision. Optimization or response surface designs enable the estimation of interactions and even quadratic-related effects, and therefore provide an idea of the (local) shape of a response surface. Among them, the 3^k factorial, the central composite design (CCD), the Box–Behnken and the Doehlert designs could be cited. It is important for optimization designs to include independently replicated runs, ideally placed at the center of the design. This enables not only an estimation of pure error for the lack-of-fit test of the candidate models, but also allows an overall decrease of the predictions variance over the complete experimental domain. Finally, optimal designs such as the D-, G- or I-optimal designs are also powerful strategies to model CQAs with a few continuous CMPs. Specifically, I-optimal designs have been developed to minimize the average variance of predictions [19], which contributes substantially to the overall uncertainty on the definition of the DS. Because of the above-described features, optimization and optimal designs are generally recommended to make a method more robust against external and non-controllable influences, especially when the experimental domain of a CMP is large or there exists no prior knowledge of the behavior of a response over the investigated domain [17–19].

5.2.5 *Statistical modeling, design space and robustness*

Once the experiments are run, appropriate statistical models must be used to analyze the data, both to understand the influence of CMPs on CQAs and to set boundaries for a DS compliant with the objectives of ICH Q8. Before discussing the various possibilities of statistically adequate models, the concepts of DS and robustness are clarified.

5.2.5.1 *Design space and robustness*

The DS concept is intimately connected with QbD approach. ICH Q8 [4] defines the DS for pharmaceutical processes as the *multi-dimensional combination and interaction of input variables that have been demonstrated to provide assurance of quality*. This definition does not explicitly apply to analytical methods. A proposed modification describes the analytical DS as the set of all combinations of input variables of a method for which assurance of the quality of the data produced by the method has been demonstrated [2, 7]. It must be emphasized that as for pharmaceutical processes, the concept of assurance of quality underscores the need for an explicit statement of the level of risk (i.e. the probability) of failing to achieve the targeted performance criteria. In other words, a key output of the analytical DS model is to provide an indication of how often the developed method will reach the desired specifications at any point of the knowledge domain [8, 11, 12, 16].

Mathematically, the DS is a subspace of the multi-dimensional experimental domain formed by the CMPs. Within this subspace, the robustness of the method and the quality of CQAs are guaranteed with a sufficient probability (Fig. 5.3). A mathematical formalism has been proposed by Peterson *et al.* [8, 20] and Lebrun [11] as

$$\text{DS} = \{\mathbf{x} \in \chi | \Pr(\mathbf{Y} \in \mathbf{A}|\mathbf{x}) \geq \pi_0\} \quad (1)$$

where $\mathbf{x} = (x_1, \ldots, x_p)'$ is a $p \times 1$ vector of CMPs, χ is the p-dimensional experimental domain, DS is the Design Space, $\Pr(\cdot)$ stands for the

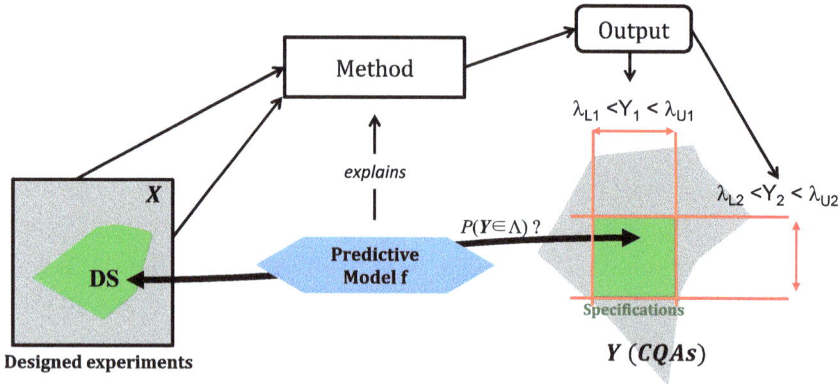

Figure 5.3: Illustration of the DS as the region of the operating conditions **x** for which there is *guarantee* that the related CQAs $\mathbf{Y} = f(\mathbf{x})$ are within acceptance limits (in red).

probability of an event, $\mathbf{Y} = (Y_1, \ldots, Y_r)'$ is a $r \times 1$ vector of CQAs, $\mathbf{A}$ is a r-dimensional subspace defined by the acceptance limits $A_1, \ldots, A_r$ of $\mathbf{Y}, \pi_0 \in [0, 1]$ is the minimum probability that the CQAs meet the specifications.

In practice, the analytical DS corresponds to a range of operating conditions where the CQAs of the analytical method meet their acceptance limits with a high probability [11, 12].

5.2.5.2 *Statistical models for the design space and robustness*

The computation of the DS from the data obtained from experimental runs may be based either on semi-empirical models such as the retention models derived from the solvophobic theory (Sec. 5.3) or on empirical chemometrics models whose equations depend on the data (Sec. 5.4). Mechanistic models are possible choices, but are not common in LC. The task of statistical data analysis consists in estimating the parameters of the chosen model if it is semi-empirical or mechanistic, or both its equation and parameters when it is empirical. In this way, predictions of chromatographic responses can be made from the estimated model. However, whatever the type of model, the requirement of level of risk inherent to DS (see Eq. (1)) implies that any method devised to fix a DS must possess certain statistical properties to be QbD-compliant [8, 9, 11, 13, 16].

First and most importantly, a QbD-compliant model should enable analysts to respond to the following question: What *probability* or *assurance* do we have that the CQAs at a given operating condition will meet the quality specifications as defined in the ATP? For such a statement to be possible, an explicit probability distribution of the future values of the CQAs that takes account of uncertainties inherent to the estimation of model parameters and unavoidable model errors is required. This distribution is known as *predictive distribution* of the CQAs (see example in Sec. 5.4.2). This requirement restricts appropriate statistical models of the DS to a group of models commonly called *probabilistic* models or *predictive* models [8, 9, 13, 17, 20]. Usually, no exact analytical expression of the predictive distribution of the CQAs is available. However, it can be approximated by stochastic simulation techniques such as the Monte Carlo-based, the bootstrap-based and the Bayesian-based methods. Conversely, *non-probabilistic* or *non-predictive* methods are statistically inadequate to compute a DS compliant with the quality assurance objectives of ICH Q8 and should be avoided [1, 9, 13]. Tools designed for statistical inference on mean responses of CQAs, such as the resolution maps or cubes methods (see Sec. 5.3.1.3) and the overlapping mean responses methods (see Sec. 5.4.3), are examples of *non-probabilistic* or *non-predictive* methods [8, 9, 12, 20]. Sadly, these confusing statistical methods have been promoted in the appendix of ICH Q8 and have been implemented by most software and used in most studies aiming to compute the DS, although the objective of assurance of quality is not guaranteed with these methods [11, 12, 14].

A second important feature required for any statistical method designed to compute the DS is the ability to model correlations among responses when several chromatographic responses are simultaneously measured and modeled. These kinds of model are conventionally referred to as *multi-response* or *multivariate* models [9, 11, 12, 20].

There is a comprehensive literature on adequate statistical models to compute the DS, and the reader is referred to previous publications for more details [8, 9, 11, 13, 17, 20]. Both theoretical and case studies of the

DoE and Bayesian models for DS computation, as a probabilistic method are also provided in Secs. 5.4 and 5.5.

5.2.6 *Validation and control strategy*

The last key step of AQbD approach is the definition of the control strategy. The goal of this step is to ensure that the method performs as intended when used routinely. Performance parameters to be monitored in routine can be derived from the outcomes of the DS analysis. These parameters are known as validity tests or system suitability tests (SST).

Statistical methods for validation of analytical methods have been extensively investigated in the scientific literature. The high performance of probabilistic or predictive methods such as tolerance intervals and accuracy profiles is widely demonstrated [21, 22]. These methods will not be discussed in this chapter though they are used in the case studies.

5.3 Statistical Methods Based on DoE and Semi-Empirical Retention Models

The ultimate objective of any chromatographic method is to achieve the complete separation of all components of a mixture in the shortest possible time. Nowadays, with the complexification of mixtures — increasing number of new substances, range of polarity and molar mass — this objective is often very tricky. It requires from analytical laboratories several time-consuming and highly expensive experiments, and advanced expertise. For instance, in many laboratories, a complex gradient elution approach is steadily replacing the generic gradient or the regular isocratic elution approaches [23]. To address this complexification, analytical scientists have investigated, early on, the possibility of using systematic, model-based, automated and, later on, computer-assisted methods to develop and optimize chromatographic methods. Recently, some of these retention models-based and automated strategies have been integrated into a systematic QbD approach, resulting in the so-called "automated QbD method development" [24–29].

In this section, we describe and discuss the advantages and limitations of the two most well-known and widely used of these computer-assisted QbD methods. They are based on the LSS and/or QSRR models.

Though we limited the discussion to these two approaches, it must be pointed out that there are similar methodologies available in many other commercial computer programs as well. Moreover, several new optimization strategies and algorithms are developed every year. Some of them are also briefly presented in the present section.

5.3.1 *The DoE and LSS models-based method*

5.3.1.1 *Overview of LSS models*

One of the earliest and most well-known models of retention behaviors is the LSS model in reversed-phase liquid chromatography (RPLC) [27,30–32]. It is a semi-empirical linear model linking the isocratic retention factor k, in RPLC mode, and the composition of a mobile phase as follows [33]:

$$\log k = a + b(\%\mathrm{B}) \tag{2}$$

where k is the retention factor and equals $(t_R - t_0)/t_0$ where t_R refers to the solute retention time and t_0 refers to the column dead time; %B is the varying percentage-volume of organic solvent in the water–organic mobile phase, and a and b are usually positive constants for a given compound and a given chromatographic condition. Equation (2) is often written as:

$$\log k = \log k_w - S\varphi \tag{3}$$

where φ is the volume-fraction of the organic modifier in the mobile phase expressed in decimal form; k_w is the extrapolated retention factor for $\varphi = 0$ (retention with water as mobile phase) and S is the solvent strength parameter, a constant for a given compound and fixed experimental conditions. In addition, in gradient mode, %B may be expressed as a function of time t after the start of the gradient; for example, $\%\mathrm{B} = c + dt$ for linear gradient elution, where c and d are constants.

For a narrow-range mobile-phase composition [34], the LSS models are generally expected to provide reliable predictions of retention times for "regular" samples, that is samples whose analytes exhibit non-intersecting retention (%B) curves and their separation order does not vary with %B [33].

The LSS models have been implemented in several commercial specialist software with ever-evolving capabilities. They provide fast computer-assisted solutions for method optimization. Some of the best-known computer programs are DryLab® (Molnár-Institute) and ChromSword®, the former claiming to be the "world standard" for chromatography modeling in both method development and training applications [26].

LSS models, as implemented in DryLab, require a low number of experiments to establish the model described in Eq. (3). For example, for predictions of gradient elution with a particular organic modifier and where only gradient time varies, two or more initial experimental runs with different gradient times are sufficient. Measured retention times are entered in the software for each analyte together with the experimental conditions — column dimensions, particle size, flow rate, and initial and final %B — for each calibration run. After calculating the coefficient values of the retention model, i.e. $\log k_w$ and S, for each analyte, the software could predict thousands of new runs for both isocratic and gradient separation as a function of the mobile phase %B or gradient conditions in a few second or minutes, each prediction corresponding to a possible chromatographic behavior [27, 33]. Optimization is then performed by searching the best predicted separation based on the critical resolution.

5.3.1.2 *DoE and modeling with LSS models*

Recently, the LSS models have been integrated within a QbD–DoE approach as the modeling tool to enhance the method understanding as requested by regulations [26–28]. In this approach, the analyst first defines the ATP and the specifications for the unique possible CQA, the critical resolution. From the program, in addition to gradient time, critical separation parameters are then selected to investigate their effects considering multifactorial experiments. A Full Factorial Design (FFD) is then generated — DryLab handles only FFD — for example 2×2 factorial design with 4 runs for gradient time and temperature, $2 \times 2 \times 3$ factorial design with 12 runs for gradient time, temperature and pH [26, 27].

Hence, these designs explicitly assume linear relationships between $\log k$ and the temperature (i.e. van't Hoff equation) and quadratic relationship between $\log k$ and the pH for each analyte [35]. For such assumptions to hold in practice, the levels of temperature and pH must be restricted to small ranges identical for all analytes. This restriction becomes theoretically infeasible for pH because the p*Ka* of investigated analytes may cover a wide range of values. Consequently, as a wide interval must be considered, the pH range must be segmented into adequate small sub-intervals, each covering at least three levels allowing to fit a quadratic model for log *k* on each sub-interval, to achieve good accuracy. This results in several models or, in other words, in several $2 \times 2 \times 3$ factorial designs. A typical example of such a multifactorial modeling approach is provided in the methodological work by Kormány *et al.* [35]. In this study aiming to optimize separation of a mixture of amlodipine and seven impurities, a pH range of 2.8–6.4 was considered with seven levels (i.e. 2.8, 3.4, 4.0, 4.6, 5.2, 5.8 and 6.4). Three quadric models of log *k* vs. pH were fitted independently for the ranges 2.8–4.0, 4.0–5.2 and 5.2–6.4.

Separation optimization may be further enhanced by extending the designs for measured factors with a "virtual" qualitative factor, the column type. Indeed, DryLab® includes a database of major types of columns and can simulate different column parameters from this database and evaluate the influence of column shifts on the quality of the separation [26, 27]. An obvious advantage of such a simulation approach is that column selection may be virtually optimized without additional expensive experimental runs.

5.3.1.3 *Design space and robustness tests with LSS models*

With the LSS approach, a "DS" is estimated by resolution maps or cubes as any condition (i.e. combination of method parameters) with critical resolution, $R_{s,crit}$, greater than the specification. A working point representing the best critical resolution or separation may also be determined.

After the DS is fixed, a robustness test is performed as follows. For each of the *p* measured and virtual method parameters, the user sets a nominal value (i.e. the working point) inside the DS and a tolerable deviation from this value (i.e. a low and a high level). Then the program simulates a 3^p factorial design and computes $R_{s,crit}$ for each combination of virtual and measured parameters (i.e. 64–729 conditions). Finally, for each $R_{s,crit}$ value, the number of conditions producing it is calculated. The proportion of $R_{s,crit}$

values falling within the specifications is computed as an indicator of the robustness. This indicator is called the "success rate" or the "probability of success" [24, 28, 29, 35].

The reader is referred to the comprehensive publications of Snyder, Molnár, Schmidt and Baczek [24,26–28,34,36] for details on mathematical development of LSS models, their combination with DoE-DS in DryLab®, the capabilities of this program, typical workflow and the so-called successful examples of DS and robustness.

5.3.2 *The DoE and QSRR models-based method*

5.3.2.1 *Overview of the QSRR models*

Another type of retention model developed by analytical scientists to model chromatographic separations of analytes in compounds is the QSRR models. These models attempt to establish mathematical relationships between retention parameters and chemical structural descriptors of analytes (i.e. variables quantifying the information encoded in the chemical structure of analytes present in a mixture). An excellent overview of most important QSRR models is provided by Kaliszan [37]. One of these QSRR models that has been implemented in ChromSword® is the model of Galushko [38, 39]. This model attempts to predict *a priori* the retention behavior of a solute from only two descriptors and without experimental runs as follows:

$$\ln k = a(V)^{2/3} + b(\Delta G) + c \tag{4}$$

where k is the solute retention factor, V is the molecular volume descriptor, ΔG is a descriptor of the energy of interaction of the solute with water, and a, b and c are parameters that are determined by the characteristics of the RPLC column in the used eluent.

In practice, the analyst must give the structural formulae of the analytes under investigation, and then ChromSword® will suggest the best possible column–solvent combinations and the optimal separation conditions for the analysis without any prior chromatographic runs [25]. It has been demonstrated that these predictions are generally not accurate, since the available molecular descriptors are of utmost importance for generating robust QSRR models. The role of descriptors is to extract relevant information about molecular shape, hydrophobic/hydrophilic volume of interactions, dipole moments, and physico-chemical descriptors (such as intrinsic solubility, $\log P$ and molecular diffusion), which are not always achieved in the case of ionized compounds or isotopomeric structures. However, these predictions may serve either as initial theoretical guess for subsequent method optimization procedures or as a screening step that reduces the number of effective chromatographic experiments [36].

5.3.2.2 *DS and robustness tests with QSRR-LSS models*

ChromSword® also offers capabilities to predict retention behaviors of analytes based on the LSS models of Eq. (3) with one or more experimental runs. The predictions accuracy is then equivalent to DryLab® [36]. The program also offers similar capabilities for a systematic approach to optimization with DoE, to testing robustness of a method and creating a DS easily and automatically. The reader is referred to Galushko *et al.* [25] for detailed information on the capabilities and workflow with this software.

5.3.3 *Other existing or newly emerging strategies*

Many new method development strategies arise every year. This reflects the ever-growing need for analytical scientists to find new tools to optimize chromatographic methods. For instance, it is possible to empirically develop QSRR models by screening descriptors through experiments and statistical modeling. In this approach, a linear regression model of retention parameters as a function of a large initial set of descriptors is used. Then, using variable selection algorithms such as a genetic algorithm, a restricted set of descriptors with better predictive abilities than the others is selected. This model-derived QSRR generally requires large numbers of experiments but can provide some useful understanding of molecular mechanisms of retention [40]. Moreover, this empirical approach to QSRR may be integrated with a DoE approach to enhance understanding of retention behaviors of analytes [41].

Another instance of these newly developed methodologies is the generic search strategy for automated method development for LC based on the predictive elution windows stretching and shifting introduced by Tyteca *et al.* [42].

5.3.4 *Limitations and pitfalls of DoE and semi-empirical retention models-based methods*

5.3.4.1 *Issues with the validity of the linearity assumption*

All fully automated solutions for method optimization and robustness building presented above are based on semi-empirical models of retention behaviors whose accuracy stringently depends on a chain of technical restrictions for linearity to be valid (i.e. narrow ranges of mobile phase concentration, temperature, pH, organic modifier other than acetonitrile) and on the hypothetical "regularity" of mixture. Theoretical and empirical evidence suggest that these assumptions are likely valid only for very specific types of chromatography such as the RPLC with organic solvent other than acetonitrile or ion-exchange chromatography (IEX), and under restrictive working conditions [36]. Particularly, the relationship $\log k$ vs. %B is often not linear for most analytes, and errors can occur especially in estimated values of $\log k$, and consequently, in predicted separations. For instance, Tyteca *et al.* [34] extensively investigated the ability to predict separation and the applicability of the LSS and two nonlinear retention-time models, namely the quadratic and the Neue models [43], for small molecules (phenol derivatives), peptides and intact proteins. They concluded that the LSS model shows poor predictions for low molecular weight analytes and peptides which exhibited moderate to pronounced nonlinear retention behaviors over the range of applicable solvent strengths. When the practically applicable window of the solvent strength is narrow, for example for intact proteins, the LSS model showed accurate predictions. Moreover, the LSS models provide poor predictions in normal-phase liquid chromatography (NPLC) or IEX, and in RPLC with acetonitrile solvent [33, 34].

Consequently, despite the fact that these retention models appear very simple and rapid, their applicability is practically limited to chromatographic conditions and molecules where the abovementioned assumptions are likely valid. When the type of chromatography changes, for example from RPLC to NPLC, hydrophilic interaction liquid chromatography (HILIC) or to supercritical fluid chromatography (SFC), these models start to show some limitations.

5.3.4.2 *Issues with the model errors and parameters uncertainties*

From a statistical perspective, the LSS model as defined in Eq. (3) does not explicitly include a random error term. This is a rather strongly deterministic assumption (i.e. deterministic models ignore random variations, and so always predict the same outcome from a given starting point) that certainly is far from valid in practice for many reasons. First, in chromatographic sciences, random errors such as instrumental errors, sample preparation errors and so on are unavoidable and will certainly affect future results of the method. Second, the LSS equation is derived by some approximation assumptions and hence is not a perfect description of the behavior of the analytes. Third, for a fully predictive approach as argued in Sec. 5.2.5.2, an error component must be included. As a result, retention predictions obtained with such models are mean predictions of what would happen "on average", assuming the model is good and hypotheses fulfilled. These predictions do not take account of model errors and measurement uncertainties affecting future individual runs.

5.3.4.3 *Issues with the flexibility of the DoE tools*

Regarding the proposed DoEs, they are only full factorial and flexibility is not left to the analysts to choose other relevant designs. It is well known that the number of experiments required by this type of design may inefficiently increase with the number of CMPs or levels by CMP. Hence, compared with the empirical risk-based methods (see example in Sec. 5.5), the LSS and full factorial designs methods may require as many experiments, despite the fact that the models of the former methods are more complex (i.e. higher order and number of model terms). This is due to the flexibility of choice of more efficient optimization designs by the empirical risk-based methods.

5.3.4.4 *Issues with the DS and robustness*

Regarding the proposed DS, it is not based on a probabilistic model, and hence, though it defines a multivariate region of operating method parameters that shows resolution (R_S) within the specification, there is no *assurance* (i.e. probability statement) that future individual analyses will show R_S values that meet the specification (see discussion in Sec. 5.2.5). The so-called "rate of success" represents the proportion of experimental conditions from a virtual 3^p factorial design which produces an acceptable R_S. A probabilistic approach would have implied stochastic simulations considering model errors and parameter uncertainties at each of the operating conditions within the experimental domain, either from a predictive distribution of R_S if its analytical form is available, or from approximation techniques such as those discussed in Sec. 5.2.5. Two examples of computing probabilistic DS with mechanistic or semi-empirical chromatographic models are provided in Close *et al.* [44] and Garcia-Muñoz *et al.* [45].

5.4 Statistical Methods Based on DoE and Risk-based Empirical Models

The possibility of simultaneous optimization of chromatographic methods and their robustness using statistical DoE and empirical models (i.e. models whose equations depend on the data) such as multivariate linear regressions (MLRs) and related techniques, was investigated in the 2000s [46, 47]. Since then, they have matured with the integration of powerful predictive tools like those mentioned in Sec. 5.2.5 and have currently resulted in powerful optimization and risked-management statistical methods. These methods are hereto referred as DoE and empirical models-based methods. Unfortunately, to our knowledge, there is no commercial software to adequately automate their implementation [48].

This section first presents an overview of the DoE and empirical models-based methods for DS in LC method development (Sec. 5.4.1). Following this, we provide a theoretical overview of a Bayesian method to compute the DS, the most appropriate of this category (Sec. 5.4.2). This method is then illustrated by two case studies (Sec. 5.5) and the reader not interested in mathematical details can skip Sec. 5.4.2.

5.4.1 *Overview of the DoE and empirical model-based methods*

Unlike the DoE and LSS models-based methods for DS, the DoE and empirical models-based methods require no explicit models of retention behaviors of analytes in a mixture. Rather, they assume that the investigated retention characteristic (e.g. retention time, time at the beginning, the apex, the end of peaks, and so on) of analytes are unknown functions of CMPs that can be approximated by truncated multivariate local Taylor polynomials. The theoretical rationale for this is the Taylor theorem that states that any function satisfying certain conditions (derivability) may be represented by a local Taylor series expansion, and hence reasonably truncating this series results in a satisfactory polynomial approximation of the function [17, 18]. "Locality" also refers to the fact that factor ranges are constrained to the experimental domain, and do not cover all real numbers. The alert reader would have noticed that this is also one of the foundation of the DoE theory.

As a result, each retention response may be satisfactorily estimated by a flexible low-order multivariate polynomial function of the CMPs

$$y_j = f_j(\mathbf{x}, \boldsymbol{\theta}_j) + \varepsilon_j \tag{5}$$

where y_j is an observed value of the jth retention response, ε_j is the zero-mean (normally distributed) error, $f_j(\mathbf{x}, \boldsymbol{\theta}_j)$ is a low-order multivariate polynomial approximating the relationship between the jth retention response and the CMPs, $\mathbf{x}$, and $\boldsymbol{\theta}_j$ is the set of parameters of the model.

The model in Eq. (5) can be estimated empirically from a series of experimental runs from an appropriate DoE (Sec. 5.2.4), using a predictive paradigm such as the Bayesian standard multivariate regression (SMR) or the parametric bootstrap on regression. The main principle of this group of statistical techniques consists in the prediction of a subspace of CMPs that will likely produce future CQAs within specifications given the observed data and, possibly, available prior information. Therefore, a core step is the determination of the joint predictive distribution of the CQAs. This represents the multivariate probability distribution of CQAs, accounting for both correlations among CQAs, unavoidable causes of method's variations and possibly uncertainties due to unknown model parameters. If the modeled responses differ from the CQAs, then the predictive distribution of the CQAs can be computed as functions of the multivariate predictive distribution of the modeled responses based on Monte Carlo samples in an error propagation scheme. This is the case for the critical resolution which is not continuous and should not be modeled directly. Considering the predictive distribution of CQAs, the probability of conformance, of future CQAs to the specifications can be easily computed at any possible operating point of the knowledge domain. The analytical DS includes any point of a grid (approximating the domain) with acceptable probability of conformance, that is greater than a predefined level, say 0.80, 0.85 or more. This probabilistic DS is usually represented as probability maps (see example in Sec. 5.5) that look very similar to but are very different from the resolution maps or overlapping mean response contour plots.

An important point to be emphasized is that the probabilistic estimation of robustness during the optimization of the separation enables rejecting solutions which offer good separation, say $R_S > 2.0$, but poor robustness, that is, poor probability that $R_S > 2.0$. Such an approach makes the key difference between the prediction of quality and the prediction of assurance of quality as advocated by the AQbD approach.

In practice, the number of experimental runs depends on the complexity of the model and may be efficiently chosen using some of the advanced and flexible designs described in Sec. 5.2.4, such as the Central Composite or I-optimal designs.

To sum up, unlike the very light DoE and LSS models-based methods that model only the retention factor k, the DoE and empirical model-based methods may satisfactorily model any relevant retention or chromatographic descriptor of the quality of separation (for instance, the time at the beginning (t_B), at the end (t_E), at the apex (t_A), the peak width, the critical resolution, and so on). A combination of these descriptors may even be modeled simultaneously through multi-response models, and various useful CQAs other than the critical resolution (R_s) can be derived [11, 49, 50]. Consequently, this group of methods is more generic and more flexible and can fit various chromatographic responses from a wider range of techniques and elution modes including NPLC, RPLC, HILIC and even the emerging hyphenated methods such as the liquid chromatography–mass spectrometry (LC-MS), and so on. The use of flexible and efficient DoEs enables the selection of reasonable numbers of experimental runs to fit more complex models.

5.4.2 *Overview of the Bayesian DS method in LC method development*

Bayesian DS is the most appropriate predictive empirical model-based method. This section summarizes the major mathematical steps to compute it, following data acquisition through an appropriate DoE. In the Bayesian context, the definition of the DS of Eq. (1) becomes

$$\text{DS} = \{\tilde{\mathbf{x}} \in \chi | \pi(\tilde{\mathbf{x}}) = \Pr(\tilde{\mathbf{Y}} \in \mathbf{A} | \tilde{\mathbf{x}}, \mathcal{D}) \geq \pi_0\} \tag{6}$$

where $\tilde{\mathbf{x}} \in \chi$ is a new point of the CMPs' domain, $\tilde{\mathbf{Y}}$ is a $r \times 1$ vector of predicted CQAs, $\mathcal{D}$ is the available data including the observed CQAs and CMPs, $\Pr(\cdot)$ stands for the probability of an event, $\mathbf{A}$ is an r-dimensional subspace defined by the acceptance limits $A_1, \ldots, A_r$ of $\mathbf{Y}$, and π_0 is the minimum probability that the CQAs meet the specifications.

Bayesian SMR or Bayesian seemingly unrelated regression (SUR) are used to determine the distribution of $\tilde{\mathbf{Y}}$ [20, 49]. The former assumes the same covariate structure for all CQAs. Locally on χ, a low-order polynomial is usually satisfactory for an accurate estimation [2, 50]. This enables a closed-form predictive distribution offering computational efficiency, though some CQAs may be over-fitted. The latter SUR model is more flexible as it enables a different model for each CQA.

For illustration purposes, the simpler case of an SMR model as described in Peterson [20] and Lebrun *et al.* [49] is considered. Denote $\mathbf{Y}_{\text{Mat}} = (\mathbf{y}'_1, \ldots, \mathbf{y}'_n)'$ the $n \times r$ matrix of observed CQAs where $\mathbf{y}_i = (y_{i1}, \ldots, y_{ir})$

with $i = 1, \ldots, n$ is the ith independent and identically distributed replicate of the $1 \times r$ vector of CQAs observed at p operating conditions $\mathbf{x}_i = (x_{i1}, \ldots, x_{ip})$. Let $z(\mathbf{x}_i) = \mathbf{z}_i$ be the $1 \times q$ vector of regressors for $\mathbf{y}_i$, and $\mathbf{Z} = (\mathbf{z}'_1, \ldots, \mathbf{z}'_q)$ the $n \times q$ model matrix. The regression model is written as

$$\mathbf{y}_i = \mathbf{z}_i\mathbf{B} + \varepsilon_i \quad \text{with} \quad \varepsilon_i \sim \mathrm{N}_r(\mathbf{0}, \Sigma) \tag{7}$$

where ε_i is the $1 \times r$ vector of errors, Σ is $r \times r$ semi-positive definite matrix; $\mathbf{B} = (\mathbf{b}'_1, \ldots, \mathbf{b}'_q)' = (\beta'_1, \ldots, \beta'_r)$ is the $q \times r$ matrix of regression coefficients, β_j is the $1 \times q$ vector of regression coefficients for the jth CQA and $\mathbf{b}_l$ is the $1 \times r$ vector of regression coefficient for a given regressor l. The likelihood of the model is

$$\mathrm{L}(\mathbf{B}, \Sigma | \mathbf{Y}_{\mathrm{Mat}}) = \prod_{i}^{n} \mathrm{N}_r(\mathbf{z}_i\mathbf{B}, \Sigma) \tag{8}$$

When no significant information or expert knowledge are available prior to the experiments, it makes sense to assume a non-informative prior distribution of the model parameters. A possible non-informative prior density is proposed by Geisser [51] and Box and Tiao [52] as

$$p(\mathbf{B}, \Sigma) \propto |\Sigma|^{-(r+1)/2} \tag{9}$$

Using the Bayes' theorem, the prior density of Eq. (9) is combined with the likelihood of Eq. (8) to obtain closed forms of the posterior distributions of the model parameters [51, 53],

$$(\Sigma | \mathcal{D}) \sim \mathrm{W}_r^{-1}(\mathbf{D}, \nu) \quad \text{and} \quad (\mathbf{B} | \Sigma, \mathcal{D}) \sim \mathrm{N}_{q \times r}(\hat{\mathbf{B}}, \Sigma, (\mathbf{Z}'\mathbf{Z})^{-1}) \tag{10}$$

and the joint predictive distribution of a future CQA vector $\tilde{\mathbf{y}}$ at a new operating point $\tilde{\mathbf{x}}$ of the experimental domain is established as a multivariate Student-t distribution [51],

$$(\tilde{\mathbf{y}} | \tilde{\mathbf{x}}, \mathcal{D}) \sim \mathrm{T}_r(\tilde{\mathbf{z}}\hat{\mathbf{B}}, (1 + \tilde{\mathbf{z}}(\mathbf{Z}'\mathbf{Z})^{-1}\tilde{\mathbf{z}})\mathbf{D}, \nu) \tag{11}$$

where T_r is the multivariate Student distribution, W_r^{-1} is the inverse Wishart distribution, $\tilde{\mathbf{z}}$ is the covariate structure for $\tilde{\mathbf{x}}$, $\hat{\mathbf{B}} = (\mathbf{Z}'\mathbf{Z})^{-1}\,\mathbf{Z}'\mathbf{Y}_{\mathrm{Mat}}$ is the least-square estimate of $\mathbf{B}$, $\mathbf{D} = (\mathbf{Y}_{\text{Mat}} - \mathbf{Z}\hat{\mathbf{B}})'(\mathbf{Y}_{\text{Mat}} - \mathbf{Z}\hat{\mathbf{B}})$, and $\nu = n - (r + q) + 1$ is the degree of freedom that must remain positive.

Given $\tilde{\mathbf{x}}$, the predictive probability $\pi(\tilde{\mathbf{x}})$ of meeting the acceptance criteria can be approximated using S independent Monte Carlo draws $\{\tilde{\mathbf{y}}^{(s)}\}_{s=1}^{S}$ from the joint predictive distribution as

$$\pi(\tilde{\mathbf{x}}) = \Pr(\tilde{\mathbf{Y}} \in \mathbf{A}|\tilde{\mathbf{x}}, \mathcal{D}) \approx \frac{1}{S}\sum_{s=1}^{S} \mathrm{I}[\tilde{\mathbf{y}}^{(s)} \in \mathbf{A}|\tilde{\mathbf{x}}] \tag{12}$$

where $\mathrm{I}(\cdot)$ denotes the indicator function taking values either 0 or 1. The probability of conformance $\pi(\tilde{\mathbf{x}})$ is computed for a set of points of a multi-dimensional grid defined over the CMPs' domain. The analytical DS includes any point of the grid with acceptable probability of conformance.

When significant information is available prior to the experiments, informative prior distributions are used to model parameters allowing for a reduction of uncertainties about the predictions. Otherwise, non-informative priors can be used, as presented above. One may use conjugate priors, for example a matrix-normal distribution for $\mathbf{B}$ and an inverse Wishart prior for $\boldsymbol{\Sigma}$. Lebrun *et al.* [49] showed that, in that case, the joint posterior predictive distribution of $\tilde{\mathbf{Y}}$ is still a multivariate Student-t distribution.

The modeling approach described above is adaptable to SUR models. However, a closed form of the joint posterior predictive distribution of the CQAs will not be available. Markov Chain Monte Carlo (MCMC) algorithms are then used to approximate this distribution [54].

It is obvious that the risk-based approach enables to overcome the flaws of the classical mean responses approach (see Sec. 5.4.3). Moreover, it enables an explicit statement of the probability of failure to meet specifications. However, the resulting DS is often smaller than that wrongly produced by the overlapping mean responses [9, 11, 13]. These latter are then generally overly optimistic.

5.4.3 *The flawed classical mean response surface methods for DS*

In the empirical model-based approach, the most commonly used but flawed statistical methods to compute the DS are the overlapping mean responses surfaces, the optimized mean responses surfaces and desirability functions. As discussed in Sec. 5.2.5.2, these methods are not predictive or probabilistic.

The overlapping mean responses surface determines the DS as the subspace of CMPs' domain where the estimated mean responses of CQAs are all within specifications. Mathematically, this is written as

$$\begin{aligned} \text{DS} &= \{\tilde{\mathbf{x}} \in \chi | \hat{\text{E}}(\mathbf{Y}|\tilde{\mathbf{x}}) \in \mathbf{A}\} \\ &= \{\tilde{\mathbf{x}} \in \chi | \hat{\text{E}}(\text{Y}_\text{j}|\tilde{\mathbf{x}}) \in \text{A}_\text{j}, \forall j = 1, \ldots, r\} \end{aligned} \tag{13}$$

where $\hat{\text{E}}(\cdot)$ denotes the expectation function and $\tilde{\mathbf{x}}$ is a new point of the experimental domain. The expected response for each CQA $\hat{\text{E}}(\text{Y}_\text{j}|\tilde{\mathbf{x}})$ is generally obtained by fitting a model in Eq. (5) by devoted statistical estimation methods, generally the least-square or maximum likelihood estimators.

If the objective is to find an optimal solution, $\hat{\mathbf{x}}$, an optimization of the mean responses is performed and the optimal condition for a single CQA becomes

$$\hat{\mathbf{x}} = \arg\max_{\tilde{\mathbf{x}}}[\hat{\text{E}}(\text{Y}_\text{j}|\tilde{\mathbf{x}})] = \arg\max_{\tilde{\mathbf{x}}}[\hat{f}_j(\tilde{\mathbf{x}}, \hat{\boldsymbol{\theta}}_\text{j})] \tag{14}$$

For multiple CQAs' optimal solutions, a desirability index is usually calculated. This index aggregates the various mean responses into one score representing the quality of the solution, which is then optimized.

Such optimization methods are implemented by most generic software devoted to DoE such as Design-Expert Software [55], JMP [56] and Minitab [57]. The reader is referred to Del Castillo [17], Khuri and Mukhopadhyay [58] and Myers *et al.* [18] for detailed information on the historical and technical development of mean responses surface methodologies, and Lebrun [11] for various applications to analytical methods.

The flaws of these methods using mean responses have been extensively demonstrated by several works [9, 11–14]. First, the models used do not account for correlations among multiple CQAs and uncertainties about unknown model parameters. Second, the predicted CQAs are mean values. It is well established that although the mean responses meet specifications, this does not necessary imply individual future runs of the method will be within acceptance limits, due to model imprecision and measurements and process uncertainties. Consequently, the DS based on mean responses may include operating conditions with quite low assurance of quality results. Obviously, these approaches do not produce DS compatible with ICH Q8's expectations.

5.5 Case Studies of Bayesian DS Methods in LC Methods Development

Since the development of the Bayesian DS method for analytical methods development [11, 49], this strategy has been successfully applied for the robust optimization of many LC methods, demonstrating its reliability. Several research papers applying the Bayesian DS method are accessible to the interested reader [2, 59–67]. This section presents two of these works as case studies.

5.5.1 *Bayesian DS applied to non-steroidal anti-inflammatory drugs*

In this case study, 18 non-steroidal anti-inflammatory drugs (NSAID), five pharmaceutical conservatives and four associated drugs were selected and pooled into 16 groups that represent real pharmaceutical formulations under tablet, capsule, syrup or suspension forms. The first objective was the robust optimization of the LC separation using a DoE and Bayesian DS approach for the establishment of the DS. The second objective was to demonstrate that the DS obtained represents a robustness area that could facilitate geometric transfer to UHPLC. Finally, the validation of the LC method was envisaged to demonstrate its quantitative performances [68].

A central composite design comprising three CMPs was selected. The CMPs were the pH of the aqueous part of the mobile phase, the gradient time to linearly modify the proportion of methanol from 15% to 95% and the temperature of the column. This design was composed of 32 experimental conditions. The measured responses were the times at the beginning, the apex and the end of each peak. The retention factors corresponding to the measured times were modeled by multivariate multiple regression models. The selected CQA was the separation criterion (S_{crit}), which is defined as the time between the end of the second peak and the beginning of the first peak of the critical pair (i.e. both closest peaks in a chromatogram).

The first advantage of using the DoE approach is that rather than injecting the 16 groups of compounds individually to perform 16 distinct optimizations, the 27 compounds were injected all together and the 16 pharmaceutical formulations were virtually optimized using the corresponding multivariate models. In this respect, the number of experiments (32) was very low, considering the number of compounds (27) and pharmaceutical formulations (16) optimized jointly. Another advantage is that, rather than defining the optimal separation based on separation criterion maps showing mean predicted conditions where $S_{\text{crit}} > 0$, probability maps showing predicted conditions with high probability of separation (i.e. $\Pr(S_{\text{crit}} > 0) \geq 0.95$) were used.

For the group of compounds containing acetaminophen, ibuprofen, nimesulide, mefenic acid, nipagin, nipasol, sodium benzoate, butylated hydroxyanisole and butylated hydroxytoluene, the optimal separation was predicted with a gradient time of 53.1 min, a temperature of 23°C and a pH of 4.05. The corresponding probability maps are shown in Fig. 5.4.

The results at the optimal predicted conditions were compared with the predicted one (see Fig. 5.5). The gradient conditions were then transposed to UHPLC using classical geometric transfer rules. The resulting UHPLC chromatograms offered a 15-fold reduction in analysis time and a 25-fold decrease of mobile phase consumption while maintaining the separation of all compounds.

Finally, the method was validated using the total error approach and accuracy profile methodology [69–71]. The method was demonstrated to be valid for the quantification of acetaminophen and ibuprofen between 200 and 600 μg/mL, as can be seen in Fig. 5.6.

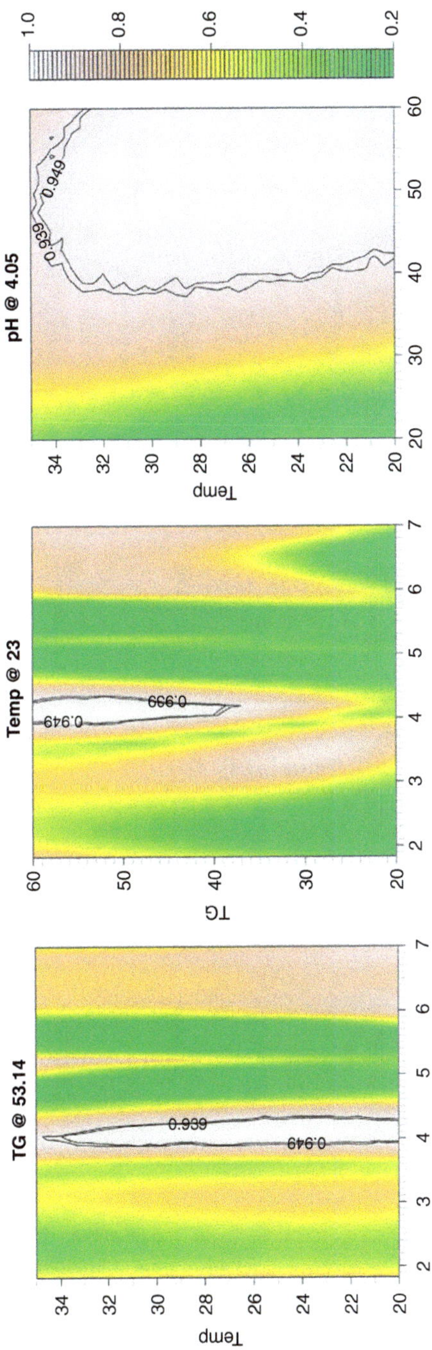

Figure 5.4: Probability maps showing predicted operating conditions and associated $\Pr(S_{\text{crit}} > 0)$. The DS is represented by the white region with minimum quality level of $\pi_0 = 0.95$, that is $\Pr(S_{\text{crit}} > 0) \geq 0.95$.

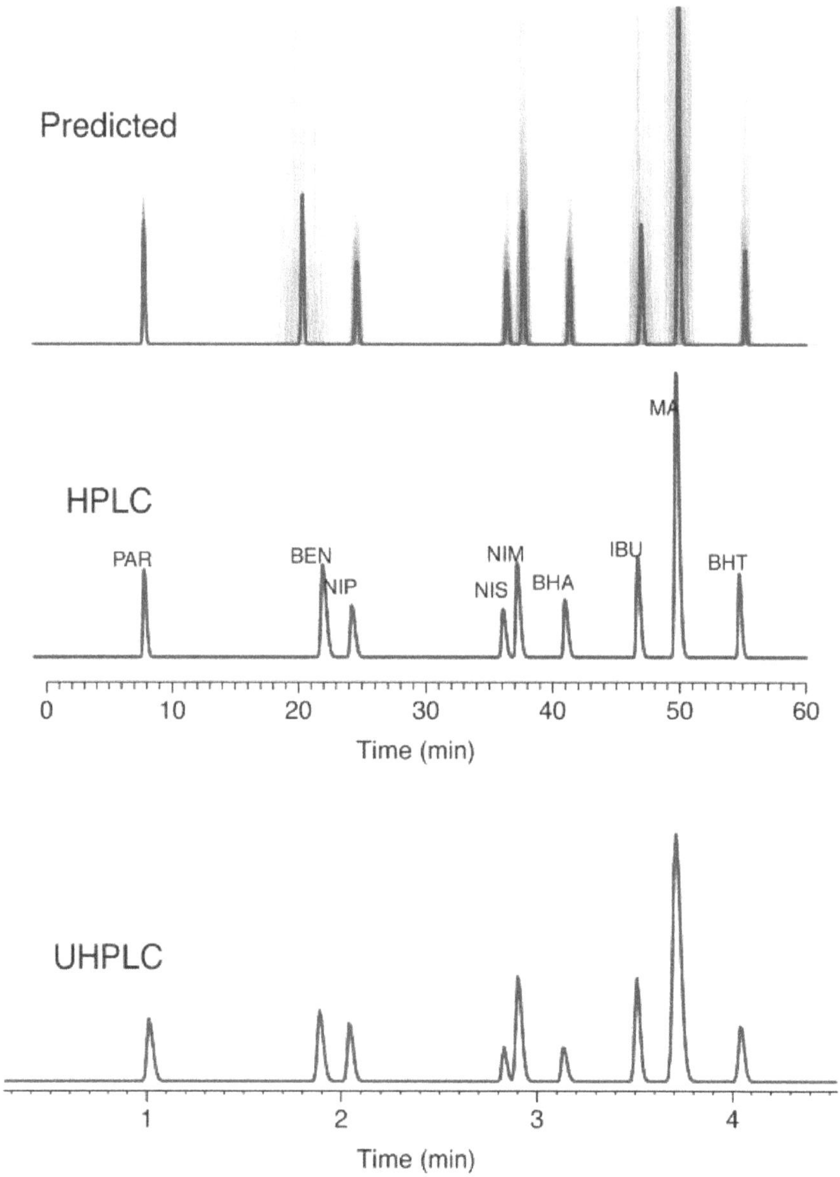

Figure 5.5: Chromatogram predicted, recorded in LC mode and in UHPLC mode at the optimal experimental conditions. Compound assignation: acetaminophen (PAR), ibuprofen (IBU), nimesulide (NIM), mefenic acid (MA), nipagin (NIP), nipasol (NIS), sodium benzoate (BEN), butylated hydroxyanisole (BHA) and butylated hydroxytoluene (BHT).

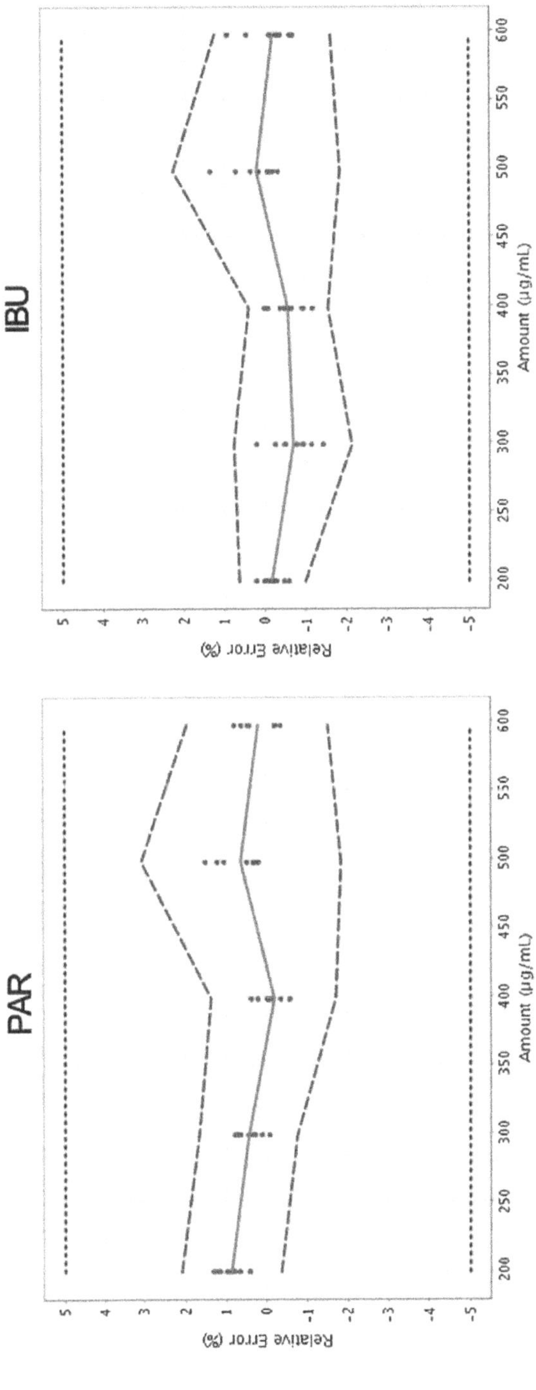

Figure 5.6: Accuracy profiles cbtained for acetaminophen (PAR) and ibuprofen (IBU).

5.5.2 *Bayesian DS for the selective determination of glucosamine and galactosamine in human plasma*

Analogous to the adjustments that may occur during a drug's life cycle, modifications of an analytical method may be needed, for example, to meet new specifications or to adapt to changes of the sample type. In this second case study, a previously developed method had to be optimized, first to enable a selective determination of glucosamine and galactosamine in another biological matrix and, second, to simultaneously optimize their chromatographic behavior as well as the sensitivity of the method. In this context, a method to determine these epimeric amino-sugars avoiding their on-column mutarotation in the presence of extracted compounds from human plasma was developed using the Bayesian DS method [72].

An initial development of the method using the QbT approach only led to an insufficient understanding of its separation performances. However, based on this acquired experience and prior knowledge about the influence of the biological sample preparation on extracted endogenous plasma compounds, several experiments were performed to confirm the chromatographic mode and select the analytical column and CMPs. Subsequently, a HILIC method coupled to tandem mass spectrometry (MS/MS) was considered. In this study, acetonitrile percentage (ACN, 80–90%) and pH (pH, 5–10) were identified as having a critical influence on both the separation and the mutarotation phenomenon. A CCD was then used and customized by adding a temperature range (T, 25–75°C) based on prior scientific knowledge of the influence of this parameter on that specific research. The custom central composite design thus obtained included 13 experimental conditions plus three repetitions at the center of the design, for a total of 15 experiments.

As in the first case study (Sec. 5.5.1), the measured responses were the retention times at the beginning, apex and end of each peak. The selected CQAs and their associated acceptance limits were the separation criterion ($S_{\text{crit}} > 0.2$ min) and the total run time (<30 min).

A multivariate linear regression was considered for the data to account for correlations among the responses as follows:

$$\begin{aligned}\mathbf{y_i} &= \mathbf{b_0} + \mathbf{b_1} \times \text{ACN}_i + \mathbf{b_2} \times \text{ACN}_i^2 \\ &\quad + \mathbf{b_3} \times \text{ACN}_i^3 + \mathbf{b_4} \times \text{pH}_i^2 + \mathbf{b_5}\end{aligned}$$

$$\times \mathrm{pH}_i^3 + \mathbf{b_6} \times \mathrm{pH}_i^4 + \mathbf{b_7} \times \mathrm{T}_i + \mathbf{b_8} \times \mathrm{ACN}_i \times \mathrm{T}_i + \mathbf{b_9}$$

$$\times \mathrm{ACN}_i \times \mathrm{pH}_i \times \mathrm{T}_i + \varepsilon_i$$

or equivalently,

$$\mathbf{y}_i = \mathbf{z}_i\mathbf{B} + \varepsilon_i \quad \text{with} \quad \varepsilon_i \sim \mathrm{N}_r(\mathbf{0}, \boldsymbol{\Sigma}) \tag{15}$$

where P is the number of peaks; $r = 3 \times P$ is the number of modeled responses; $q = 10$ is the number of regressors; $\mathbf{y_i}$ is the $1 \times r$ vector of log-transformed measured responses at run $i = 1, \ldots, N = 15$; $\mathbf{z}_i$ is the $1 \times q$ vector of regressors at run i; ε_i is the $1 \times r$ vector of errors; $\boldsymbol{\Sigma}$ is $r \times r$ semi-positive definite matrix; $\mathbf{B} = (\mathbf{b}_1', \ldots, \mathbf{b}_q')'$ is the $q \times r$ matrix of regression coefficients and $\mathbf{b}_l$ is the $1 \times r$ vector of regression coefficients for the lth regressor.

The predictive distribution of the modeled responses is given by Eq. (11). The predictive distribution of each CQA is computed as a function of the Monte Carlo samples of the modeled responses, thus propagating model errors and parameter uncertainties. From the predictive distribution of each CQA, the DS for each CQA is then computed using Eq. (12) as any point of the experimental domain where the CQAs meet their acceptance limits with a probability of at least 0.85. The resulting probability surfaces are plotted (Figs. 5.7 and 5.8).

Temperature shows a positive and homogenous influence on the chromatographic separation for values above 50°C, as illustrated in Fig. 5.7. Considering a temperature of 50°C (i.e. the softest condition for equipment robustness), Fig. 5.8 shows a computed 2D plot representation of the probability of meeting the CQAs' acceptance limits for varying acetonitrile percentage and pH. The level of quality obtained was acceptable and relatively constant between 84% and 88% of acetonitrile and between 5.25 and 6.5 for pH. Within this area, two DS with a quality level of 0.83 were identified and delimited by dark lines.

These DS represent the set of conditions where an acceptable separation of both compounds and endogenous plasma compounds was obtained within a maximum run time of 30 min. A working condition within the DS (i.e. ACN = 86%, pH = 6 and T = 50°C) was selected and validated using the accuracy profile approach, which is based on statistical tolerance intervals [69–71]. The accuracy profile obtained for the

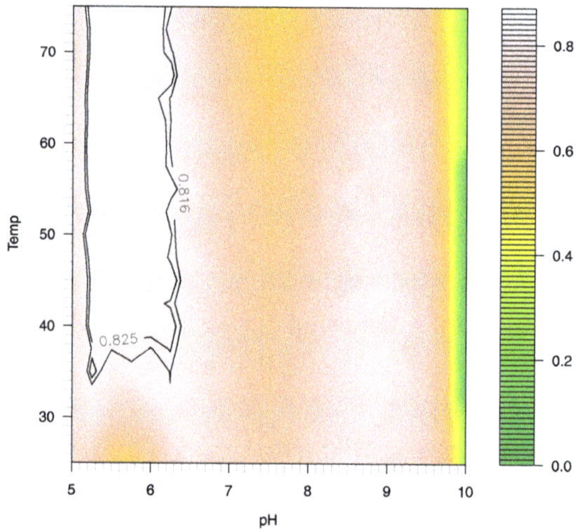

(a)

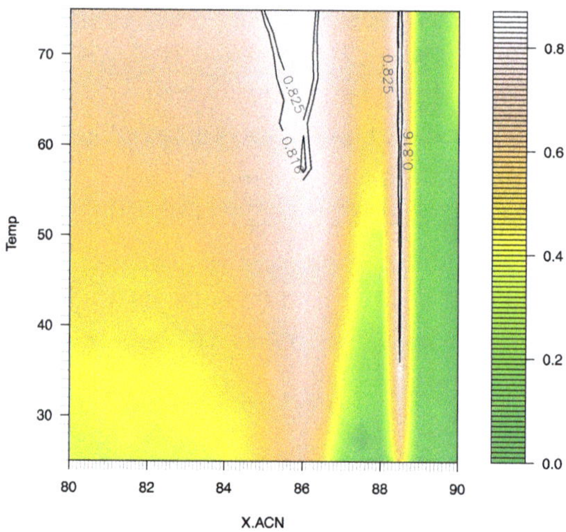

(b)

Figure 5.7: Two-dimensional probability surfaces (i.e. $P(\text{CQAs} > \lambda)$ with their DS defined by dark lines. (a) T and pH for ACN fixed at 88.5%. (b) T and ACN for a pH fixed at 5.75.

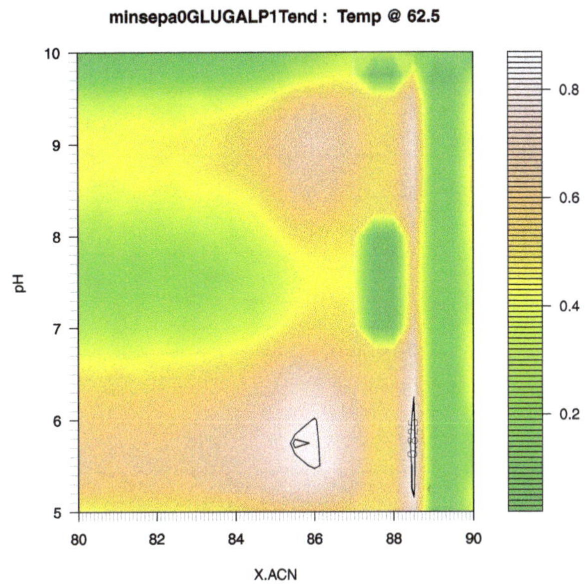

Figure 5.8: Two-dimensional probability surfaces (i.e. $P(\text{CQAs} > \lambda)$ for pH and ACN with T fixed at 50°C. The DS are defined by a dark line.

validation of the working condition for glucosamine and galactosamine are presented in Fig. 5.9. These profiles illustrate the quantitative performances of the method for a specific working condition where the separation of both compounds contained in a human plasma matrix is guaranteed.

As previously stated, the ATP defined the expectations of a method in terms of chromatographic separation and robustness but also in terms of quantitative performance. Consequently, trueness, precision or accuracy could represent a CQA of the method defining the minimal quantitative performance requirement. In this context, a first demonstration of the possibility to compute a quantitative DS representing the probability of success of the validation throughout an operational space was done as part of this case study [72].

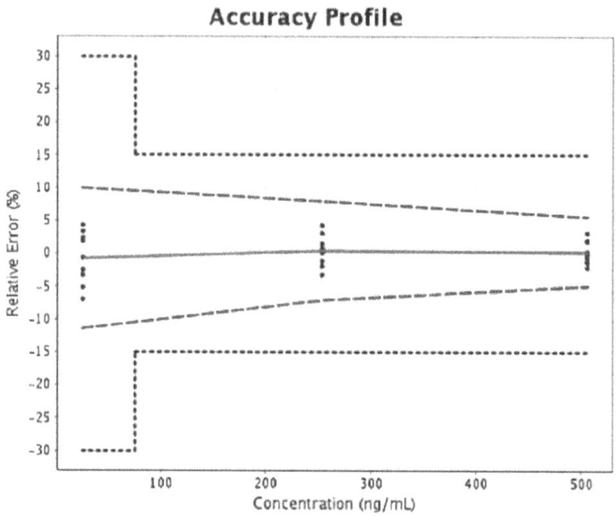

(a)

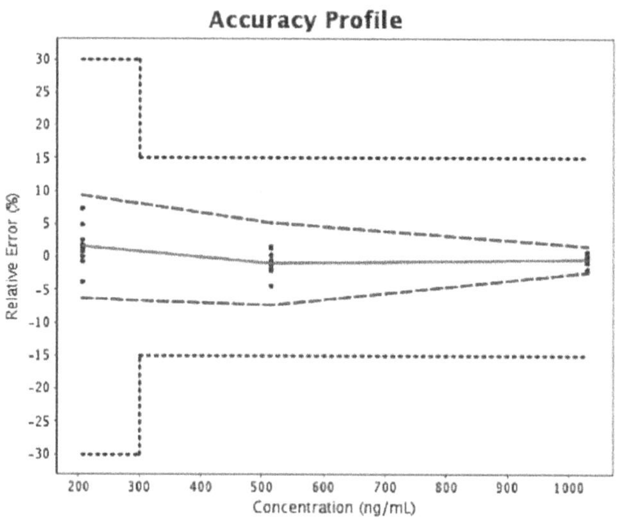

(b)

Figure 5.9: Accuracy profile of the validation of the selected working conditions (i.e. ACN = 86%, pH = 6 and T = 50°C) for (a) glucosamine and (b) galactosamine.

5.6 Conclusions

Since the earlier stages of chromatography, analytical scientists have been investigating the possibility of using mathematical models to optimize the development of chromatographic methods. These investigations first resulted in semi-empirical models mostly applied to reverse-phase chromatography, such as the popular LSS and the QSRR models. Later, with the advances in computer science, these models have been implemented in computer software, enabling the model-based automation of the optimization of chromatographic methods. These commercial solutions undoubtedly demonstrate significant achievements in optimization of separation methods as they largely substitute costly and time-consuming experimentations and calculations by model-based computer simulations.

Nonetheless, these first models lack enough versatility to precisely simulate the existing diversity of chromatographic elution modes such as normal-phase, HILIC, or emerging techniques such as SFC. Furthermore, the past decade has seen the enforcement of new, compelling quality regulations, resulting in a shift of paradigm from the unstructured Quality by Testing (QbT) approach to the systematic QbD approach to method development. A key outcome of this later paradigm is the concept of DS, which defined the method operating conditions that are supposed to guarantee a delivery of quality results by the method. A critical point in the estimation of this DS is the requirement of selecting experimental conditions of the studied domain with high probability of delivering results that meet a set of specifications routinely, rather than simply fixing a subspace of the experimental domain where specifications are met.

Given these new challenges, improvements of semi-empirical models have been considered by developing and integrating capabilities to implement key QbD statistical tools such as DoE, DS and robustness. Unfortunately, such models still lack the flexibility to fit the existing diversity of chromatographic techniques. Moreover, the concept of DS, as implemented, does not comply with this critical criterion of quality assurance of predictions.

As an alternative, empirical (data-driven) statistical methods for robust optimization have been developed. These methods combine the advantages of great flexibility for precise modeling of a wide range of chromatographic modes and effective risk management, leading to powerful optimization tools that fully comply with the new quality regulations. For now, the implementation of such tools requires programming skills and the expertise of statisticians. Nonetheless, integration of empirical models into analytical software could be greatly beneficial for the analytical chemistry community. This will enable the popularization of QbD-compliant methods for optimization of chromatographic methods, with regard to the quality assurance objectives of the ICH Q8 guidelines.

References

[1] E. Rozet, P. Lebrun, B. Debrus, Ph. Hubert, New methodology for the development of chromatographic methods with bioanalytical application, *Bioanalysis* 4(7) (2012) 755–758.

[2] C. Hubert, P. Lebrun, S. Houari, E. Ziemons, E. Rozet, Ph. Hubert, Improvement of a stability-indicating method by Quality-by-Design versus Quality-by-Testing: A case of learning process, *J. Pharm. Biomed. Anal.* 88 (2014) 401–409.

[3] U.S. Pharmacopeial Convention, new chapter 1224, 1225, 1226, USP panel expert.

[4] ICH, Q8(R2), Pharmaceutical development. International Conference on Harmonization on Technical Requirements on Registration of Pharmaceuticals for Human Use, Geneva, Switzerland, (2009).

[5] ICH, Q9, Quality Risk Management. International Conference on Harmonization on Technical Requirements on Registration of Pharmaceuticals for Human Use, Geneva, Switzerland, (2005).

[6] ICH, Q10, Pharmaceutical Quality System. International Conference on Harmonization on Technical Requirements on Registration of Pharmaceuticals for Human Use, Geneva, Switzerland, (2008).

[7] P. Borman, K. Truman, D. Thompson, P. Nethercote, M. Chatfield, The application of quality by design to analytical methods, *Pharm. Technol.* 31(10) (2007) 142–152.

[8] J.J. Peterson, R.D. Snee, P.R. McAllister, T.L. Schofield, A.J. Carella, Statistics in pharmaceutical development and manufacturing, *J. Qual. Technol.* 41(2) (2009) 111–134.

[9] J.J. Peterson, What your ICH Q8 design space needs: A multivariate predictive distribution, *Pharm. Manufact.* 8(10) (2010) 23–28.

[10] C.F. Poole, Editorial on "Chemometrics-assisted method development in reversed-phase liquid chromatography" by R. Cela, E.Y. Ordonez, J.B. Quintana, R. Rodil, *J. Chromatogr. A* 1287 (2013) 1.

[11] P. Lebrun, Bayesian Design Space applied to Pharmaceutical Development, Ph.D. thesis (2012), University of Liège, Belgium.

[12] E. Rozet, P. Lebrun, Ph. Hubert, D. Debrus, B. Boulanger, Design spaces for analytical methods, *Trends Anal. Chem.* 42 (2013) 157–167.
[13] J.J. Peterson, K. Lief, The ICH Q8 definition of design space: A comparison of the overlapping means and the Bayesian predictive approaches. *Stat. Biopharm. Res.* 2(2) (2010) 249–259.
[14] F.G. Vogt, A.S. Kord, Development of quality-by-design analytical methods, *J. Pharm. Sci.* 100(3) (2011) 797–812.
[15] B.D. Hibbert, Experimental design in chromatography: A tutorial review, *J. Chromatogr. B* 910 (2012) 2–13.
[16] J.J. Peterson, S. Altan, Overview of drug development and statistical tools for manufacturing and testing, In: *Nonclinical Statistics for Pharmaceutical and Biotechnology Industries*, Springer International Publishing, Switzerland, 2016, pp. 383–414.
[17] E. Del Castillo, *Process Optimization: A Statistical Approach*, International Series in Operations Research & Management Science, Vol. 5, Springer US, New York, USA, 2007.
[18] R.H. Myers, D.C. Montgomery, C.M. Anderson-Cook, *Response Surface Methodology: Process and Product Optimization Using Designed Experiments*, 4th Edition, Wiley series in probability and statistics, John Wiley & Sons, New Jersey, USA, 2016.
[19] P. Goos, B. Jones, *Optimal Design of Experiments: A Case Study Approach*, John Wiley & Sons, Chichester, UK, 2011.
[20] J.J. Peterson, A Bayesian approach to the ICH Q8 definition of design space, *J. Biopharm. Stat.* 18(5) (2008) 959–975.
[21] E. Rozet, R.D. Marini, E. Ziemons, B. Boulanger, Ph. Hubert, Advances in validation, risk and uncertainty assessment of bioanalytical methods, *J. Pharm. Biomed. Anal.* 55(4) (2011) 848–858.
[22] A. Dispas, P. Lebrun, Ph. Hubert, Validation of supercritical fluid chromatography methods. In: C.F. Poole (ed.) *Supercritical Fluid Chromatography, Handbooks in Separation Science*, Amsterdam, The Netherlands, 2017, pp. 317–344.
[23] I. Molnár, Computerized design of separation strategies by reversed-phase liquid chromatography: Development of DryLab software, *J. Chromatogr. A* 965 (2002) 175–194.
[24] L.R. Snyder, L. Wrisley, Computer-facilitated HPLC method development using DryLab® Software. In: *HPLC Made to Measure: A Practical Handbook for Optimization*, WILEY-VCH Verlag GmbH & Co. KGaA, Weinheim, Germany, 2006, pp. 567–586.
[25] S. Galushko, V. Tanchuk, I. Shishkina, O. Pylypchenko, W.-D. Beinert, ChromSword® software for automated and computer-assisted development of HPLC methods. In: *HPLC Made to Measure: A Practical Handbook for Optimization*, WILEY-VCH Verlag GmbH & Co. KGaA, Weinheim, Germany, 2006, pp. 587–600.
[26] I. Molnár, H.J. Rieger, K.E. Monks, Aspects of the "Design Space" in high pressure liquid chromatography method development, *J. Chromatogr. A* 1217 (2010) 3193–3200.
[27] I. Molnár, H.-J. Rieger, R. Kormány, Modeling of HPLC methods using QbD principles in HPLC. In: *Advances in Chromatography*, Vol. 53, CRC Press, USA, 2016, pp. 331–350.
[28] A.H. Schmidt, I. Molnár, Using an innovative Quality-by-design approach for development of a stability indicating UHPLC method for ebastine in the API and pharmaceutical formulations, *J. Pharm. Biomed. Anal.* 78–79 (2013) 65–74.

[29] S. Fekete, R. Kormány, D. Guillarme, Computer-assisted method development for small and large molecules, *LC-GC Europe* 30(6) (2017) 14–21.
[30] L.R. Snyder, Linear elution adsorption chromatography: VII. gradient elution theory, *J. Chromatogr. A* 13 (1964) 415–434.
[31] L.R. Snyder, H.D. Warren, Linear elution adsorption chromatography: VIII. gradient elution practice. the effect of alkyl substituents on retention volume, *J. Chromatogr. A* 15 (1964) 344–360.
[32] L.R. Snyder, J.W. Dolan, The linear-solvent-strength model of gradient elution, *Adv. Chromatogr.* 38 (1998) 115–187.
[33] L.R. Snyder, J.W. Dolan, *High Performance Gradient Elution: The Practical Application of The Linear-Solvent-Strength Model*, John Wiley & Sons Inc., Hoboken, New Jersey, USA, 2007.
[34] E. Tyteca, J. De Vos, N. Vankova, P. Cesla, G. Desmet, S. Eeltink, Applicability of linear and nonlinear retention-time models for reversed-phase liquid chromatography separations of small molecules, peptides, and intact proteins, *J. Sep. Sci.* 39(7) (2016) 1249–1257.
[35] R. Kormány, J. Fekete, D. Guillarme, S. Fekete, Reliability of simulated robustness testing in fast liquid chromatography, using state-of-the-art column technology, instrumentation and modelling software, *J. Pharm. Biomed. Anal.* 89 (2014) 67–75.
[36] T. Baczek, Computer-assisted optimization of liquid chromatography separations of drugs and related substances, *Curr. Pharm. Anal.* 4(3) 2008 151–161.
[37] R. Kaliszan, QSRR: Quantitative structure-(chromatographic) retention relationships, *Chem. Rev.* 107(7) (2007) 3212–3246.
[38] S.V. Galushko, Calculation of retention and selectivity in reversed-phase liquid chromatography, *J. Chromatogr. A* 552 (1991) 91–102.
[39] S.V. Galushko, A.A. Kamenchuk, G.L. Pit, Calculation of retention in reversed-phase liquid chromatography: IV. ChromDream software for the selection of initial conditions and for simulating chromatographic behaviour, *J. Chromatogr. A* 660 (1994) 47–59.
[40] S. Schefzick, C. Kibbey, M.P. Bradley, Prediction of HPLC conditions using QSPR techniques: An effective tool to improve combinatorial library design, *J. Comb. Chem.* 6(6) (2004) 916–927.
[41] M. Taraji, P.R. Haddad, R.I.J. Amos, M. Talebi, R. Szucs, J.W. Dolan, C.A. Pohl, Rapid method development in hydrophilic interaction liquid chromatography for pharmaceutical analysis using a combination of quantitative structure-retention relationships and design of experiments, *Anal. Chem.* 89(3) (2017) 1870–1878.
[42] E. Tyteca, A. Liekens, D. Clicq, A. Fanigliulo, B. Debrus, S. Rudaz, D. Guillarme, G. Desmet, *Anal. Chem.* 84(18) (2012) 7823–7830.
[43] U.D. Neue, H.-J. Kuss, Improved reversed-phase gradient retention modeling, *J. Chromatogr. A* 1217 (2010) 3794–3803.
[44] E.J. Close, J.R. Salm, D.G. Bracewell, E. Sorensen, A model based approach for identifying robust operating conditions for industrial chromatography with process variability, *Chem. Eng. Sci.* 116 (2014) 284–295.
[45] S. García-Muñoz, C.V. Luciani, S. Vaidyaraman, K.D. Seibert, Definition of design spaces using mechanistic models and geometric projections of probability maps, *Org. Process Res. Dev.* 19(8) (2015) 1012–1023.

[46] W. Dewé, R.D. Marini, P. Chiap, Ph. Hubert, J. Crommen, B. Boulanger, Development of response models for optimizing HPLC methods, *Chemometr. Intell. Lab.* 74(2) (2004) 263–268.
[47] P. Lebrun, B. Govaerts, B. Debrus, A. Ceccato, G. Caliaro, Ph. Hubert, B. Boulanger, *Chemometr. Intell. Lab.* 91(1) (2013) 4–16.
[48] B. Debrus, D. Guillarme, S. Rudaz, Improved quality-by-design compliant methodology for method development in reversed-phase liquid chromatography, *J. Pharm. Biomed. Anal.* 84 (2013) 215–223.
[49] P. Lebrun, B. Boulanger, B. Debrus, P. Lambert, Ph. Hubert, A Bayesian design space for analytical methods based on multivariate models and predictions, *J. Biopharm. Stat.* 23(6) (2013) 1330–1351.
[50] B. Debrus, P. Lebrun, A. Ceccato, G. Caliaro, E. Rozet, I. Nistor, R. Oprean, F.J. Rupérez, C. Barbas, B. Boulanger, Ph. Hubert, Application of new methodologies based on design of experiments, independent component analysis and design space for robust optimization in liquid chromatography, *Anal. Chim. Acta* 691(1–2) (2011) 33–42.
[51] S. Geisser, Bayesian estimation in multivariate analysis, *Ann. Math. Statist.* 36(1) (1965) 150–159.
[52] G.E.P. Box, G.C. Tiao, *Bayesian Inference in Statistical Analysis*, Wiley Classic Library, New York, USA, 1973.
[53] S.J. Press, *Applied Multivariate Analysis: Using Bayesian and Frequentist Methods of Inference*, Holt, Rinehart and Winston, New York, USA, 1972.
[54] J.J. Peterson, G. Miró-Quesada, E. Del Castillo, A Bayesian reliability approach to multiple response optimization with seemingly unrelated regression models, *Qual. Technol. Quant. M.* 6(4) (2009) 353–369.
[55] Stat-Ease, Inc., Design-Expert® Software Version 10, from www.stat-ease.com, 01 September 2017.
[56] SAS Institute, Inc., JMP® Software from www.jmp.com, 01 September 2017.
[57] Minitab Ltd, Minitab Software Version 18 from www.minitab.com, 01 September 2017.
[58] A.I. Khuri, S. Mukhopadhyay, Response surface methodology, *WIREs: Comp. Stat.* 2(2) (2010) 128–149.
[59] B. Debrus, P. Lebrun, J. Mbinze Kindenge, F. Lecomte, A. Ceccato, G. Caliaro, J. Mavar Tayey Mbay, B. Boulanger, R.D. Marini, E. Rozet, Ph. Hubert, Innovative high-performance liquid chromatography method development for the screening of 19 antimalarial drugs based on a generic approach, using design of experiments, independent component analysis and design space, *J. Chromatogr. A* 1218 (2011) 5205–5215.
[60] A. Dispas, P. Lebrun, B. Andri, E. Rozet, Ph. Hubert, Robust method optimization strategy — A useful tool for method transfer: The case of SFC, *J. Pharm. Biomed. Anal.* 88 (2014) 519–524.
[61] B. Debrus, P. Lebrun, A. Ceccato, G. Caliaro, E. Rozet, I. Nistor, R. Oprean, F.J. Rupérez, C. Barbas, B. Boulanger, Ph. Hubert, Application of new methodologies based on design of experiments, independent component analysis and Design Space for robust optimization in liquid chromatography, *Anal. Chim. Acta* 691(1–2) (2011) 33–42.
[62] M.H. Rafamantanana, B. Debrus, G.E. Raoelison, E. Rozet, P. Lebrun, S. Uverg-Ratsimamanga, Ph. Hubert, J. Quetin-Leclercq, Application of design of experiments

and design space methodology for the HPLC-UV separation optimization of aporphine alkaloids from leaves of *Spirospermum penduliflorum* Thouars, *J. Pharm. Biomed. Anal.* 62 (2012) 23–32.
[63] C. Lamalle, R.D. Marini, B. Debrus, P. Lebrun, J. Crommen, Ph. Hubert, A.-C. Servais, M. Fillet, Development of a generic micellar electrokinetic chromatography method for the separation of 15 antimalarial drugs as a tool to detect medicine counterfeiting, *Electrophoresis* 33(11) (2012) 1669–1678.
[64] B. Andri, P. Lebrun, A. Dispas, R. Klinkenberg, B. Streel, E. Ziemons, R.D. Marini, Ph. Hubert, Optimization and validation of a fast supercritical fluid chromatography method for the quantitative determination of vitamin D3 and its related impurities, *J. Chromatogr. A* 1491 (2017) 171–181.
[65] A. Dispas, V. Desfontaine, B. Andri, P. Lebrun, D. Kotoni, A. Clarke, D. Guillarme, Ph. Hubert, Quantitative determination of salbutamol sulfate impurities using achiral supercritical fluid chromatography, *J. Pharm. Biomed. Anal.* 134 (2017) 170–180.
[66] A. Vemic, T. Rakić, A. Malenović, M. Medenica, Chaotropic salts in liquid chromatographic method development for the determination of pramipexole and its impurities following quality-by-design principles, *J. Pharm. Biomed. Anal.* 102 (2015) 314–320.
[67] J. Pantović, A. Malenović, A. Vemić, N. Kostić, M. Medenica, Development of liquid chromatographic method for the analysis of dabigatran etexilate mesilate and its ten impurities supported by quality-by-design methodology, *J. Pharm. Biomed. Anal.* 111 (2015) 7–13.
[68] J.K. Mbinze, P. Lebrun, B. Debrus, A. Dispas, N. Kalenda, J. Mavar Tayey Mlbay, T. Schofield, B. Boulanger, E. Rozet, Ph. Hubert, R.D. Marini, Application of an innovative Design Space optimization strategy to the development of liquid chromatographic methods to combat potentially counterfeit nonsteroidal anti-inflammatory drugs, *J. Chromatogr. A* 1263 (2012) 113–124.
[69] R.K. Budrick, D.J. LeBlond, D. Sandell, H. Yang, H. Pappa, Statistical methods for validation of procedure accuracy and precision, *Pharm. Forum* 39 (2013).
[70] E. Rozet, A. Ceccato, C. Hubert, E. Ziemons, R. Oprean, S. Rudaz, B. Boulanger, Ph. Hubert, Analysis of recent pharmaceutical regulatory documents on analytical method validation, *J. Chromatogr. A* 1158 (2007) 111.
[71] E. Rozet, V. Wascotte, N. Lecouturier, V. Preat, W. Dewé, B. Boulanger, Ph. Hubert, Improvement of the decision efficiency of the accuracy profile by means of a desirability function for analytical methods validation. Application to a diacetyl-monoxime colorimetric assay used for the determination of urea in transdermal iontophoretic extract, *Anal. Chim. Acta* 591 (2007) 239.
[72] C. Hubert, S. Houari, E. Rozet, P. Lebrun, Ph. Hubert, Towards a full integration of optimization and validation phases: An analytical-quality-by-design approach, *J. Chromatogr. A* 1395 (2015) 88–98.

Chapter 6

Optimization of Peak Capacity

Krisztián Horváth
Department of Analytical Chemistry,
University of Pannonia, Egyetem u. 10,
8200 Veszprém, Hungary
raksi@almos.uni-pannon.hu

6.1 Introduction

The ultimate goal of analytical liquid chromatography is to provide high separation power in the shortest time possible. In HPLC, performance means peak width. The higher the performance of a chromatographic method, the narrower the peaks are on the chromatogram. Several measures exist for the quantification of quality of a separation or of a chromatogram. The most commonly used one is the number of theoretical plates, N, which is considered as a benchmark measure. The use of plate count, however, has disadvantages. Although it can estimate widths of peaks in isocratic measurements, it cannot be used in gradient separations directly, nor can it tell anything about the overall separation power of the chromatographic method. A column even with the highest plate count ever is useless if all the compounds elute together in a very narrow time range. Resolution, on the other hand, can be used for the characterization of separation quality of neighboring compounds in both isocratic and gradient runs. However, resolution does not serve any information on the general column performance. Peak capacity, a concept introduced by Giddings [1] in 1967, is a very intuitive and, at the same, time much more general measure than plate count and resolution. Peak capacity is the maximum number of components resolvable by HPLC with a unity resolution [2]. It combines the entire chromatographic space with the variability of the peak widths over the chromatogram. While the number of the actually resolved peaks depends on the nature of solutes existing in a particular mixture, peak capacity can be used to approximate the overall separation power of a given column. Since the introduction of the peak capacity concept, it has been used widely in chromatography, both in theoretical studies and method developments.

In method development, peak capacity has a significant importance for the analysis of complex samples (e.g. protein tryptic digests). Complete resolution of all components in these samples is often impossible by uni-dimensional chromatography due to the large number of compounds, even if it is smaller than the peak capacity offered by the method. In that case, the analyst should focus on decreasing the degree of overlap of the components by maximizing the peak capacity of the system. When simpler mixtures containing much fewer components are analyzed, the optimization of resolutions of pairs of compounds by adjusting the selectivities through the variation of separation conditions is a suitable approach. The effectiveness of this concept, however, is limited as the number of components becomes much larger than 15–20 [3].

Comparison of separations is not always a straightforward task. Giddings introduced [4] the concept of kinetic plots to compare the theoretical limit of separating speed of gas and liquid chromatography by plotting the logarithm of analysis time against the logarithm of plate count. This approach was used and extended by Knox and Saleem [5] and Guiochon [6]. In 1997, Poppe [7] proposed to plot plate time, t_0/N, against N to obtain a clearer comparison of chromatographic columns. Desmet *et al.* [8] extended Gidding's concept and generated a broad family of kinetic plots that allow the direct comparison of the performance of different LC supports. Since its introduction, applications of Poppe plots have become widespread in development and evaluation of stationary phases and efficient chromatographic methods. Even if Poppe plots were constructed for isocratic separations originally, they were extended for gradient [9] chromatography as well. In these approaches, the gradient times are used instead of t_0 to generate the Poppe plots.

In this chapter, possible concepts are presented for the optimization of chromatographic peak capacities. The majority of the results are dedicated to reversed-phase gradient chromatography, or at least separations modes where the linear solvent strength model [10] applies.

The most important algorithms used for the calculation of results of this chapter are presented in Python programming language.[a] The reader can use, modify and share it freely without the permission of the author. The main reasons of using Python in optimization of chromatographic separations are the following:

- Python is free and open source, whereas other closed-source commercial products can sometimes be very expensive.
- Python is easy to read and has relatively short learning curve.
- Python integrates well with other languages (e.g. C/C++, Fortran).
- A large number of general-purpose or more specialized libraries exists for Python.
- A huge scientific community is built up around Python. It is easy to find help and information from other scientists.

Python has impressive libraries applicable in everyday scientific tasks. The codes shared in this chapter are based on NumPy[b] and SciPy[c] libraries. These two libraries together cover most of MATLAB's basic functionality and form parts of many of the toolkits. Additionally, they have great documentation and an active community. The figures were generated with the Matplotlib[d] plotting library, which is able to produce publication-quality figures in a variety of formats and interactive environments.

As of the writing of this chapter (autumn of 2017), Python 3.6, NumPy 1.13, SciPy 1.0, and Matplotlib 2.0 are the actual versions of the language and libraries. The codes shared in this chapter were tested and worked with these versions. Even if Python language and its libraries are evolving gradually, the codes can be used directly or with slight modification for at

[a] https://www.python.org/.
[b] http://www.numpy.org/.
[c] https://www.scipy.org/.
[d] http://matplotlib.org/.

```
import matplotlib.pyplot as plt     #import of Matplotlib library
import numpy as np                  #import of NumPy library
#import functions from SciPy library
from scipy.optimize import brentq, minimize, curve_fit

def trgrad(tg, fi0, dfi, t0, S, k0):
    """
    Calculation of gradient retention time.
    For meaning of parameters, see Eqs. (12)-(21)
    """
    b = S*dfi/tg*t0
    kfi0 = k0*np.exp(-S*fi0)
    return t0*(1 + np.log(1+kfi0*b)/b)

def peakcap(tg, fi0, dfi, t0, H, L, S, k0):
    """
    Calculation of gradient peak capacity.
    For meaning of parameters, see Eqs. (12)-(21)
    """
    N = L/H
    b = S*dfi/tg*t0
    kfi0 = k0*np.exp(-S*fi0)
    kL = kfi0/(1+kfi0*b)
    p = b*kfi0/(1+kfi0)
    Theta = np.sqrt((1+p+1/3*p**2))/(1+p)
    Q = np.sqrt(1 + b + 1/3*b**2)
    tau = 1+kfi0*b
    return 1 + np.sqrt(N)/4/Q*np.log((b/6 + 1/b*(Q**2*tau-1) + Theta*Q*tau*(1+kL))/(1+b/2+Q))

def peakcapProduction(tg, fi0, t0, H, L, S, k0):
    """
    Calculation of gradient peak capacity production, (tG+t0)/n.
    For meaning of parameters, see Eqs. (12)-(21)
    """
    dfi = brentq(lambda dfi: trgrad(tg, fi0, dfi, t0, S, k0)-t0-tg, 1e-6, 1-fi0)
    return (tg+t0)/peakcap(tg, fi0, dfi, t0, H, L, S, k0)

def h(nu, A, B, C):
    """
    Calculation of height equvalent to a theoretical plate.
    For meaning of parameters, see Eqs. (31) and (32)
    """
    return A*nu**(1/3)+B/nu+C*nu

A, B, C = 1.0, 1.5, 0.05        #parameters of Knox equation Eq. (31)
fi = 1e3                        #column resistance factor
eta = 1e-3                      #dynamic viscosity (Pa sec),
dm = 1e-9                       #diffusion coefficient, m^2/sec
```

Listing 6.1: Libraries, constants and function definitions necessary to run Listings 6.3–6.6.

least a decade after publishing this book, most probably. Note that Python uses indentation to structure its programs and scripts into blocks. When using the codes presented in Listings 6.1–6.6, please pay careful attention to the leading spaces at the beginning of lines.

The author of this chapter recommends the installation of a Python distribution. In 2017, the two most popular and complete distributions aimed at the need of scientific community are Anaconda Python

Distribution[e] and Enthought Python Distribution.[f] These distributions contains all the necessary tools and libraries required to run the Python codes presented in Listings 6.1–6.6.

A part of the Python codes used during the construction of figures presented in this chapter were common in each program. In Listing 6.1, imports of the libraries, definitions of functions and constants that are necessary to run all the other codes are presented. The content of Listing 6.1 should be copied before the codes presented in Listings 6.2–6.6.

6.2 Theory

Peak capacity is the measure of the number of peaks that can fit into an elution time window t_1 to t_n with a fixed — usually unity — resolution [2]. There are several approaches for the derivation of peak capacity. Originally, it was defined by Giddings for isocratic chromatography, [1] and subsequently extended by Horváth and Lipsky [11] to gradient elution chromatography. Grushka [12] later also derived an equation for computing peak capacity in gradient elution.

Here, we follow Grushka's approach that is general enough to apply for both isocratic and gradient separations as well. According to this approach, peak capacity of a chromatographic separation can be calculated by the solution of an ordinary differential equation.

$$\frac{\mathrm{d}n}{\mathrm{d}t} = \frac{1}{w(t)} \tag{1}$$

with the following initial condition:

$$n(t_1) = 1 \tag{2}$$

where w is peak width generally referred to as four times standard deviation of a chromatographic peak ($w = 4\sigma$), t is time, and t_1 the retention time of the first eluting compound.

[e] http://www.anaconda.com/distribution/.

[f] http://www.enthought.com/product/enthought-python-distribution.

The solution of Eq. (1) requires knowledge of the peak widths as a function of retention time, $w(t)$. The general solution can be written as

$$n = 1 + \int_{t_1}^{t_n} \frac{1}{w(t)}\, dt \tag{3}$$

where t_n is the retention time of the last peak. Accordingly, the width of the accessible separation window is $t_n - t_1$.

6.2.1 *Peak capacity in isocratic elution*

Under isocratic elution conditions, the velocity of the sample bands are constant throughout the column. Widths of peaks are affected solely by kinetic processes.[g] The dependency of peak widths on retention time can be written as

$$w(t) = \frac{4}{\sqrt{N}}\, t \tag{4}$$

Therefore, the solution of Eq. (1) is

$$n = 1 + \frac{\sqrt{N}}{4} \ln \frac{t_n}{t_1} \tag{5}$$

t_n can be rewritten as the sum of t_1 and the relative retention window

$$t_n = t_1\,(1 + \delta) \tag{6}$$

where δ is the width of retention window relative to the retention time of the first compound

$$\delta = \frac{t_n - t_1}{t_1} \tag{7}$$

Note that δ is the retention factor, k, when t_1 equals to the column hold up time, t_0.

By combining Eqs. (5)–(7), peak capacity can be expressed as

$$n = 1 + \frac{\sqrt{N}}{4} \ln(1 + \delta) \tag{8}$$

[g]This statement is strictly true only under linear conditions when the isotherms of compounds are linear. Under nonlinear conditions, thermodynamic processes also influence peak shapes.

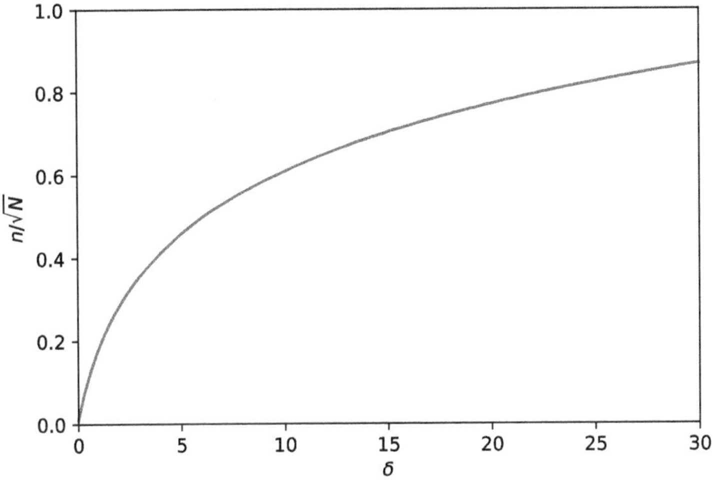

Figure 6.1: Isocratic peak capacity relative to the square root of N as a function of relative retention window, δ.

In Fig. 6.1, the isocratic peak capacity relative to the square root of N can be seen as a function of δ. The figure shows that the wider the retention window, the higher the achievable peak capacity is. The increase of n, however, is less remarkable at larger δ values.

It is important to note that peak capacity of isocratic separations is not the ratio of the retention window and average peak width. That would be

$$\frac{t_n - t_1}{\overline{w}} = \frac{t_n - t_1}{\frac{1}{t_n - t_1}\int_{t_1}^{t_n} w(t)\mathrm{d}t} = \frac{\sqrt{N}}{2}\frac{t_n - t_1}{t_n + t_1} \tag{9}$$

In the equations above, the extra-column band broadening was not taken into account. In several cases, however, extra-column processes have a large impact on the width of peaks. Assuming that the extra-column variance is σ_{ext}^2, peak widths and peak capacities can be rewritten as

$$w(t) = 4\sqrt{\frac{t^2}{N} + \sigma_{\mathrm{ext}}^2} \tag{10}$$

and

$$n = 1 + \frac{\sqrt{N}}{4} \ln \frac{1 + \delta + \sqrt{(1+\delta)^2 + \frac{\sigma_{\text{ext}}^2}{\sigma_1^2}}}{1 + \sqrt{1 + \frac{\sigma_{\text{ext}}^2}{\sigma_1^2}}} \tag{11}$$

where σ_1^2 is the variance of the first eluting peak.

6.2.2 *Peak capacity in gradient elution*

In gradient chromatography, eluent composition is varied during the separation in order to gradually decrease the retention of solutes. Estimation of retention times and peak widths requires the solution of two ordinary differential equations [13] and the knowledge of the change of eluent composition as a function of time, $\varphi(t)$ and the relationship between the retention factor and eluent composition, $k(\varphi)$. In gradient chromatography, it is impossible to derive general equations for the estimation of peak capacity due to the wide variety of the parameters that affect peak shapes. Several assumptions have to be defined regarding the shape of gradient and retention behavior of compounds.

Poppe *et al.* [13] derived simplified equations for the calculation of retention times and peak variances in the case of linear gradients and linear solvent strength (LSS) behavior which means that the composition of stronger eluent component was a linear function of time and that the isocratic retention of a solute ($\ln k$) was assumed to be a linear function of the volume fraction of the stronger eluent modifier (φ)

$$k = k_0 \exp(-S\,\varphi) \tag{12}$$

where k_0 the retention factor of the compounds for $\varphi = 0$, $-S$ the slope of $\ln k[\varphi]$ vs. φ plot. S is a practical measure of the retention sensitivity of a compound toward the change of eluent composition.

Under these conditions, the retention time of a compound, t_R, and the width of its peak can be calculated by the following set of equations [13]:

$$t_R = t_0 \left(1 + \frac{\ln(k_{\varphi_0} b + 1)}{b}\right) \tag{13}$$

$$w = 4\,\frac{L}{\sqrt{N}}\,\frac{1 + k_L}{u_0}\,\Theta \tag{14}$$

with b the gradient steepness

$$b = S\,t_0\,\frac{\Delta\varphi}{t_G} \tag{15}$$

where $\Delta\varphi$ is the change of stronger eluent component in t_G gradient time, k_L the retention factor of the compound at the column outlet. Θ represents the band compression [14, 15] that arises from the rear part of the band migrating at a velocity higher than the front part.

$$k_L = \frac{k_{\varphi 0}}{1 + k_{\varphi 0}\,b} \tag{16}$$

and

$$\Theta = \frac{\sqrt{1 + p + \frac{1}{3}p^2}}{1 + p} \tag{17}$$

where

$$p = b\frac{k_{\varphi 0}}{1 + k_{\varphi 0}} \tag{18}$$

and $k_{\varphi 0}$ is the retention factor of solute at the beginning of analysis $(\varphi = \varphi_0)$

By rearranging Eq. (13) for $k_{\varphi 0}$ and substituting it into Eq. (14), $w(t)$ can be generated. It still cannot be integrated since the value of b is different from solute to solute. Accordingly, an additional assumption has to be made regarding the constant values of S for all the sample compounds. In that case, peak capacity of linear gradients in case of LSS behavior becomes

$$n = 1 + \frac{\sqrt{N}}{4}\frac{1}{Q}\ln\left(\frac{\frac{b}{6} + \frac{1}{b}\left(Q^2\tau_n - 1\right) + \Theta_n\,Q\,\tau_n(1 + k_{L,n})}{1 + \frac{b}{2} + Q}\right) \tag{19}$$

with

$$Q = \sqrt{1 + b + \frac{1}{3}b^2} \tag{20}$$

and

$$\tau_n = \exp\left[b\left(\frac{t_n}{t_0} - 1\right)\right] \tag{21}$$

where $k_{L,n}$ and Θ_n refers to the last eluting compounds (see Eqs. (16) and (17)). Note that Eqs. (19)–(21) are essentially the same as Eqs. (14)–(17) of Ref. [16] derived by Gritti and Guiochon. However, here, they are presented in a different grouping of parameters.

In Fig. 6.2, gradient peak capacities as a function of t_G can be seen at different combinations of S and k_0 parameters. As opposed to isocratic separations (see Fig 6.1), peak capacities approach a maximal peak capacity as analysis time increases in gradient separations. The maximal peak capacity that can be achieved with gradient elution can be determined as the limit of Eq. (19) as t_G approaches infinity.

$$n_{\max} = 1 + \frac{\sqrt{N}}{4} \ln\left(1 + k_{\varphi 0,n}\right) \tag{22}$$

where $k_{\varphi 0,n}$ is the retention factor of the last eluting compound at the beginning of the analysis.

Peak widths and peak capacities calculated by Eqs. (14) and (19) are valid in the ideal case. In practice, several effects cause peak broadening downstream of the column (e.g. peak spreading in tubings, connections, detector cell, etc.). Since the retention factor of compounds are large at

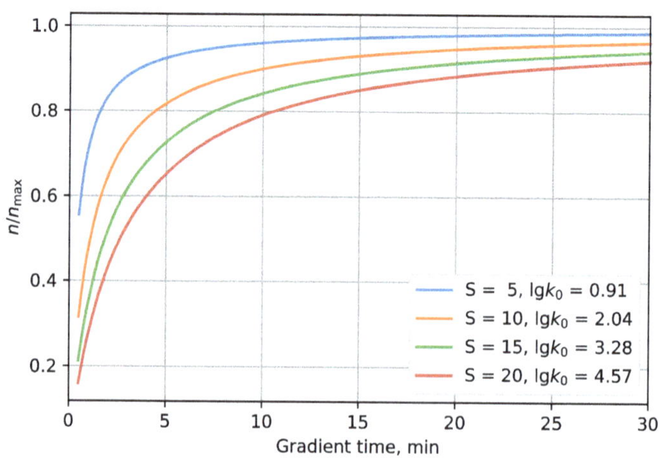

Figure 6.2: Gradient peak capacity relative to $n_{\max}$ as a function of gradient time, t_G. Parameters of calculation: column length $L = 10\,\text{cm}$, particle diameter $d_p = 1.7\,\mu\text{m}$, pressure drop $\Delta P = 1200\,\text{bar}$, plate count $N = 25{,}000$, column hold-up time $t_0 = 28.8\,\text{s}$.

the beginning of analysis, solutes are focused in narrow bands at the head of column after injection. Therefore, the pre-column effects usually do not affect final peak shapes. Thus, the peak variance has to be completed with the contributions of post-column broadening. Accordingly, widths of detected peaks can be calculated as

$$w \simeq 4\sqrt{\frac{L^2}{N}\left(\frac{1+k_L}{u_0}\right)^2 \Theta^2 + \sigma_{t,\text{pc}}^2} \tag{23}$$

where $\sigma_{t,\text{pc}}^2$ represents the contribution of the post-column processes to the variance of the peak.

6.3 Optimization of Peak Capacity

In practice, optimization of peak capacities is necessary when the sample contains a large number of compounds. In that case, the goal usually is not the complete baseline separation of all compounds, but the spreading of analytes as much as possible before introduction into mass spectrometer or a second chromatographic dimension. The analyst usually has two different goals: (1) achieving a given target peak capacity within as short a time as possible or (2) reaching the highest possible peak capacity in a given analysis time. In the following, general concepts and considerations are presented that are applicable to fulfill both goals of peak capacity optimization.

6.3.1 *Optimization of isocratic separations*

Close examination of Eq. (11) highlights that peak capacity in isocratic mode can be optimized by

- minimizing the extra-column band broadening (σ_{ext}^2),
- maximizing the retention window (δ),
- maximizing column efficiency (N).

6.3.1.1 *Extra-column band broadening*

It is well known that extra-column band broadening has a deteriorating effect on separation performance. The relative decrease of apparent column

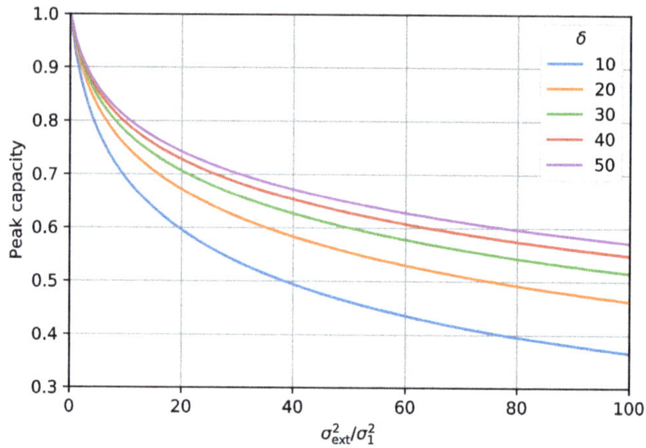

Figure 6.3: Relative peak capacity as a function of ratio of extra-column and column variance at different widths of retention window δ.

plate count depends on the relation of column- and extra-column variances.

$$\frac{\Delta N}{N} = -\frac{\sigma^2_{\text{ext}}}{\sigma^2_{\text{col}} + \sigma^2_{\text{ext}}} = -\frac{\sigma^2_{\text{ext}}}{\frac{t_R^2}{N} + \sigma^2_{\text{ext}}} \tag{24}$$

Typical values of contributions of UHPLC, optimized HPLC and non-optimized HPLC systems to the peak variances are 5, 25 and 100 μL^2 [17, 18], respectively.[h] Note, however, that the actual values vary from instrument to instrument. Since column variance is proportional to the square of retention time, the deteriorating effect of extra-column band broadening is more significant for the early eluting compounds. In Fig. 6.3, the decrease of peak capacities can be seen as the $\sigma^2_{\text{ext}}/\sigma^2_1$ ratio varies. The volumetric variance of a peak eluted at the hold-up time from a 150 × 4.6 mm column packed with 5 μm particles is typically larger than 175 μL^2. Therefore, $\sigma^2_{\text{ext}}/\sigma^2_1$ is lower than 0.5 when an older HPLC

[h] For practical reasons, volumetric variances are usually used for the quantification of extra-column band broadening (see Refs. [17, 18]). By dividing it with the square of flow rate, volumetric variances can be converted to temporal variances. Similarly, temporal variances can be converted to volumetric ones by multiplying with the square of flow rate. Therefore, volumetric variance of a chromatographic peak is the ratio of square of retention volume to the plate number, V_R^2/N.

instrument is used. Less than 5% decrease of peak capacity should be expected in that case. Using a 50 × 2.1 mm column packed with 1.7 μm particles, however, the variance of the peak eluting with the column dead volume can be less than 1 μL^2. Even a state-of-the-art instrument with a 5 μL^2 extra-column variance can decrease the peak capacity by more than 10%. Using these columns in outdated, non-optimized instrument, the peak capacity might be lower than half of that provided by the column itself. The effect of σ^2_{ext} is more significant when the retention window is narrow. Note that the decrease of peak capacity is unaffected by the actual column plate number in practice when N is sufficiently large.

It can be concluded that extra-column processes might have a significant effect on the achievable peak capacity depending on the type of column and instrument used. Accordingly, the extra-column effects cannot be neglected in the optimization of isocratic peak capacities, especially when ultra-high performance columns are used. The minimization of σ^2_{ext} by reducing volumes of capillaries and detector is necessary when the goal of separation is to achieve high peak capacities in a reasonable analysis time.

6.3.1.2 *Width of retention window*

Equation (11) clearly shows that the wider the retention window, the more powerful the separation is. After some simple considerations, the relative retention window, δ, can be rewritten as

$$\delta = \begin{cases} (\alpha - 1)\dfrac{k_1}{1 + k_1} & \text{if } k_1 > 0 \\ k_n & \text{if } k_1 = 0 \end{cases} \tag{25}$$

where α is the selectivity of the last and first eluting solutes, $\alpha = k_n/k_1$.

Equation (25) has similar form than the simplified resolution equation, which is one of the most important relationships in development and optimization of isocratic separations (see e.g. Eq. (2.24) of Ref. [19]). Equation (25) leads to the important conclusion that peak capacity of isocratic separations can be improved with the increase of α.

The selectivity between the first and last eluting solutes can be adjusted by the proper choice of stationary phase and separation conditions. Variation of mobile phase composition can be a suitable strategy if the compounds respond differently to the change of eluent composition. In reversed-phase and hydrophilic interaction modes, this condition usually, applies especially in the case of analysis of biological samples. In ion chromatography, however, only selectivities of ions having different charge can be modified by the change of the concentration of the electrolyte. When the ions have the same charge, other approaches should be applied for the variation of α (e.g. addition of organic modifier to the eluent).

The change of separation temperature can also be a suitable option for the selectivity improvement, especially when the difference between the adsorption enthalpies, ΔH, of the first and last eluting compounds is large. Since late eluting solutes usually have higher affinities toward the stationary phase, the decrease of temperature might improve α. Note, however, that it increases the pressure drop of the column and might decreases the overall column efficiency significantly.

Equation (25) shows that by increasing the retention factor of the first eluting peak, k_1, peak capacity of isocratic separation can also be improved supposing that α remains constant. Note, however, that this scenario is rather theoretical, it does not have any significance in practice. Increasing the retention of the first eluting compound while α is kept constant would increase the analysis time so drastically that the cost of analysis in time and solvent consumption would be too much for the extra information gained by the improved peak capacity. Instead, during the optimization of isocratic separations, α should be maximized by increasing k_n and decreasing k_1 as low as possible, ideally until zero.

It is worth studying the rate of peak capacity production, ν_n, with the increase of analysis time. Rate of peak capacity production can be defined as

$$\nu_n = \frac{\mathrm{d}n}{\mathrm{d}t_n} = \frac{1}{4}\frac{\sqrt{N}}{t_n} \tag{26}$$

Equation (26) reveals that ν_n is the highest at the beginning of chromatogram ($t_n = t_0$), and it decreases gradually as the time passes. The

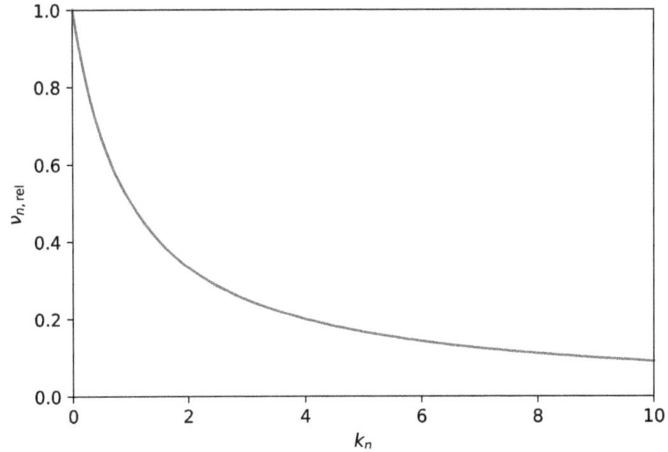

Figure 6.4: Rate of peak capacity production relative to the initial ν_n.

rate of peak capacity production relative to the initial ν_n is given by

$$\nu_{n,\mathrm{rel}} = \frac{\frac{1}{4}\frac{\sqrt{N}}{t_n}}{\frac{1}{4}\frac{\sqrt{N}}{t_0}} = \frac{1}{1+k_n} \qquad (27)$$

In Fig. 6.4, $\nu_{n,\mathrm{rel}}$ can be seen as a function of retention factor of the last eluting component. It is obvious that most of the peak capacities are gained close to the hold up time of the column. As the analysis time increases, the rate of peak capacity production decreases remarkably. This is due to the fact that peaks become wider as their retention times increase, as is shown by Eq. (4). As the retention of compounds increases, their migration velocity decreases, and so more and more time is necessary to elute one band from the chromatographic column.

A 20-peak-capacity chromatogram can be seen in Fig. 6.5. The figure clearly shows the peak generation phenomenon discussed above. Most of the eluted peaks appear at the beginning of the chromatogram. Late eluting peaks are wider, and more time is necessary to ensure unity resolution between them. Half of the total peak capacities are generated in the first third of the analysis time. Figures 6.4 and 6.5 emphasize the importance of reducing the retention of the first eluting peak. These figures also highlight the necessity of reducing extra-column broadening, since most of the

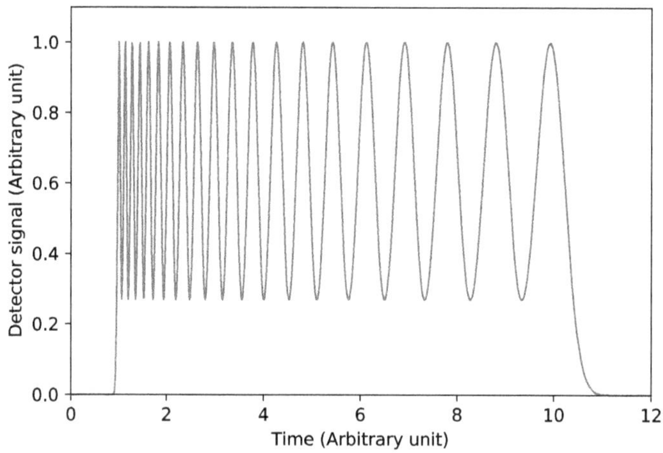

Figure 6.5: Chromatogram of a 20-peak-capacity separation in the case of isocratic elution.

peak capacities are generated in the beginning of analysis time. The narrow, early eluting peaks are more sensitive toward the detrimental effect of extra-column processes.

Specific peak capacity production, n', shows how much total peak capacities are generated in a given unit of time. It can be defined as

$$n' = \frac{n}{t_n} \tag{28}$$

Specific peak capacity production has an optimum at

$$t_{n,\mathrm{opt}} = t_1 \exp\left(1 - \frac{4}{\sqrt{N}}\right) \simeq 2.718\, t_1 \tag{29}$$

with a maximal value of

$$n'_{\mathrm{opt}} = \frac{1}{t_{n,\mathrm{opt}}} \frac{\sqrt{N}}{4} \simeq 0.092\, \frac{\sqrt{N}}{t_1} \tag{30}$$

In Fig. 6.6, specific peak capacity production can be seen as a function of analysis time. Note that the time axis is scaled for $t_{n,\mathrm{opt}}$, and the y axis is scaled for n'_{opt}. The figure clearly represent that the specific gain of peak capacity decreases with the increase of analysis time. Accordingly, the

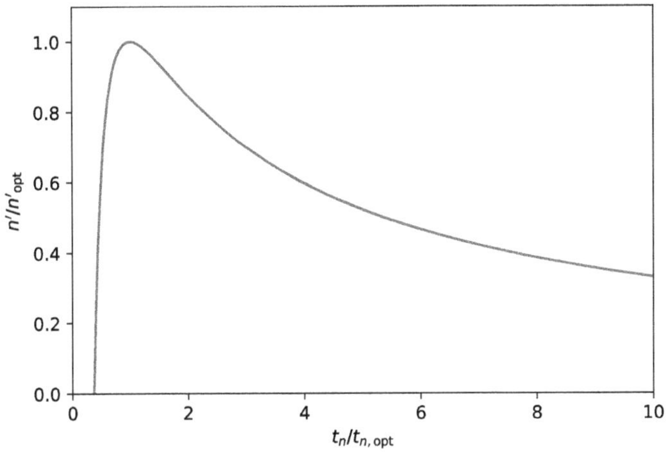

Figure 6.6: Specific peak capacity production, Eq. (30), as a function of analysis time.

analyst has to make a trade-off between the analysis time and separation power of the chromatographic system based on the knowledge based on the sample composition analyzed.

6.3.1.3 *Plate number*

A third option for increasing the of peak capacity is the optimization of number of theoretical plates, N. In order to estimate column efficiency under different separation conditions, one should choose a proper plate height or plate number model. The most accurate equation for describing the dependence of the plate height on mobile phase linear velocity is offered by the general rate model. It is, however, so complex that it is rarely used in method development. The van Deemter and the Knox equations are the most widely used plate height equations in practice. The simple form of the analytical solution for the minimum plate height obtained from van Deemter's equation allows one to locate the optimum velocity and minimum plate height and to obtain insight into the contribution of kinetic processes to it. The Knox equation, however, provides a better fit to liquid chromatography data than van Deemter equation. Therefore, it will be used in the following.

Knox equation can be defined as

$$h = A\,\nu^{\frac{1}{3}} + \frac{B}{\nu} + C\,\nu \tag{31}$$

where A, B and C are dimensionless constant parameters. Their typical values are 0.8–1.0, 1.5 and 0.02–0.05, respectively; h is the reduced plate height that is the ratio of the height equivalent to a theoretical plate and the particle diameter, $h = H/d_p$, and ν is the reduced velocity.

$$\nu = \frac{u_0\,d_p}{D_m} \tag{32}$$

where u_0 is the linear velocity of the mobile phase, d_p, the particle size and D_m, the diffusion coefficient of the solute molecules. Note that the reduced velocity is the same as the Péclet number used in study of transport phenomena.

An obvious way for peak capacity optimization in isocratic chromatography is the minimization of Eq. (31). Unfortunately, the exact solution of the minimum plate height based on Knox equation is too complicated to be informative. In Listing 6.2, however, a simple Python code is presented for obtaining the parameters of Knox plate height equation and the optimal eluent velocity.

As was discussed in the introduction, a popular approach of characterization and comparison of column performances is the use of Poppe plots. In a single graph, it includes information on the plate number, the analysis time, and the maximum pressure that can be applied with the chromatographic system. Therefore, it is more capable for optimization purposes than a single plate height equation. In a Poppe plot, the plate time (H/u_0 or N/t_0) is plotted against separation efficiency. During the construction of these plots, it is assumed that the chromatographic system is operated at the maximum allowed pressure, ΔP. The latter can be described by the Kozeny–Carman equation

$$\Delta P = \frac{\phi\,u_0\,\eta\,L}{d_p^2} \tag{33}$$

where L is the column length, η the dynamic viscosity of the eluent and ϕ the column resistance factor, which is in the range of 500–1000 (1000 is assumed in this work).

```
    #-#-#-#    Use with Listing 6.1    #-#-#-#

#Experimental data. Replace with actual data!
#eluent linear velocities, m/s
u0s = np.array([9.62e-05, 1.443e-04, 1.924e-04, 2.405e-04, 3.849e-04,\
         4.811e-04, 9.623e-04, 1.4435e-03, 1.9247e-03, 2.4059e-03,\
         2.8871e-03, 3.3683e-03, 3.8495e-03, 4.3307e-03, 4.8119e-03,\
         5.2931e-03, 5.7743e-03, 6.2555e-03, 6.7367e-03, 7.2179e-03,\
         7.6991e-03, 8.1802e-03])
#number of theoretical plates at u0s eluent velocities
Ns = np.array([3377.5, 4773.7, 6232.9, 7110.8, 9453.8, 11123.4,\
      14019.4, 14868.8, 14336.9, 13828.5, 13666.4, 12903.1, 12553.7,\
      12279.4, 11725.8, 11183.2, 11100.6, 10710.1, 10196.1, 10023.4,\
      9899.3, 9699.0])

dp = 1.7e-6          #particle diameter, m
L = 0.05             #column length, m
nus = u0s*dp/dm      #eluent reduced velocities
Hs = L/Ns            #heights equivalent to a theoretical plate, m


fp, fcv = curve_fit(h, u0s*dp/dm, Hs/dp)  #fitting Knox equation on experimental data
A, B, C = fp[0], fp[1], fp[2]             #parameters of Knox equation
print('A = {0:.3f}, B = {1:.3f}, C = {2:.3f}'.format(A, B, C)) #print results

opt = minimize(h, 2, args=(A, B, C))    #minimize Knox equation with parameters A, B, C
print('Minimum reduced HETP: {0:.3f}, minimum HETP: {1:.3f} um'\
    .format(opt.fun, opt.fun*dp*1e6))         #print minimum HETP
print('Optimal reduced velocity: {0:.3f}, optimal velocity: {1:.3f} cm/min'\
    .format(opt.x[0], opt.x[0]*dm/dp*100*60)) #print optimal velocity
```

Listing 6.2: Python code for determination of parameters of Knox plate equation.

Considering that the column length, L, is the product of the required plate count, N_{req}, and the plate height (that is $h\, d_p$), Eq. (33) can be rewritten as

$$\Delta P = \frac{\phi\, u_0\, \eta\, N_{\text{req}}\, h}{d_p} \tag{34}$$

From Eq. (34), it is possible to calculate the maximal plate number generated when the column works at the optimal eluent velocity.

$$N_{\text{opt}} = \frac{d_p^2\, \Delta P}{\nu_{\min}\, D_m\, \eta\, \phi\, h_{\min}} \tag{35}$$

where $h_{\min}$ is the minimum reduced plate height obtained at $\nu_{\min}$ optimal reduced eluent velocity.

The column length necessary to generate N_{opt} plate numbers is

$$L_{\text{opt}} = N_{\text{opt}}\, h_{\min}\, d_p = \frac{d_p^3\, \Delta P}{\nu_{\min}\, D_m\, \eta\, \phi} \tag{36}$$

with a dead time of

$$t_{0,\mathrm{opt}} = \frac{N_{\mathrm{opt}}^2\, h_{\min}^2\, \eta\, \phi}{\Delta P} = \frac{d_p^4\, \Delta P}{\nu_{\min}^2\, D_m^2\, \eta\, \phi} \tag{37}$$

Note that Eq. (35) is not the maximum plate number that can be generated by the given stationary phase. N_{opt} is the maximum achievable plate count when the column is operated at the optimal eluent velocity and the column length is maximized in order to reach ΔP. The overall maximum of plate number is

$$N_{\max} = \frac{d_p^2\, \Delta P}{B\, D_m\, \eta\, \phi} \tag{38}$$

where B is a parameter of the plate height Eq. (31). If van Deemter equation is used to estimate column efficiency, $N_{\max}$ becomes the same mathematically as in case of Knox equation. Considering the typical values of B, $\nu_{\min}$ and $h_{\min}$, it can be predicted that $N_{\max}$ is $\sim$3 times larger than N_{opt}. This observation serves the important conclusion that plate number can be further increased by using longer columns even if the eluent velocity becomes smaller than the optimal one.

The minimum reduced plate height of a well-packed column is $\sim$2.0 in the case of fully porous and $\sim$1.7 in case of core–shell phases with optimal reduced flow rate 2.0–3.0. Assuming that $\nu_{\min}$ is 2.8, N_{opt} of a column packed with 1.7 μm fully porous particles operated at 1200 bar pressure is 62,000 for small molecules ($D_m = 10^{-9}$ cm^2/s). A 21-cm long column with a slightly more than 2 min dead time is necessary to obtain this separation performance. For 5 μm particles and 400 bar pressure drop, N_{opt} is larger, $\sim$180,000. This efficiency can be generated with a 1.8-m long column and 53 min dead time.

Equations (35)–(38) serve some important conclusions. The achievable plate number is directly proportional to the applied pressure and to the square of particle diameter. Accordingly, the larger the particles used, the higher the plate number that can be generated by the chromatographic system. A two-fold increase of d_p can produce 4 times more plates. It was shown in Eq. (8) that the peak capacity is proportional to the square root of N. Therefore, n is directly proportional to the particle diameter and to the square root of pressure of separation. One can generate more peak capacities by using larger particles and higher operating pressures. At the same time, however, the time taken for analysis increases with the fourth power of d_p. It means that a two-fold increase of peak capacity requires 16 times more analysis time and an 8 times more longer column if the peak capacity is increased by the duplication of particle size. The same improvement can be achieved by quadruplication of the pressure. In that case, both the analysis time and column length are quadrupled. Note that these conclusions are valid numerically only for columns operated at the optimal eluent velocity. The tendencies, however, are valid in any eluent conditions.

In Eq. (35), there is no column length. The idea behind Eq. (35) is that the column works at the optimal flow rate that produces the minimal plate height. The column length is adjusted in order to generate the maximal pressure drop, ΔP. In the construction of Poppe plots, the same approach can be used with the difference that the eluent velocity is varied in order to generate the required plate number. The following equation is solved for ν:

$$\frac{\Delta P\, d_p}{\phi\, \eta\, N_{\text{req}}} - \frac{\nu\, D_m}{d_p}\, h = 0 \tag{39}$$

Note that h is a function of ν.

The plate height, plate number, column length and dead time are calculated by appropriate substitutions into Eqs. (31) and (35)–(37). By plotting H/u_0 or N/t_0 against N_{req}, the Poppe plot can be constructed.

A simplified approach for the construction of Poppe plot is to vary column length in a wide range. It defines the value of u_0 from Eq. (33). u_0 allows the calculation of plate height by Eq. (31), then the plate number as L/H, and the dead time as L/u_0. By plotting N/t_0 against N_{req}, the Poppe plot can be constructed. In Listing 6.3, this simplified approach is presented for the construction of Poppe plot. It can be seen that this approach does not need any numerical optimization algorithm.

In Fig. 6.7, the Poppe plot of different diameter column packings can be seen. The pressure drop is varied according to the typical maximum pressure used with these particles. Since square root of N is required for

```
    #-#-#-#     Use with Listing 6.1     #-#-#-#

ls = np.logspace(-4, 2, 1000)        #definition of range of column lengths

#------------CONSTRUCTION OF POPPE PLOTS------------#
#iterate through  d_p - ΔP data pairs,
#dp - particle diameter, dP - max pressure drop
for dp, dP in [(1.3e-6, 1200e5), (1.7e-6, 1200e5), (1.7e-6, 400e5), (3e-6, 600e5), \
        (5e-6, 400e5)]:
    u0s = dP*dp**2/(fi*eta*ls)       #eluent linear velocities, see Eq. (33)
    nus = u0s*dp/dm                  #reduced velocities,  Eq. (32)
    Ns = ls/(h(nus, A, B, C)*dp)     #number of theoretical plates
    t0s = ls/u0s                     #dead times
    plt.loglog(np.sqrt(Ns), t0s/np.sqrt(Ns))    #plot results

#------------PLOTTING OF CONSTANT t0 LINES------------#
Ns = np.logspace(2,6,10)             #definition of range of plate numbers
#iterate through different t0s
for t0 in [1e-1, 1e0, 1e1, 1e2, 1e3, 1e4]:
    #at each t0, plot t0/√N
    %%with black dashed line
    plt.loglog(np.sqrt(Ns),  t0/np.sqrt(Ns), 'k--', linewidth=0.5)

plt.xlim(1e1,1e3)                    #set x limits
plt.ylim(1e-3,1e2)                   #set y limits
plt.xlabel(r'$\lg \sqrt{N}$')        #set label of x axis
plt.ylabel(r'$\lg t_0 / \sqrt{N}$') #set label of x axis
plt.show()                           #show plot
```

Listing 6.3: Python code for construction of Poppe plot.

the estimation of isocratic peak capacities, $t_0/\sqrt{N}$ is plotted against $\sqrt{N}$ in the figure. Since the value of $\frac{\ln(1+\delta)}{4}$ in Eq. (8) is close to unity under most of the practically relevant conditions ($\delta > 15$), Fig. 6.7 can be considered as a kinetic plot of isocratic peak capacities. The diagonal lines represent zones of constant analysis times.

Figure 6.7 demonstrates clearly that, in the practically relevant range of analysis times ($t_0 = 10 - 100$ s), higher peak capacities can be generated with columns packed by smaller packing material than by larger ones, provided that each column is operated at the highest allowed pressures. Similarly, it is possible to achieve the same peak capacity in significantly shorter analysis times by applying ultra-high performance stationary phases. The advantages of larger particle sizes arose when the goal was to produce very high peak capacities (>500). The vertical asymptotes correspond to the square root of maximal plate counts as is calculated by Eq. (38). As can be seen, with the use of larger particles higher peak capacities can be achieved. The cost of this separation power is the

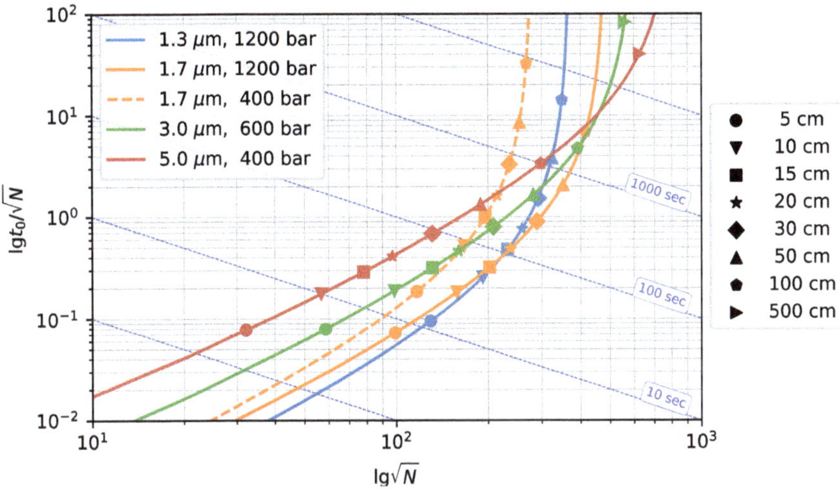

Figure 6.7: Poppe plot of isocratic peak capacity. Parameters of calculations: maximal pressure drop $\Delta P = 400$ bar, viscosity $\eta = 0.001$ Pa s, flow resistance factor $\phi = 1000$, diffusion coefficient $D_m = 10^{-9}$ m/s, reduced plate height expression Eq. (31) with parameters $A = 1.0$, $B = 1.5$, $C = 0.05$ [7].

extremely large analysis time, however. The figure also emphasize that the use of 1.7 μm particles in a 400-bar HPLC system does not offer significant improvement over larger particles in the practically relevant range of analysis times.

In general, it can be concluded that when time is not a limiting factor, the peak capacity of an isocratic separation can be maximized by the increase of retention window and the use of large particles and the longest possible columns consistent with the pressure limit of the instrument.

Even if Poppe plot allows for a detailed comparison of different stationary phases and separation strategies, most of the points on the curves do not have any practical relevance. No one has, e.g. a 17.9-cm long column packed with 5 μm particles to generate 8000 plates. Instead, there are one or more 5, 10, and 15-cm long columns in the drawer. In Fig. 6.7, points calculated for column lengths that are possible to combine from commercially available columns are also presented. These points present the practically relevant separation conditions. By using

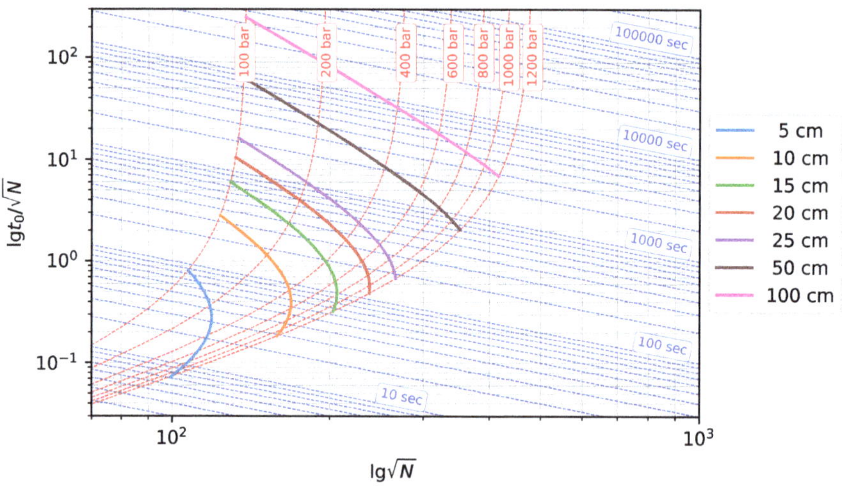

Figure 6.8: Nomogram for the design and optimization of isocratic separation with 1.7 μm particles. Parameters for calculation can be found in the caption of Fig. 6.7.

these points, one can easily compare different separation strategies and decide on the most appropriate one considering the required peak capacity and the available instrumentation and consumables present in the analytical lab.

A more complete design and optimization of isocratic separation can be achieved by constructing nomogram-like Poppe plots, as is shown in Fig. 6.8. This figure is calculated for 1.7 μm particles. Red dashed lines represents pressure drops, blue dashed lines the column hold-up times and the thick color lines some typical column lengths that can be combined by connecting commercially available columns. Figure 6.8 gives a deep insight into the influence of chromatographic conditions on the achievable separation power. It can be concluded that by increasing the column length, the plate number can be improved even if ΔP remains constant. The increase of ΔP always decrease the plate time, even if N might decrease since the eluent velocity exceeds the optimal one. When the columns are short, it is advantageous to operate the column at the optimal flow rate. For large columns, maximal ΔP is smaller than that necessary to produce that eluent flow rate.

Nomograms such as Fig. 6.8 can be used in method development directly. First, the analyst should define the maximal pressure drop applicable. Then, moving along the "isobar", the column length that produces the desired plate count in an acceptable analysis time can be found. One can generate nomograms like Fig. 6.8 for any phases available in the lab. The optimal column dimensions, stationary phases and operating conditions can be selected directly by the comparison of these nomograms.

In Listing 6.4, a Python code is presented for the generation of nomogram-like Poppe plots. Note that the import of NumPy and Matplotlib packages, the definition of reduced plate height equation and some constant parameters are not included in Listing 6.4. Those can be found in Listing 6.1. Therefore, the two codes should be used together in order to generate the nomogram.

```
    #-#-#-#    Use with Listing 6.1    #-#-#-#

dp = 1.7e-6        #definition of particle diameter, m

#------------CONSTRUCTION OF "ISOBARS"------------#
ls = np.logspace(-4,2, 1000)       #definition of range of column lengths
#iterate through different pressure drops
for dP in [100e5, 200e5, 400e5, 600e5, 800e5, 1000e5, 1200e5]:
    u0s = dP*dp**2/(fi*eta*ls)             #eluent velocity, m/sec, see  Eq. (33)
    Ns = ls/(h(u0s*dp/dm, A, B, C)*dp)     #plate numbers
    t0s = ls/u0s                           #column hold up times, sec
    plt.loglog(np.sqrt(Ns), t0s/np.sqrt(Ns), 'r--', linewidth=0.5)  #plot isobars

#------------PLOTTING OF CONSTANT t0 LINES------------#
Ns = np.logspace(2,6,1000)          #definition of range of plate numbers
#iterate through different t0s
for i in [1e-1, 1e0, 1e1, 1e2, 1e3, 1e4, 1e5]:
    #at each t0, plot t0/√N
    %%with blue dashed line
    plt.loglog(np.sqrt(Ns),  i/np.sqrt(Ns), 'b--', linewidth=0.5)

#------------CONSTRUCTION OF POPPE CURVES FOR DIFFERENT COLUMN LENGTHS------------#
dP = np.logspace(7,np.log10(1200e5), 1000)    #definition of pressure range, Pa
#iterate through different column lengths that have practical relevance
for length in [0.05, 0.1, 0.15, 0.2, 0.25, 0.5, 1]:
    u0s = dP*dp**2/(fi*eta*length)         #eluent velocity, m/sec, see  Eq. (33)
    Ns = length/(h(u0s*dp/dm, A, B, C)*dp)     #plate numbers
    t0s = length/u0s                       #column hold up times, sec
    plt.loglog(np.sqrt(Ns), t0s/np.sqrt(Ns)) #plot Poppe curves in log-log scale

plt.xlim(70,1e3)                      #set x limits
plt.ylim(1e-2,1e3)                    #set y limits
plt.xlabel('$\lg \sqrt{N}$')          #label of x axis
plt.ylabel(r'$\lg t_0 / \sqrt{N}$') #label of y axis
plt.show()                            #show plot
```

Listing 6.4: Python code for construction of nomogram-like isocratic Poppe plot.

6.3.2 *Optimization of gradient separations*

6.3.2.1 *Extra-column broadening*

It was shown previously that extra-column band broadening has a detrimental effect on achievable peak capacities in isocratic elution. In Fig. 6.9, a typical 20-peak-capacity chromatogram of gradient separation is shown. The timescale is the same as in Fig. 6.5. It can be seen that the same peak capacity could be generated in much less time. The peaks in Fig. 6.9 remain narrow throughout the whole separation range. Accordingly, in gradient elution the extra-column effects should be more significant than in isocratic runs.

In Fig. 6.10, the relative decrease of peak capacities are shown for a 50×2.1-mm column packed with 1.7 μm particles and a 150×4.6-mm column packed with 5 μm particles as a function of volumetric extra-column variance. Note that typical values of extra-column variances of UHPLC, optimized HPLC and non-optimized HPLC systems are 5, 25 and 100 μL^2 [17, 18], respectively. Figure 6.10 emphasize the necessity of minimizing extra-column volumes of chromatographic system. Even a state-of-the-art

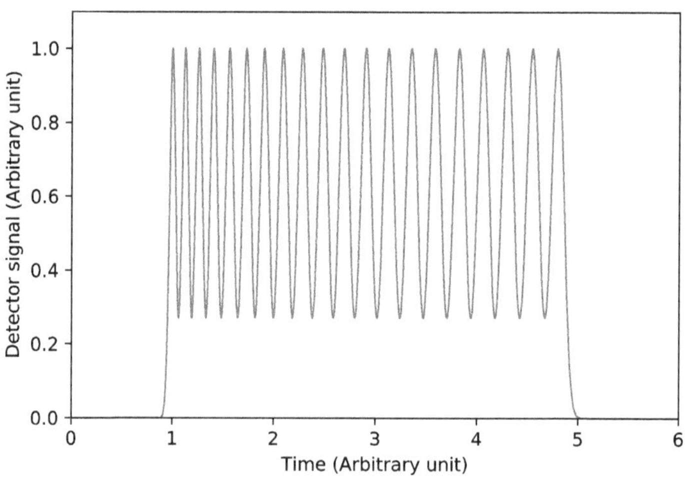

Figure 6.9: Chromatogram of a 20-peak-capacity separation in the case of gradient elution.

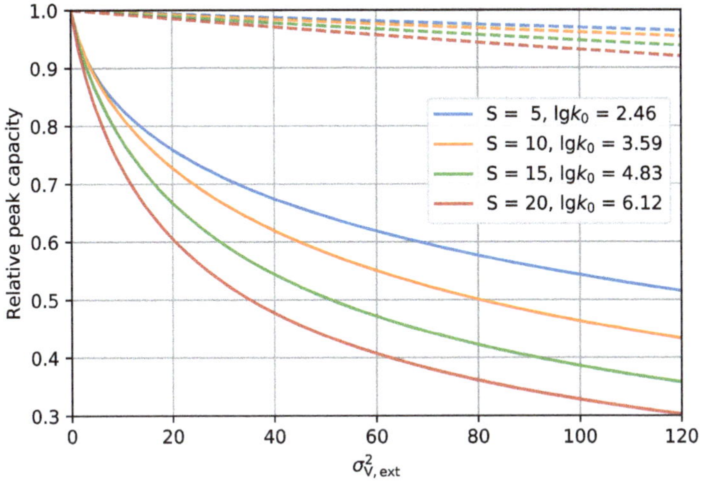

Figure 6.10: Relative decrease of gradient peak capacity as a function of extra-column variance. Solid lines: 50 × 2.1 mm column packed with 1.7 μm particles ($H = 2.81$), dashed lines: 150 × 4.6 mm column packed with 5 μm particles ($H = 2.97$).

chromatograph can decrease the peak capacity of an ultra-high performance column by 10%. The use of these columns in an obsolete hardware is senseless practically. Even a system with 20 μL^2 extra-column variance — that corresponds to a well-optimized conventional HPLC or even some UHPLC systems — might decrease n by 20–40%. For large columns, with large dead volumes, the effect of system volume is less detrimental.

6.3.2.2 *Gradient conditions*

Optimization of conditions of gradient separations is a much more complex task than that of isocratic separations. Some of the parameters affecting peak capacity of gradient runs are not mutually independent. Therefore, numerical algorithms should be used in order to find the optimal separation conditions.

Since Poppe plot can be applied directly in isocratic method development (see Fig. 6.8), it would be useful to apply the same concept in gradient runs as well. There are several approaches to construct gradient kinetic plots [9,20–24]. Here, we use Eq. (19) as the basis for calculations. Since gradient peak capacity is calculated by integrating $1/w(t)$ between t_0 and the retention time of the last eluting compound, t_n, application of Eq. (19) requires that t_n be equal to the sum of gradient and hold up times, $t_n = t_0 + t_G$. It ensures that the last compound elutes exactly at the time when the gradient leaves the column. This scenario can be called as an "utterly utilized gradient". By rearranging Eq. (13), t_G can be calculated by the following equation:

$$t_G = k_{\varphi 0} \frac{S\, t_0\, \Delta\varphi}{\exp\left(S\, \Delta\varphi\right) - 1} \tag{40}$$

A simple strategy to construct gradient Poppe plot (Listing 6.5) is varying column length in a relatively wide range while particle diameter of

```
    #-#-#-#     Use with Listing 6.1     #-#-#-#

fi0, dfi = 0.05, 0.7        #initial eluent composition, change of eluent composition
k0, S = 1e6, 20             #parameters of linear solvent strenth model, Eq. (12)
#retention factor of the eluting compound at the beginning of analysis
kfi0 = k0*np.exp(-S*fi0)

#------------PLOTTING OF CONSTANT tg+t0 LINES------------#
pcs = np.logspace(1,3,10)           #definition of range of peak capacities
#iterate through different t0+tG values
for tgt0 in [1e1, 1e2, 1e3, 1e4, 1e5]:
    #at each t0+tG, plot (t0+tG)/n with blue dashed line
    plt.loglog(pcs, tgt0/pcs, 'b--', linewidth=0.5)

#------------CONSTRUCTION OF GRADIENT POPPE CURVES FOR DIFFERENT COLUMN LENGTHS------------#
ls = np.logspace(-3,1,100)    #definition of range of column lengths
#iterate through different  dp - ΔP data pairs,
#dp - particle diameter, dP - max pressure drop, c - color, m - linestyle
for dp, dP, c, m in [(1.7e-6, 1200e5, 'C0', '-'), (1.7e-6, 400e5, 'C0', '--'), \
        (3e-6, 600e5, 'C1', '-'), (3e-6, 400e5, 'C1', '--'), (5e-6, 400e5, 'C2', '-')]:
    u0s = dP*dp**2/(fi*eta*ls)      #eluent linear velocities
    Hs = h(u0s*dp/dm, A, B, C)*dp   #plate heights
    t0s = ls/u0s                    #column hold-up times
    tgs = kfi0*S*dfi*t0s/(np.exp(S*dfi)-1)    #gradient times, see Eq. (40)
    pcs = peakcap(tgs, fi0, dfi, t0s, Hs, ls, S, k0)  #peak capacities, see
    Eq. (19)
    #plot curves in log-log scale with 'm' linestyle, 'c' color. Add label to legend.
    plt.loglog(pcs, tgs/pcs, m, color=c, \
            label=r'{0:1.1f} $\mu$m, {1:4.0f} bar'.format(dp*1e6, dP*1e-5))

plt.xlim(30, 1000)                          #set x limits
plt.ylim(0.1,500)                           #set y limits
plt.xlabel('Gradient peak capacity, $n$') #label of x axis
plt.ylabel(r'$(t_G+t_0)/n$')                #label of y axix
plt.legend()                                #show legend
plt.show()                                  #show plot
```

Listing 6.5: Python code for construction of gradient Poppe plot shown in Fig. 6.11.

the stationary phase, d_p, initial eluent concentration, φ_0, and the change of eluent composition, $\Delta\varphi$, are set constant. At each column length, the maximal eluent velocity, $u_{0,\max}$, is calculated. It defines the values of plate height, H, and column hold-up time, t_0. The gradient time, t_G is determined by Eq. (40). Finally, the peak capacity of the separation is calculated using Eq. (19) at each column length. By plotting the peak time, t_{peak}, against peak capacity, one can construct the gradient Poppe plot. Peak time is the ratio of total analysis time and peak capacity.

$$t_{\text{peak}} = \frac{t_G + t_0}{n} \tag{41}$$

Note that in the construction of gradient Poppe plot, the column hold-up time should be taken into consideration.

Figure 6.11 shows the gradient Poppe plots of columns operated at different pressures and packed with particles of different sizes. For the sake of comparability, the same viscosity and diffusion coefficient were used for the calculations as in Figs. 6.7 and 6.8, even if the applied k_0 (10^6) and S

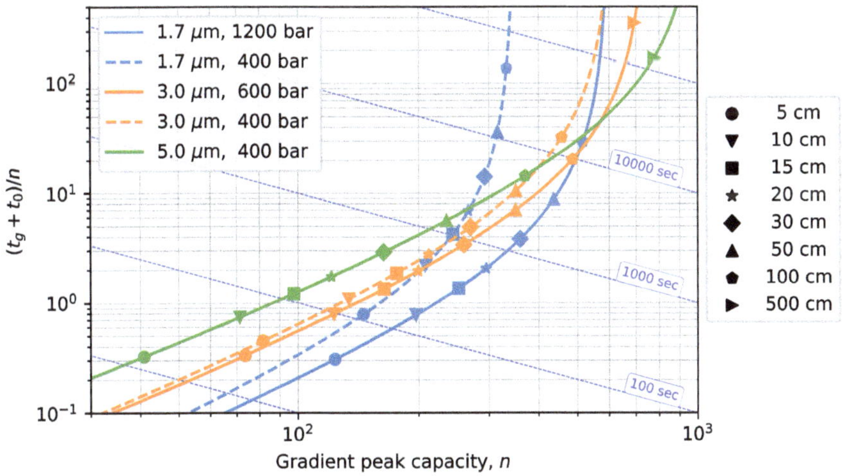

Figure 6.11: Poppe plot of gradient peak capacity for columns operated at different pressures and packed with particles of different sizes. Parameters of calculations: $k_0 = 10^6$, $S = 20$, $\varphi = 0.05$, $\Delta\varphi = 0.7$, $D_m = 10^{-9}$ m^2/s, $\eta = 0.001$ Pa s, $\phi = 1000$.

values (20) suggest a large molecule, such as a large peptide. The trends shown in Fig. 6.11 are similar to the plots in Figs. 6.7. In the practical range of analysis times and column lengths, higher peak capacities can be achieved by using columns packed by smaller particles so long as the maximum operating pressure applicable to the phase is applied. Even if low pressure is used, ultra-high-performance particles can provide higher rate of peak capacity production and faster analysis than larger particles. The vertical asymptotes of curves presented in Fig. 6.11 correspond to the maximal achievable peak capacities as they are calculated by Eq. (22). It can be seen that long columns packed by large particles can provide very high peak capacities, even if it takes more analysis times. The application of large pressures provides higher peak capacities and faster separations. It is desirable to use a column at the highest applicable flow rate in order to generate the highest peak capacity possible. These conclusions are in agreement with the isocratic Poppe plots.

Comparison of Figs. 6.7 and 6.11 emphasize the obvious conclusion that gradient separations are superior to isocratic ones. It was shown that $\sqrt{N}$ in Fig. 6.7 corresponds to the isocratic peak capacity. Therefore, the figures can be compared directly. It can be seen that a $\sim$1000 s separation can generate 200–400 peak capacities in gradient run. The dead-time required to achieve the same order of n in isocratic separation is also $\sim$1000 s. Considering, however, that the total analysis time of an isocratic run is 20–40 times larger than the t_0 when the goal is to reach high separation power, it is indisputable that much higher peak capacities can be generated in much shorter time by gradient separation than by isocratic mode.

Equation (19) shows that gradient steepness, b, is an important factor that influences separation power significantly. b consists of four parameters. S is fixed in the approach used here. t_0 is defined by the column length and pressure drop (through $u_{0,\max}$). The gradient time and change of eluent composition are not mutually independent parameters. Constrain shown in Eq. (40) defines their strict relationship. In Fig. 6.11, value of $\Delta\varphi$ was set to 0.7. It is obvious that this artificially chosen parameter cannot serve with the optimal peak capacities and peak times. The proper choice of t_G and $\Delta\varphi$ is essential in the optimization of gradients. Both too

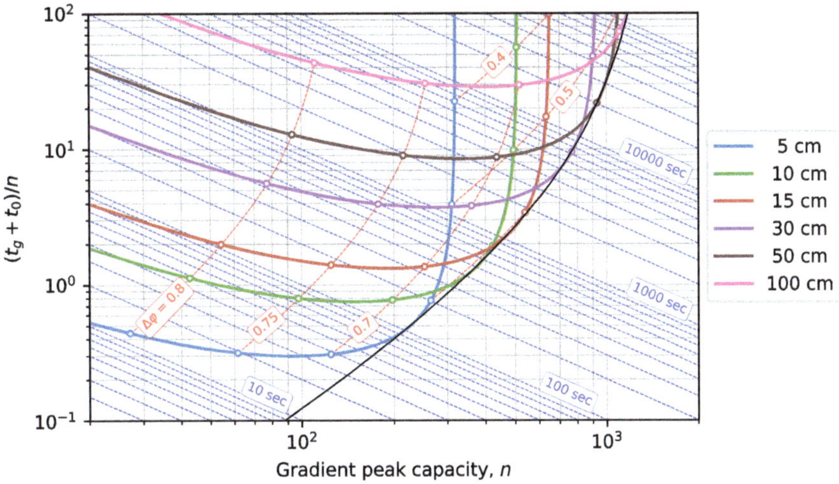

Figure 6.12: Nomogram for the support of optimization of gradient peak separation. For parameters of calculations, see Fig. 6.7.

steep and too shallow gradients are detrimental to the achievable peak capacity. Therefore, it is necessary to apply an optimization method for the determination of t_G and $\Delta\varphi$.

Figure 6.12 presents a nomogram-like gradient Poppe plot constructed for 1.7 μm particles, 1200 bar max. pressure drop, and column lengths that have practical relevance. The figure is similar to Fig. 6.8. It can also be used directly in method development. By using Fig. 6.12, one can find the separation conditions that (1) offer the highest peak capacity in a given analysis time or (2) requires the shortest time to generate a given peak capacity. In the first scenario, the analyst should move on the straight line of target analysis time (dashed blue lines on the figure) to find the column length, L, that offer the highest peak capacity. The dead time can be determined from the column length and pressure drop applied in the analysis by rearranging Eq. (33) for u_0. Gradient time, t_G is given as the difference of total analysis time and t_0. The required change of eluent composition can be determined either by interpolating between the iso-$\Delta\varphi$ lines (dashed red lines on the figure) or by calculating it from Eq. (40). Since Eq. (40) cannot be rearranged to calculate $\Delta\varphi$ directly, a proper root finding algorithm, such as the following, is necessary for the calculation of its value:

```
dfi = brentq(lambda dfi: trgrad(tg, fi0, dfi, t0, S, k0)-(t0+tg), 1e-6, 1-fi0)
```

Here, we took Brent's method provided by SciPy scientific computing library to find $\Delta\varphi$ where the retention time of the last eluting compound is equal to the sum of column hold-up time and gradient time. The bracketing values of $\Delta\varphi$ required by Brent's method were chosen as 10^{-6} and $1 - \varphi_0$ since $\Delta\varphi$ should be larger than zero and smaller than or equal to $1 - \varphi_0$.

In the second scenario, the analyst should first find the column that offers the target peak capacity in the shortest analysis time. It can be determined from Fig. 6.12 directly. The eluent velocity is given by Eq. (33). The dead time can be determined as L/u_0. The total analysis time is given as the product of n and peak time, $(t_G + t_0)/n$. Then, gradient time is given as the difference of total analysis time and t_0. $\Delta\varphi$ can be determined as was shown in the first scenario. Alternatively, t_G can be determined by Brent's method after estimating $\Delta\varphi$ from the nomogram:

```
H = h(u0*dp/dm, A, B, C)*dp    #calculation of plate height
tg = brentq(lambda tg: peakcap(tg, fi0, dfi, t0, H, L, S, k0)-pctarget, 1e-6, 1e5)
```

Here, the Brent's method is used to find the t_G that produces the target peak capacity (`pctarget` in the code).

In Fig. 6.12, the thin black envelope shows the overall optimum of gradient separation. The points of envelope represents the optimal column length that produces the highest peak capacity and the lowest peak time at a given analysis time. The envelope demonstrates the limit of achievable separation power by a given type of particle.

Listing 6.6 shows a Python code that allows the construct of nomogram-like Poppe plot for the optimization of gradient separations. In order to be able to construct nomograms such as Fig. 6.12, the analyst has to determine or at least estimate k_0 of the last eluting compound, a nominal S value that represents the overall sample compounds and the parameters of plate height equation. By generating nomograms like Fig. 6.12 for any phases available in the lab, one can compare different scenarios for the

```
    #-#-#-#    Use with Listing 6.1    #-#-#-#

fi0 = 0.05          #initial eluent composition, φ0
k0, S = 1e6, 20     #parameters of linear solvent strength model
dp, dP = 1.7e-6, 1200e5     #particle diameter and max. pressure drop
#retention factor of the last eluting compound at the beginning of
%analysis
kfi0 = k0*np.exp(-S*fi0)

#------------CONSTRUCTION OF ISO-Δφ LINES------------#
ls = np.logspace(np.log10(0.05),0,100)          #definition of range of column lengths
#iterate through different Δφ values
for dfi in [0.4, 0.5, 0.6, 0.7, .75, 0.8]:
    u0s = dP*dp**2/(fi*eta*ls)        #eluent linear velocities
    Hs = h(u0s*dp/dm, A, B, C)*dp     #plate heights
    t0s = ls/u0s                      #column hold-up times
    tgs = kfi0*S*dfi*t0s/(np.exp(S*dfi)-1)    #gradient times, see Eq. (40)
    pcs = peakcap(tgs, fi0, dfi, t0s, Hs, ls, S, k0)   #peak capacities, see Eq. (19)
    plt.loglog(pcs, (tgs+t0s)/pcs, 'r--', linewidth=0.5)      #plot iso-Δφ lines in log-log scale

#------------PLOTTING OF CONSTANT tg+t0 LINES------------#
pcs = np.logspace(1,3.4,100)          #definition of range of peak capacities
#iterate through different t0+tG values
for tgt0 in [1e1, 1e2, 1e3, 1e4, 1e5]:
    #at each t0+tG, plot (t0+tG)/n with blue dashed line
    plt.loglog(pcs, tgt0/pcs, 'b--', linewidth=0.5)

#------------CONSTRUCTION OF GRADIENT POPPE CURVES FOR DIFFERENT COLUMN LENGTHS------------#
dfis = np.linspace(0.05, 1-fi0, 100)          #definition of range of column lengths
#iterate through different column lengths that have practical relevance
for length in [0.05, 0.1, 0.15, 0.3, 0.5, 1.0]:
    u0 = dP*dp**2/(fi*eta*length)     #eluent linear velocities
    H = h(u0*dp/dm, A, B, C)*dp       #plate heights
    t0 = length/u0                    #column hold-up times
    tgs = kfi0*S*dfis*t0/(np.exp(S*dfis)-1)   #gradient times, see Eq. (40)
    pcs = peakcap(tgs, fi0, dfis, t0, H, length, S, k0)  #peak capacities, see Eq. (19)
    plt.loglog(pcs, (tgs+t0)/pcs)     #plot curves in log-log scale

plt.xlim(2e1, 2e3)                        #set x limits
plt.ylim(1e-1,1e2)                        #set y limits
plt.xlabel('Gradient peak capacity, $n$') #label of x axis
plt.ylabel(r'$(t_G + t_0) / n$')          #label of y axix
plt.show()                                #show plot
```

Listing 6.6: Python code for generating nomogram-like gradient Poppe plot.

analysis of the given sample. The optimal column dimensions, stationary phases and operating conditions can be determined directly by the use of these nomograms, as was shown in the earlier paragraphs.

6.4 Conclusions

Proper optimization of peak capacities of analytical HPLC methods is unavoidable in the analysis of samples containing a large number of compounds. A well-optimized method can offer the same peak capacity in much less analysis time, consuming much less solvents than a non-optimized procedure. Before any method optimization, the analyst should minimize extra-column volumes by changing connection capillaries and detector cell, especially if ultra-high-performance columns are used. In this chapter, the construction and application of Poppe plots were demonstrated in analytical method development. Poppe plots are suitable tools in optimization of peak capacities. In isocratic runs, one can optimize the width of retention window and column plate count separately. At the same time, gradient elution needs a holistic optimization. The parameters affecting the peak capacity generated by the chromatographic system are not mutually independent. Change of one parameter changes the optimal value of other parameters as well. Fortunately, Poppe plots offer a general approach for the optimization of both isocratic and gradient separations. By the use of Python codes shared in this chapter, nomograms can be constructed that allow the determination of most of the optimal separation conditions. The use of Poppe plots in method development provides the analyst a simple and effective tool for optimization of HPLC analyses.

Acknowledgment

The author acknowledges the financial support of the János Bolyai Research Scholarship of the Hungarian Academy of Sciences.

References

[1] J.C. Giddings, Maximum number of components resolvable by gel filtration and other elution chromatographic methods, *Anal. Chem.* 39 (1967) 1027–1028.
[2] U.D. Neue, Peak capacity in unidimensional chromatography, *J. Chromatogr. A* 1184 (2008) 107–130.
[3] J.W. Dolan, L.R. Snyder, N.M. Djordjevic, D.W. Hill, T.J. Waeghe, Reversed-phase liquid chromatographic separation of complex samples by optimizing temperature and gradient time: I. Peak capacity limitations, *J. Chromatogr. A* 857 (1999) 1–20.
[4] J.C. Giddings, Comparison of theoretical limit of separating speed in gas and liquid chromatography, *Anal. Chem.* 37 (1965) 60–63.
[5] J.H. Knox, M. Saleem, Kinetic conditions for optimum speed and resolution in column chromatography, *J. Chromatogr. Sci.* 7 (1969) 614–622.
[6] G. Guiochon, Comparison of the theoretical limits of separating speed in liquid and gas chromatography, *Anal. Chem.* 52 (1980) 2002–2008.
[7] H. Poppe, Some reflections on speed and efficiency of modern chromatographic methods, *J. Chromatogr. A* 778 (1997) 3–21.

[8] G. Desmet, D. Clicq, P. Gzil, Geometry-independent plate height representation methods for the direct comparison of the kinetic performance of lc supports with a different size or morphology, *Anal. Chem.* 77 (2005) 4058–4070.
[9] X. Wang, D.R. Stoll, P.W. Carr, P.J. Schoenmakers, A graphical method for understanding the kinetics of peak capacity production in gradient elution liquid chromatography, *J. Chromatogr. A* 1125 (2006) 177–181.
[10] L.R. Snyder, Linear elution adsorption chromatography. VII. Gradient elution theory, *J. Chromatogr.* 13 (1964) 415–434.
[11] Cs. Horváth, S.R. Lipsky, Peak capacity in chromatography, *Anal. Chem.* 39 (1967) 1893.
[12] E. Grushka, Chromatographic peak capacity and the factors influencing it, *Anal. Chem.* 42 (1970) 1142–1147.
[13] H. Poppe, J. Paanakker, M. Bronckhorst, Peak width in solvent-programmed chromatography. i. general description of peak broadening in solvent-programmed elution, *J. Chromatogr. A* 204 (1981) 77–84.
[14] L.R. Snyder, D.L. Saunders, Optimized solvent programming for separations of complex samples by liquid–solid adsorption chromatography in columns, *J. Chromatogr. Sci.* 7 (1969) 195–208.
[15] L.R. Snyder, J.W. Dolan, J.R. Gant, Gradient elution in high-performance liquid chromatography: I. theoretical basis for reversed-phase systems, *J. Chromatogr. A* 165 (1979) 3–30.
[16] F. Gritti, G. Guiochon, Performance of columns packed with the new shell kinetex-c18 particles in gradient elution chromatography, *J. Chromatogr. A* 1217 (2010) 1604–1615.
[17] F. Gritti, C.A. Sanchez, T. Farkas, G. Guiochon, Achieving the full performance of highly efficient columns by optimizing conventional benchmark high-performance liquid chromatography instruments, *J. Chromatogr. A* 1217 (2010) 3000–3012.
[18] S. Fekete, J. Fekete, The impact of extra-column band broadening on the chromatographic efficiency of 5cm long narrow-bore very efficient columns, *J. Chromatogr. A* 1218 (2011) 5286–5291.
[19] L.R. Snyder, J.J. Kirkland, J.W. Dolan, *Introduction to Modern Liquid Chromatography*, John Wiley & Sons, Inc., Hoboken, NJ, USA, 2010.
[20] K. Horváth, F. Gritti, J.N. Fairchild, G. Guiochon, On the optimization of the shell thickness of superficially porous particles, *J. Chromatogr. A* 1217 (2010) 6373–6381.
[21] T.J. Causon, E.F. Hilder, R.A. Shellie, P.R. Haddad, Probing the kinetic performance limits for ion chromatography. ii. gradient conditions for small ions, *J. Chromatogr. A* 1217 (2010) 5063–5068.
[22] X. Wang, W.E. Barber, W.J. Long, Applications of superficially porous particles: High speed, high efficiency or both? *J. Chromatogr. A* 1228 (2012) 72–88.
[23] S. Fekete, D. Guillarme, Possibilities of new generation columns packed with $1.3\,\mu$m core–shell particles in gradient elution mode, *J. Chromatogr. A* 1320 (2013) 86–95.
[24] S. Fekete, J.-L. Veuthey, D. Guillarme, Achievable separation performance and analysis time in current liquid chromatographic practice for monoclonal antibody separations, *J. Pharmaceut. Biomed.* 141 (2017) 59–69.

Chapter 7

"HPLC Teaching Simulator": A Simple Excel Tool for Teaching Liquid Chromatography

Davy Guillarme* and Jean-Luc Veuthey

School of Pharmaceutical Sciences,
University of Geneva, University of Lausanne,
CMU — Rue Michel Servet 1, 1211 Geneva 4, Switzerland

**davy.guillarme@unige.ch*

7.1 Introduction

High-performance liquid chromatography (HPLC) is currently one of the main analytical techniques in the industry, widely used in areas ranging from research and development to quality control laboratories. HPLC is also taught at the university in the analytical chemistry program, for students of chemistry, biology and pharmacy. In HPLC, there is a complex interplay among the solute contained within the mixture to be analyzed, the mobile phase and the stationary phase. There are a lot of chemical interactions that take place between these three partners, and this is why the technique is particularly difficult to master. It is indeed important to keep in mind that there are a significant number of parameters (e.g. physico-chemical properties and molecular weight of the solutes; nature, composition, temperature, pH and flow rate of the mobile phase; chemical nature and dimensions of the stationary phase, etc.) influencing the quality of the separation in terms of retention time, selectivity, efficiency, pressure drop, peak area, etc.

Various commercial HPLC simulators are available on the market, including DryLab (Molnar-Institute) [1], ChromSword® (Iris Tech) [2],

LC & GC simulator (Advanced Chemistry Development) and Osiris (Datalys) [3]. These software are particularly useful to efficiently develop HPLC methods based on a limited number of initial experiments, and they can also be used to better understand the principles of HPLC. However, these tools remain relatively difficult to use for students and, above all, expensive to purchase; this is why they are only scarcely used in education program.

To master the principles of chromatography in a relatively cheap way, several free or low-cost computer-based HPLC simulators have also been proposed in the past [4–7]. As shown in Ref. [8], there are currently six HPLC simulators available, but most of them are not available anymore, or not fully compatible with contemporary computers. To the best of our knowledge, the most interesting HPLC simulator was released in 2013 and is completely free of charge [8]. The software interface is relatively easy to use and a simulated chromatogram is redrawn when an experimental parameter is changed.

In comparison to this software named "HPLC simulator" developed by Prof. Dwight Stoll from the Gustavus Adolphus College (USA), the philosophy of our program called "HPLC teaching assistant" is quite different, but certainly also complementary. First of all, this is a simple Excel spreadsheet, which can be easily used on any computer, without installing Java. Second, our tool allows to easily link compound's physico-chemical properties (log P, pK_a) and chromatographic behavior. Finally, each spreadsheet of our Excel tool describes one given concept (e.g. the effect of mobile phase pH on the chromatogram). The goal of this chapter is to provide the theoretical background of our Excel tool, and also show some practical examples to illustrate its utility for teaching chromatography. Among the available features, it allows to (i) visualize the change in resolution when modifying retention, selectivity and efficiency, (ii) understand the van Deemter equation and kinetic performance in HPLC, (iii) illustrate the importance of analytes lipophilicity on retention in reversed-phase liquid chromatography (RPLC), (iv) handle the RPLC retention, taking into account the compounds pK_a and mobile phase pH, (v) simulate the impact of mobile phase temperature on RPLC separations, (vi) understand the chromatographic behavior in isocratic and gradient elution modes, (vii) show the influence of the instrument (injected volume and tubing geometry) on kinetic performance and sensitivity and (viii) demonstrate the impact of analyte molecular weight on thermodynamic and kinetic performance in RPLC mode.

7.2 Chromatographic Resolution — Impact of Retention, Selectivity and Efficiency

In liquid chromatography (LC), the separation of two peaks is usually described by the concept of chromatographic resolution (R_s), which represents the difference in retention times (t_R) between two consecutive peaks, divided by the average peak widths at the baseline of both peaks (w). The following equation can be used to experimentally measure the chromatographic resolution:

$$Rs = \frac{2 \times (tR_2 - tR_1)}{w_1 + w_2} \tag{1}$$

To better highlight the impact of analytical conditions on resolution, the fundamental equation of resolution (Eq. (2)) can be transformed into an expression where the retention factor (k), selectivity (α) and plate number (N) values appear:

$$Rs = \frac{\sqrt{N}}{4} \times \frac{k}{k+1} \times \frac{\alpha - 1}{\alpha} \tag{2}$$

In the first spreadsheet of our Excel tool (see Fig. 7.1), the impact of k, α and N on R_s can be graphically visualized. For this purpose, a chromatogram has been simulated and the chromatographic resolution of two molecules when modifying the values of k, α and N is shown. In the proposed example, shown in Fig. 7.1, the column has dimensions of 150×4.6 mm and is operated at 1 mL/min (dead time of 1.74 min, highlighted with the red line on the chromatogram).

In addition to the chromatogram, three different graphics were located at the bottom of the spreadsheet to illustrate the resolution change when modifying the three individual variables (k, α and N). As illustrated, the impact of α on resolution is extremely strong: when α values vary between 1 and 4, the resolution is drastically enhanced and a plateau is attained for very high α values ($\alpha > 4$), but such values are far from the usual range

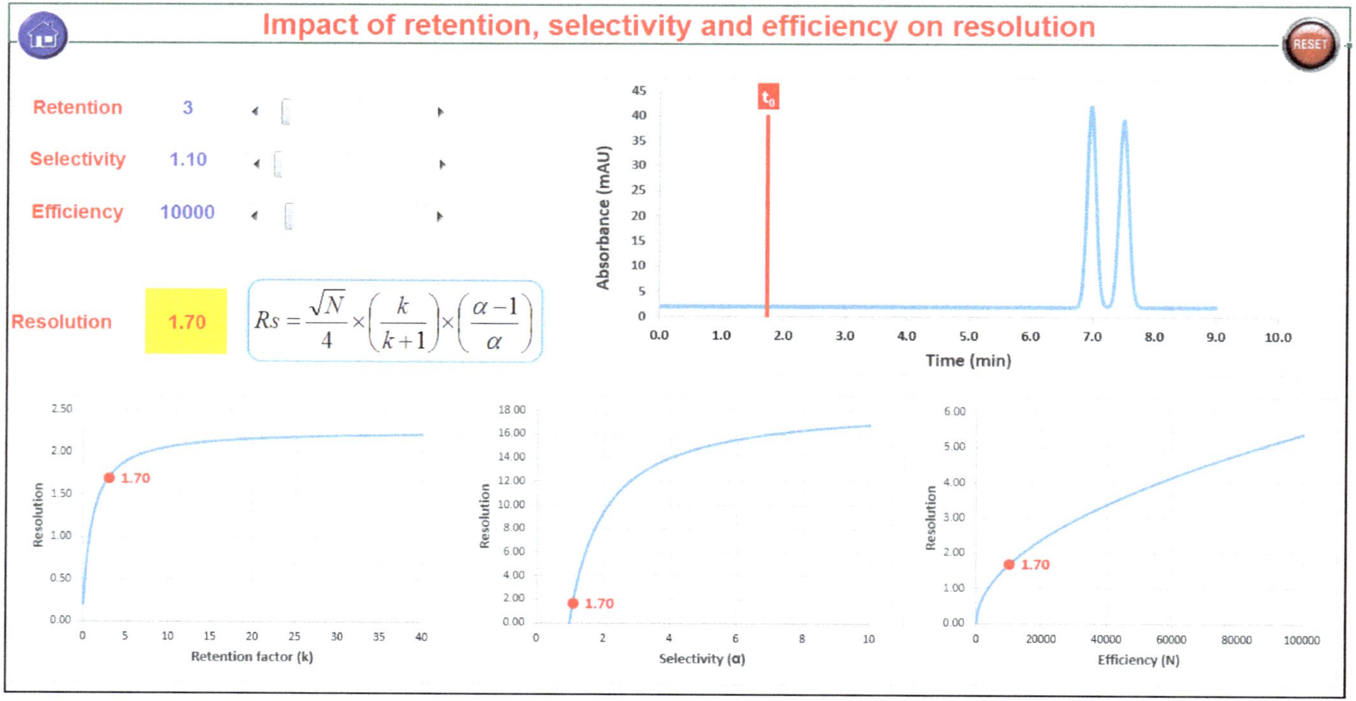

Figure 7.1: Excel spreadsheet highlighting the impact of retention, selectivity and efficiency on resolution.

of selectivity in HPLC. Retention factor (k) also plays a key role to improve resolution for k values below 3. However, its impact becomes modest for k values between 3 and 10, and very low for k higher than 10 (a plateau is observed). Indeed, for $k > 10$, the analysis time becomes extremely long, and the improvement in resolution is minor (this can be easily checked by simulating a chromatogram obtained for very high k values). Finally, because efficiency (N) is expressed as the square root in Eq. (2), its impact on resolution is limited, and N should be drastically increased to have a clear effect on R_S.

Based on these observations, a workflow can be described for method development, that includes three main steps once the best stationary phase and mobile phase conditions have been selected: The steps are as follows:

(i) Select column dimensions with a sufficient plate count, taking into account the sample complexity (a column able to produce 10,000 plates may be a good starting point in HPLC).
(ii) Adjust the solvent strength (percentage of organic modifier) to attain a reasonable retention factor (comprised between 1 and 10).
(iii) Optimize selectivity by tuning other chromatographic parameters (pH, temperature, etc.). In some cases, it is important to keep in mind that the retention and efficiency can also be influenced during the selectivity optimization step.

7.3 Chromatographic Efficiency and van Deemter Curves — Impact of Column Dimensions

In HPLC, the plate number (N) and column backpressure (ΔP) may be strongly influenced by the column dimensions (length, internal diameter and particle size) and mobile phase (flow rate and viscosity). In the second spreadsheet of our Excel tool, the kinetic performance was assessed for a mixture of three compounds having a molecular weight of about 100 Da. For all these calculations, a generic HPLC column having a porosity (ε) of 0.7 and a flow resistance (Φ) of 500 was considered. The column temperature was also fixed at 30°C, and the mobile phase was composed of 30% ACN and 70% aqueous buffer.

The following equation was used to calculate the plate number (N) taking into account the column dimensions:

$$N = \frac{L}{H} \tag{3}$$

where L is the column length (mm) and H is the height equivalent to a theoretical plate (μm).

To estimate the H value, the van Deemter equation has to be considered:

$$H = A + \frac{B}{u} + Cu \tag{4}$$

In this equation, the A, B and C terms correspond to eddy dispersion, longitudinal diffusion and mass transfer, respectively. These values depend on the solute, column and mobile phase conditions. The u value corresponds to the linear velocity (mm/s), which is estimated by taking into account the mobile phase flow rate (F), column porosity (ε) and column internal diameter (d_c), with the following equation:

$$u = \frac{4 \times F}{\pi \times d_c^2 \times \varepsilon} \tag{5}$$

In our case, we wanted to use some generic a, b and c terms ($a = 1$, $b = 4$, $c = 0.05$). This is only possible if these parameters are independent of the analytical conditions. Therefore, Eq. (4) was transformed into its reduced form:

$$h = a + \frac{b}{\nu} + c\nu \tag{6}$$

where h is the reduced height equivalent to a theoretical plate and ν is the reduced linear velocity, which could be expressed according to the following equations:

$$h = \frac{H}{d_p} \tag{7}$$

$$\nu = \frac{u \times d_p}{D_m} \tag{8}$$

Here, d_p represents the column particle diameter (μm) and D_m is the diffusion coefficient of the compound in the mobile phase (m^2/s), which can be estimated using the Wilke–Chang equation [9].

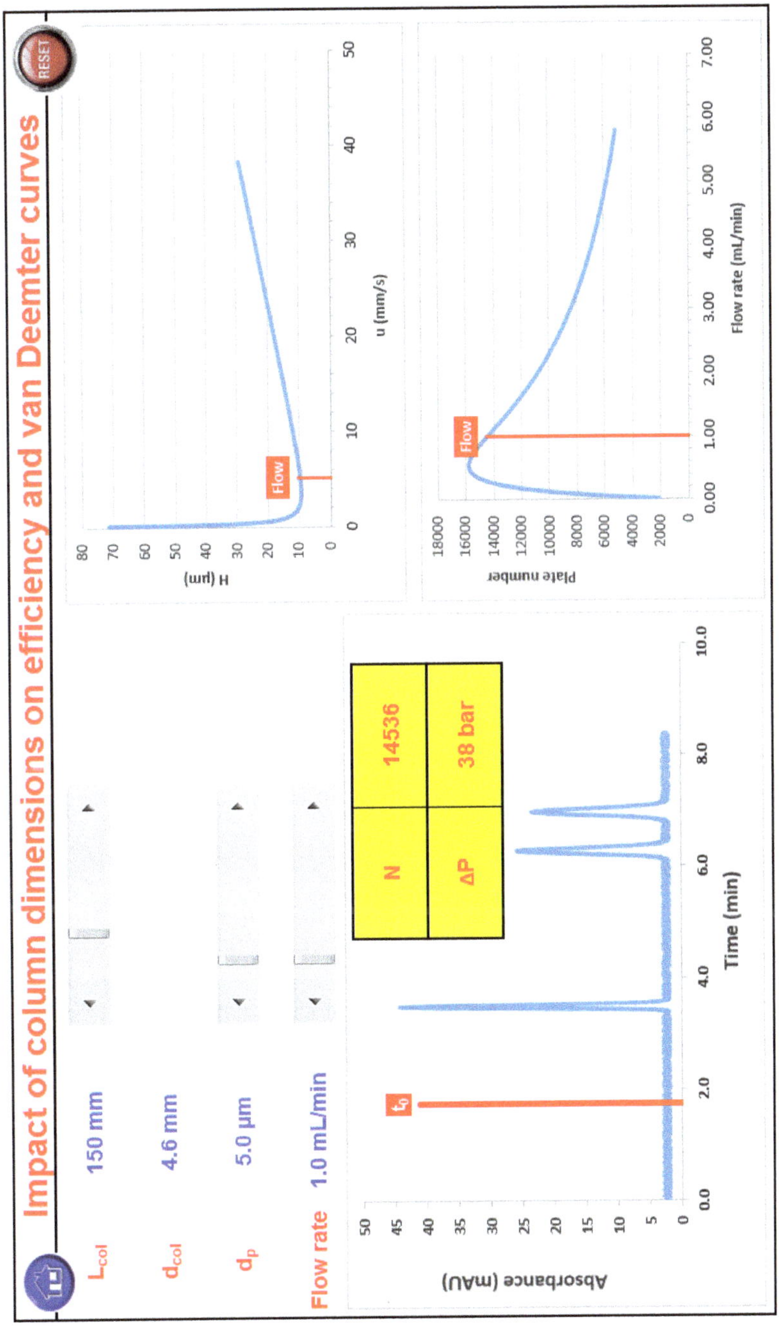

Figure 7.2: Excel spreadsheet showing the impact of column dimensions on efficiency and van Deemter curves.

Finally, the column pressure drop was calculated thanks to the Darcy's law, with η being the mobile phase viscosity (cP):

$$\Delta P = \frac{\eta \times L \times u \times \Phi}{d_p^2} \tag{9}$$

As shown in Fig. 7.2, the impact of column dimensions (L_{col}, d_{col} and d_p) and mobile phase flow rate (F) on N and ΔP can be directly assessed. Indeed, a computer-generated chromatogram with three compounds ($k =$ 1.0, 2.6 and 3.0) was added and shows the performance when altering column dimensions and flow rate. Besides the simulated chromatogram, the van Deemter curve, $H = \mathrm{f}(u)$ and the more practical curve representing $N =$ $\mathrm{f}(F)$ were also drawn for the tested set of conditions. In this spreadsheet, the user is free to modify the column dimensions (L_{col}, d_{col} and d_p) and mobile phase flow rate to see the impact on the simulated chromatogram located at the bottom of the spreadsheet. The corresponding plate count and column pressure drop are also provided. In addition, the user is able to visualize the corresponding van Deemter curve and evaluate whether the employed conditions are far from the optimal linear velocity (or flow rate), by taking into account the red line drawn on the van Deemter curve. This could help the user to assess the maximal plate number achievable on the selected column geometry under optimal flow rate conditions.

7.4 Retention in RPLC Conditions — The Importance of Lipophilicity

Under RPLC conditions, the retention on alkyl stationary phases (e.g. C4, C8, C18) is mostly driven by the compound lipophilicity, which can be expressed as the partition coefficient (P). P is defined as the ratio of compound concentrations found in two immiscible phases (generally 1-octanol and water) at equilibrium. To have reasonable values of partition coefficients, the logarithm of P is preferentially considered ($\log P$), and is defined by the following expression:

$$\log P = \log\left(\frac{C_{\text{octanol}}}{C_{\text{aqueous}}}\right) \tag{10}$$

When $\log P$ values are lower than 0, the molecules are considered as hydrophilic (e.g. the molecules have more affinity for the hydrophilic mobile phase rather than for the hydrophobic stationary phase) and will be poorly retained under RPLC conditions. On the other hand, if the $\log P$ values are superior to 0, the molecules are lipophilic, and preferably interact with the hydrophobic stationary phase, leading to enhanced retention. In generic RPLC conditions, using a C18 stationary phase and a mixture of MeOH and water as mobile phase, only substances having $\log P$ values between -1 and $+6$ can be satisfactorily analyzed.

In this third Excel spreadsheet, a simulated chromatogram including three compounds with different $\log P$ values (values have to be set by the user) shows the chromatographic behavior for a given mobile phase composition (%MeOH). To construct this spreadsheet, the transformation of $\log P$ values into retention factors and retention times was performed, based on a previous study from our group [10], and considering a C18 column and a mobile phase containing MeOH and water.

The following empirical equation was employed to calculate $\log k_w$, which is defined as the extrapolated retention factor to pure water, allowing to perfectly mimic 1-octanol/water partitioning [11]) based on the $\log P$ value set by the user [10]:

$$\log k_w = 0.83 \times \log P + 0.21 \tag{11}$$

Then, the $\log k_w$ value was transformed into the $\log k$ value at the mobile phase composition set by the user, using the following equation coming from the linear solvent strength (LSS) theory from Snyder [12]:

$$\log k = \log k_w - S\Phi \tag{12}$$

where Φ is the fraction of organic solvent in the mobile phase (value comprised between 0 and 1) and S is a parameter of a given solute, corresponding to the elution strength of the organic modifier (i.e. the slope of the logarithmic plot: $d(\log k)/d\Phi$). Under RPLC conditions, S values generally vary from three (compounds of about 100 Da) to more than 100 for large proteins ($>$50,000 Da). In this Excel spreadsheet, a generic value of 4 was considered for S, as it is typical of small molecules ($<$300 Da). Finally, the $\log k$ values obtained from Eq. (12) were transformed into t_r

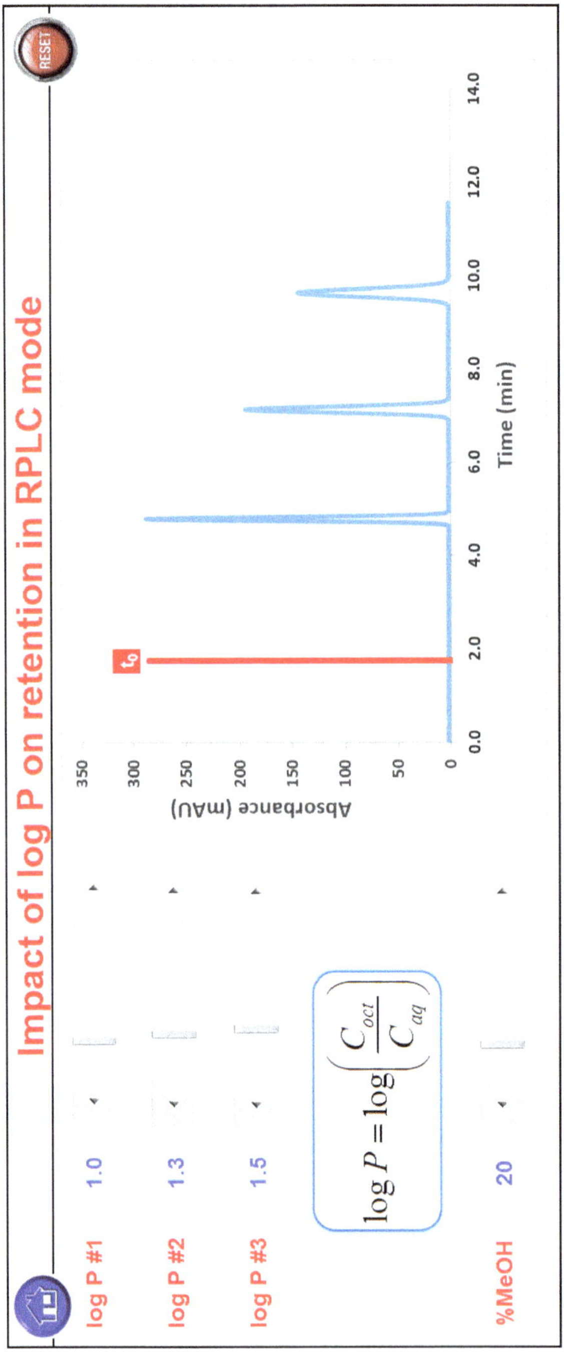

Figure 7.3: Excel spreadsheet showing the impact of log P on retention under RPLC conditions.

to construct the final chromatogram reported in Fig. 7.3, considering a column of 150 × 4.6 mm, 5 μm used at a flow rate of 1 mL/min (column dead time of 1.74 min). As shown in Fig. 7.3, the user can modify the log P values of three representative compounds, to directly observe the impact on retention for a simulate chromatogram. In addition, the user also has the possibility to tune the mobile phase composition, by adjusting the %MeOH. Various log P values can be tested to highlight the versatility of RPLC, since this mode of chromatography is compatible with a wide range of compounds lipophilicity, when using a generic C18 material. The retention can then be easily tuned by adjusting the proportion of organic modifier contained in the mobile phase. In the case where compounds of very different lipophilicity have to be analyzed simultaneously, gradient elution has to be preferentially used.

7.5 Impact of Compound Ionization on Retention and Selectivity in RPLC Mode

One of the most important parameters for tuning retention and selectivity in RPLC is the mobile phase pH. Indeed, when analyzing ionizable compounds, the pH may strongly impact the relative amount of neutral and ionized forms. In RPLC conditions, the retention of basic analytes will decrease at low mobile phase pH (pH $< \mathrm{p}K_a$ of the molecule), since the basic molecule will be mostly under an ionized form. On the other hand, the retention of acidic substances will increase at low pH (pH $< \mathrm{p}K_a$ of the molecule), as the acidic molecule will be mostly under a neutral form. It is indeed well established that ionized compounds are more hydrophilic than neutral ones, and therefore, the retention and selectivity can be easily altered when tuning the mobile phase pH since it modifies the percentages of ionized/neutral forms. When dealing with ionizable substances, the partition coefficient (log P) cannot be used and should be replaced by the distribution coefficient (log D), expressed with the following expression for an acidic molecule [13]:

$$\log D = \log\left(\frac{[AH]_{\text{octanol}}}{[A^-]_{\text{aqueous}} + [AH]_{\text{aqueous}}}\right) \tag{13}$$

For an acid, the percentages of neutral and ionized forms at a given pH are obtained from the following equations [14]:

$$\% \text{ ionized} = \frac{100}{1 + 10^{\text{p}Ka - \text{pH}}} \tag{14}$$

$$\% \text{ neutral} = \frac{100}{1 + 10^{\text{pH} - \text{p}Ka}} \tag{15}$$

For a base, the percentages of neutral and ionized forms at a given pH are obtained from the following equations [14]:

$$\% \text{ ionized} = \frac{100}{1 + 10^{\text{pH} - \text{p}Ka}} \tag{16}$$

$$\% \text{ neutral} = \frac{100}{1 + 10^{\text{p}Ka - \text{pH}}} \tag{17}$$

All the calculations in the Excel spreadsheet were made for a column of $150 \times 4.6\,\text{mm}$, $5\,\mu\text{m}$ at a flow rate of 1 mL/min and using MeOH as the organic modifier. As shown in Fig. 7.4, only two compounds were selected for the simulation to limit the complexity of the chomatogram, and a color code was employed to distinguish the two substances (green and blue, for compound 1 and 2, respectively). For these two model compounds, users can set $\log P$ and $\text{p}K_a$ values, and also the chemical nature of the compound (acidic or basic). In addition, the mobile phase pH should also be indicated, as well as the %MeOH in the mobile phase. Based on these inputs, the $\log D$ is calculated for the two compounds, at the pH indicated by the user, using the following two equations that have to be applied for an acid (Eq. (18)) and a base (Eq. (19)), respectively [13]:

$$\log D = \log P - \log\left(1 + 10^{\text{pH} - \text{p}Ka}\right) \tag{18}$$

$$\log D = \log P - \log\left(1 + 10^{\text{p}Ka - \text{pH}}\right) \tag{19}$$

For (pH–pK_a) values higher than $|3.5|$, $\log D$ becomes a constant value, since the pH is too far from the pK_a of the compound to have an impact on the retention in RPLC. The $\log D$ values were then easily transformed into retention factors using Eqs. (11) and (12).

The impact of mobile phase pH and compounds pK_a can be directly visualized in Fig. 7.4. Except the simulated chromatogram, the ionization

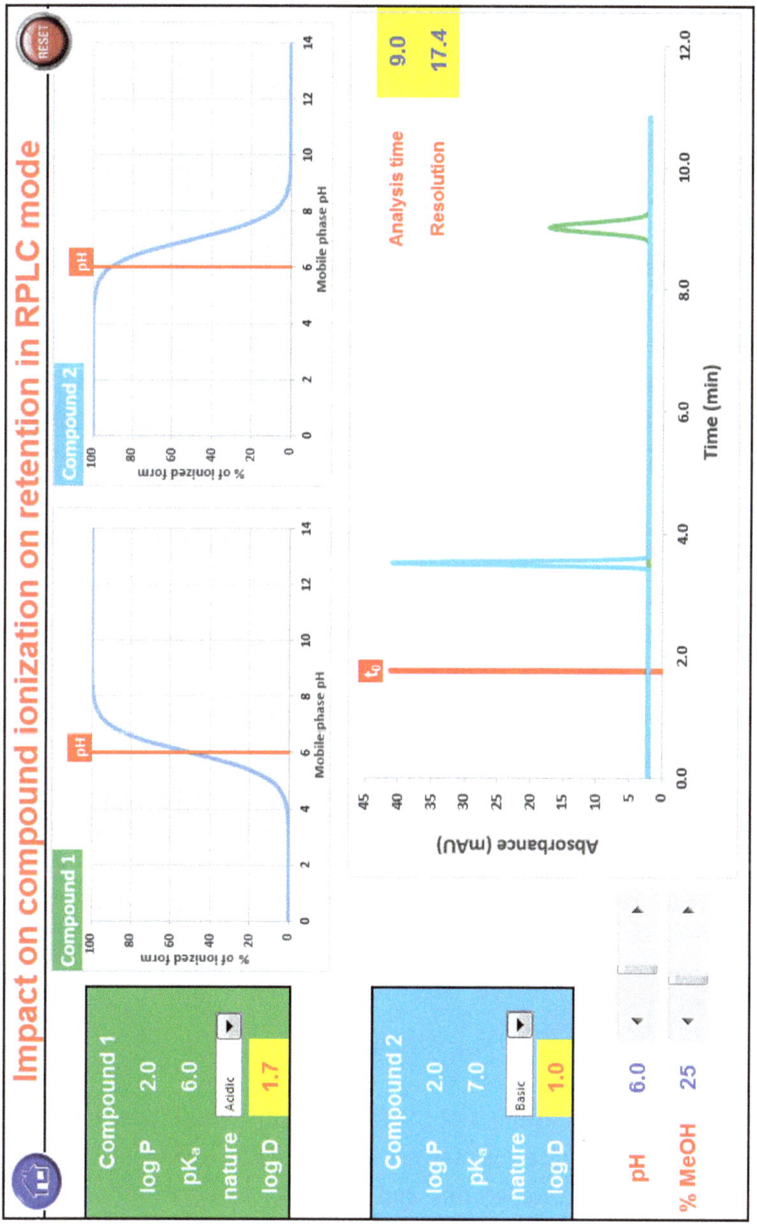

Figure 7.4: Excel spreadsheet highlighting the impact of compound ionization on retention in RPLC mode.

profiles of the two simulated compounds is also presented in the upper part of the spreadsheet. It provides the amount of ionized form as a function of mobile phase pH. A sigmoidal shape is observed when using Eqs. (14)–(17).

As example, the impact of mobile phase pH (modified in a narrow range around the pK_a of the substances) on the chromatographic separation of two species (acid and base) having identical log P values can be directly visualized, by appropriately selecting pH, pK_a and log P values in the Excel spreadsheet. Because the mobile phase pH is very close to the pK_a of both substances, the impact of pH change on retention will be significant on both compounds and, in some cases, the elution order can even be reversed. This example illustrates the impact of pH on retention, selectivity and resolution and the importance of adequately controlling this variable under RPLC conditions.

7.6 Impact of Mobile Phase Temperature in RPLC Mode

Mobile phase temperature has an impact on kinetic performance (shape of the van Deemter curve, optimal linear velocity and column pressure drop) and also thermodynamic performance (change in retention and selectivity) in RPLC conditions.

In terms of kinetic performance, the optimal linear velocity (u_{opt}) visualized on the van Deemter curve is given by the following equation:

$$u_{\text{opt}} = \frac{\nu_{\text{opt}} \times D_m}{d_p} \tag{20}$$

In this expression, the diffusion coefficient, D_m of a solute A in a solvent B can be expressed using the Wilke–Chang equation:

$$D_m = 7.4 \times 10^{-8} \frac{(\Phi_B M_B)^{1/2} T}{\eta_B V_A^{0.6}} \tag{21}$$

where M_B is the molecular weight of solvent B, T is the absolute temperature (K), η_B is the solvent B viscosity (cP) at T, V_A is the molar volume of solute A at its normal boiling temperature and Φ_B, is the association factor [15] of solvent B (dimensionless). Wilke and Chang recommended a value of Φ_B equal to 2.6 when the solvent is water, 1.9 when the solvent is MeOH, and

1.0 when the solvents are unassociated (for example, ACN). The selected organic solvent for the calculation was ACN. The mobile phase viscosity was determined from Ref. [16] and depends on the nature of organic solvent, mobile phase composition and its temperature.

The column pressure drop reported in this Excel spreadsheet was calculated using the Darcy's law (Eq. (9)). It depends on the mobile phase viscosity and is therefore also impacted by mobile phase temperature.

In terms of thermodynamic performance, $\log k$ decreases linearly as a function of $1/T$(1/K), according to the van't Hoff equation:

$$\log k = \frac{\Delta S^0}{R} - \frac{\Delta H^0}{RT} \tag{22}$$

Depending on the compound, the slopes of the van't Hoff curves could be different, since the enthalpy (ΔH) and entropy (ΔS) might vary. Based on our experience, when the temperature increases by 30–40°C, the retention factor is generally divided by a factor of 2, depending on the compound (this is an empirical value, observed in Ref. [16]). Three different compounds were arbitrarily selected. For the first compound ($\log P$ of 2.0), the k value was divided by a factor 2 each 40°C increase; for the second compound ($\log P$ of 2.5), k was divided by a factor 2 each 30°C increase; for the third compound ($\log P$ of 2.8), k was divided by a factor 2 each 38°C increase.

In this Excel spreadsheet (see Fig. 7.5), the impact of column dimensions (L_{col}, d_{col} and d_p), mobile phase flow rate (F), percentage of organic solvent, compound molecular weight (MW) and mobile phase temperature on kinetic and thermodynamic performance can be directly visualized. A first representation shows the kinetic performance (N vs. F), while a second one highlights the thermodynamic behavior ($\log k$ vs. $1/T$) of the three compounds. In addition, a simulated chromatogram at the bottom of Fig. 7.5 illustrates the chromatographic behavior of the three compounds, when modifying mobile phase temperature. As expected, the retention decreases at elevated temperature, while t_0 obviously remains constant. This confirms that the percentage of organic solvent has to be adjusted to achieve similar retention factors at different temperatures. For example, at 1% ACN, the retention at 150°C becomes equivalent to the retention at

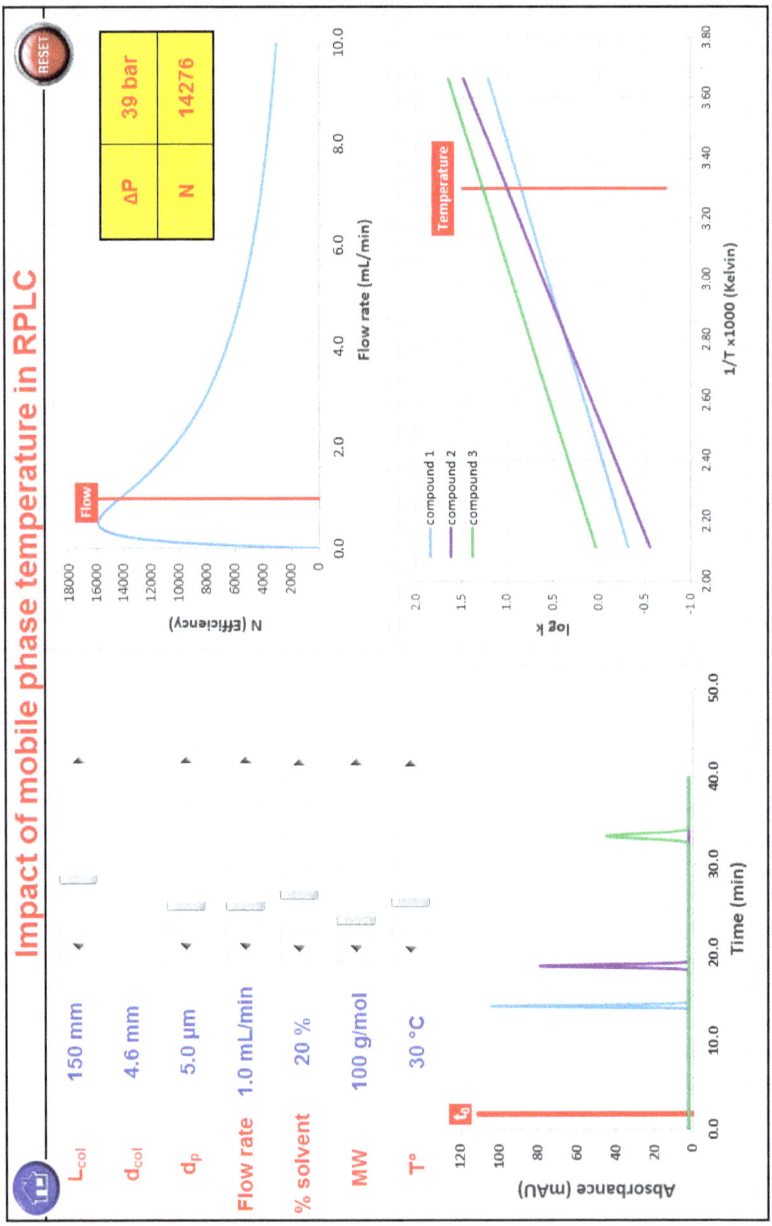

Figure 7.5: Excel spreadsheet highlighting the impact of mobile phase temperature, under RPLC conditions.

30°C and 20% ACN. Besides retention, the selectivity can be modified and the elution order can even be reversed at high temperature. This observation proves that temperature can be considered as an effective parameter for tuning selectivity, provided that temperature is investigated in a sufficiently large range and solute structures vary significantly.

7.7 Chromatographic Optimization in RPLC Isocratic Mode

In RPLC, the retention model can be described by the LSS theory (Eq. (12)) [17]. This theory shows that the $\log k$ value of a given compound varies linearly with the amount of organic modifier in the mobile phase. This linear behavior can be successfully used for developing RPLC method and optimizing resolution. To efficiently develop a method, the retention models ($\log k = f(\%\text{MeOH})$) can be drawn for all the compounds contained within a mixture. Then, a sufficient selectivity can be visually observed for a composition of MeOH where the curves do not cross each other. In this Excel spreadsheet (Fig. (7.6)), the retention models of five compounds having $\log P$ values between 2.2 and 2.8 and S values between 4 and 6.3 were represented. Then, the minimal resolution for each %MeOH was plotted as a function of %MeOH, and the corresponding chromatogram was also shown. This type of procedure can be used for optimizing a RPLC method. Indeed, since the LSS curves are linear, only two initial runs with different %MeOH are required to find out the intercepts and slopes of the curves for each compound and optimize the chromatographic separation under isocratic conditions. This type of calculation is included in the commercially available optimization software, such as DryLab (Molnar-Institute), ChromSword® (Iris Tech), LC & GC simulator (Advanced Chemistry Development) and Osiris (Datalys). However, in these software, the preliminary experiments are performed in the gradient mode and calculations are performed using a more complex iterative procedure.

A chromatographic separation of five model substances can be simulated for any mobile phase composition. As shown in Fig. 7.6, the retention is decreasing at high percentage of MeOH, which is in line with the LSS theory. In addition, the selectivity is also altered, and even if the analysis

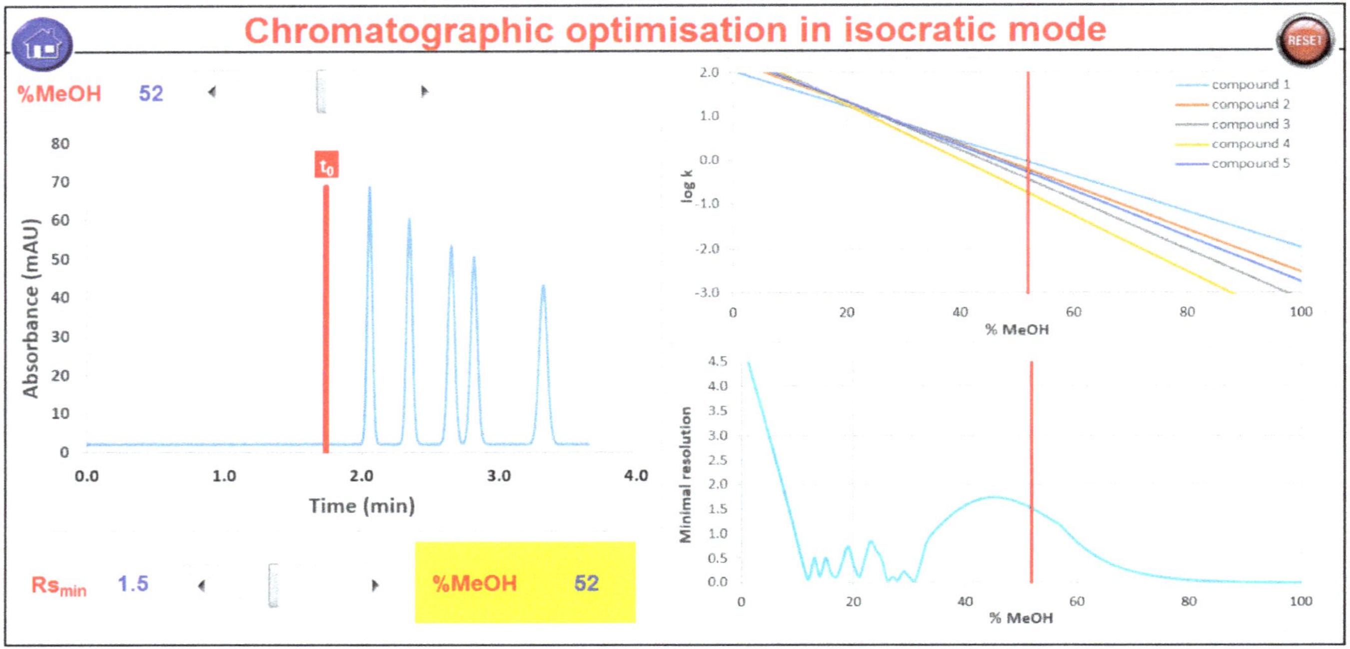

Figure 7.6: Excel spreadsheet showing how to perform chromatographic optimization in isocratic mode.

time was longer at 35% MeOH, the separation was not better, compared to the one obtained at 52% MeOH. However, the separation obtained at 70% MeOH was clearly the worst one in terms of selectivity, due to the too low retention of the five substances under these conditions ($k \ll 1$). In this Excel spreadsheet, the user can also set a minimal resolution value (as an example 1.5, which corresponds to a baseline separation of the five compounds) and the optimal corresponding % MeOH value is calculated, based on the representation of the minimal resolution as a function of % MeOH.

7.8 Understanding the Gradient Elution Mode in RPLC Conditions

The gradient mode is often used in RPLC to analyze compounds having too diverse lipophilicity. A chromatogram of five substances eluted under gradient conditions was simulated in this Excel spreadsheet (see Fig. 7.7), and the following procedure was employed to construct the chromatogram.

First of all, five substances of different lipophilicity ($\log P$ values from 1.4 to 3.4) were initially selected. The S values of these five analytes were arbitrarily fixed between 4.0 and 6.3, respectively.

Then, based on the gradient profile (% initial, % final and gradient time), column dimensions (L_{col}, d_{col}, d_p) and mobile phase flow rate (F) set by the user (see Fig. 7.7), the corresponding elution composition (C_e) of each compound was calculated. Then, the C_e values were transformed into t_r under the employed gradient conditions.

Next, the peak widths (w) in gradient mode were simulated using the following expression:

$$w = \frac{t_0 \times (1 + k_e)}{\sqrt{N}} \tag{23}$$

where N and t_0 were calculated from the column dimensions and mobile phase flow rate. k_e is the elution retention factor, calculated from the elution composition (C_e), the $\log k_w$ value obtained from the $\log P$ value (using Eq. (11)) and the S value, using the following equation:

$$\log k_e = \log k_w - S \times C_e \tag{24}$$

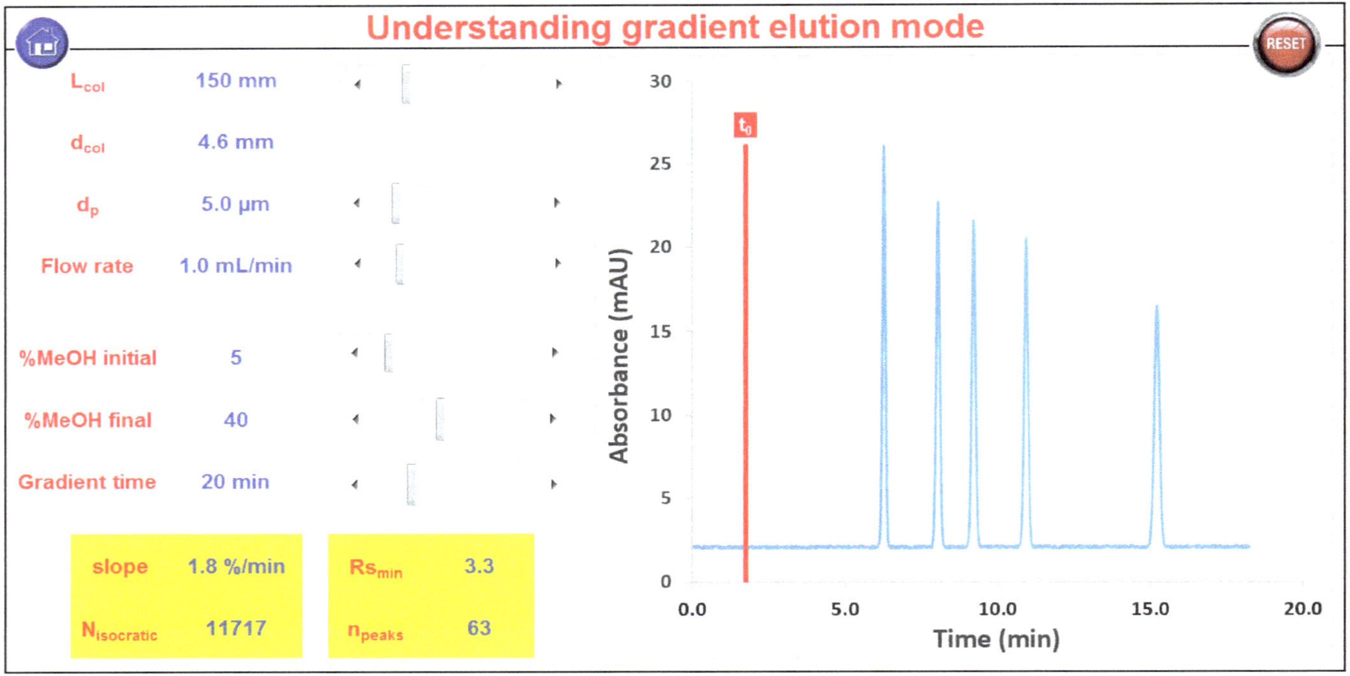

Figure 7.7: Excel spreadsheet highlighting the gradient elution concept, and the impact of various parameters on the final chromatogram.

Finally, the corresponding peak capacity (n_{peaks}) was calculated using the Neue equation [18]:

$$n_{\text{peaks}} = 1 + \frac{\sqrt{N}}{4} \times \frac{1}{b+1} \ln\left(\frac{b+1}{b} e^{S\times\Delta\Phi} - \frac{1}{b}\right) \tag{25}$$

where b is the gradient steepness, expressed as:

$$b = \frac{t_0 \cdot \Delta\Phi \cdot S}{tG} \tag{26}$$

where tG is the gradient time, and $\Delta\Phi$ is the change in solvent composition during the gradient, ranging from 0 to 1.

The gradient Excel spreadsheet allows the simulation of gradient chromatograms for various column dimensions, mobile phase flow rates and gradient conditions. The user can modify only the gradient profile (simplest case), but it is also possible to easily visualize the impact of column dimensions and mobile phase flow rate on the chromatogram obtained under gradient conditions. As an example, for a fixed gradient time, a higher mobile phase flow rate often improves resolution and peak capacity, while simultaneously reducing the elution time. This behavior can be explained by the lower column dead time at elevated mobile phase flow rate. Then, the k_e values are increased, leading to better overall performance [19].

7.9 The Impact of Injected Volume in RPLC Conditions

The injected volume influences the chromatographic separation in RPLC conditions. This behavior was simulated in isocratic conditions in this Excel spreadsheet (see Fig. 7.8). First of all, a higher sensitivity (C_{max}) is expected when increasing the injected volume (V_{inj}), taking into account the following equation [20]:

$$C_{\text{max}} \propto \frac{\sqrt{N} \times V_{\text{inj}}}{L \times d_c^2 \times (1+k)} \tag{27}$$

However, a very large volume cannot be injected in RPLC, and so a compromise has to be found. Indeed, peaks become much broader and plate count is reduced at high V_{inj}, due to the contribution of injected volume to extra-column band broadening. Some more explanations are provided below.

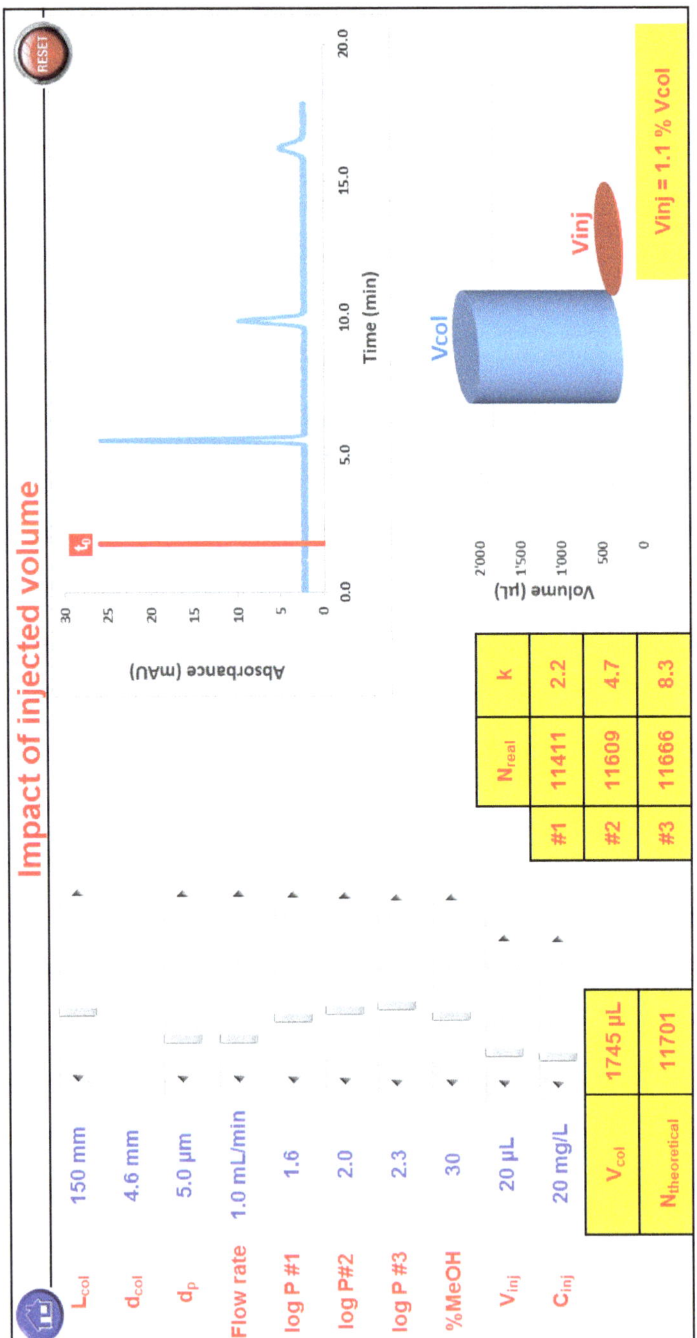

Figure 7.8: Excel spreadsheet showing the impact of injected volume.

In HPLC, the observed peak variance (σ^2_{tot}) is the sum of the chromatographic column dispersion (σ^2_{col}), the dispersion related to the injection system (σ^2_{inj}) and the dispersion related to the rest of the equipment (tubing and detector), σ^2_{ext}. It can be expressed using the following equation:

$$\sigma^2_{\text{tot}} = \sigma^2_{\text{col}} + \sigma^2_{\text{inj}} + \sigma^2_{\text{ext}} \tag{28}$$

For the sake of simplicity, the σ^2_{ext} was neglected in this Excel spreadsheet.

Dispersion linked to the chromatographic column itself (σ^2_{col}) can be expressed by the following equation, when a limited amount of sample is injected:

$$\sigma^2_{\text{col}} = \frac{V_R}{\sqrt{N}} = \frac{V_0 \cdot (1+k)}{\sqrt{N}} \tag{29}$$

where σ^2_{col} is the column variance (in μL^2), N is the plate count and V_R is the retention volume, which is a function of the column dead volume V_0 and retention factor, k.

The dispersion related to the injection (σ^2_{inj}) can be expressed as [21]:

$$\sigma^2_{\text{inj}} = K_{\text{inj}} \cdot \frac{V^2_{\text{inj}}}{12} \tag{30}$$

where K_{inj} is a constant (generally comprised between 1 and 3) depending on the injection mode. In our case, this number was set to a generic value of 2.

The observed plates number (N_{obs}) could then be estimated by the following expression, accounting for the loss of efficiency due to the injector:

$$N_{\text{obs}} = N_{\text{col}} \cdot \frac{1}{1 + \frac{\sigma^2_{\text{inj}}}{\sigma^2_{\text{col}} + \sigma^2_{\text{inj}}}} \tag{31}$$

where N_{col} is the theoretical plate number of the chromatographic support.

The impact of V_{inj} on chromatographic performance (sensitivity and plate count) can therefore be assessed using Eqs. (27) and (31). As illustrated in Fig. 7.8, the user should set the column dimensions (L_{col}, d_{col}, d_p), mobile phase flow rate, log P values of three model compounds, percentage of MeOH in the mobile phase and injected volume, as well as compound concentration. Then, the chromatogram is simulated under the suggested conditions, and a graph representing the column volume vs. injected volume is also shown at the bottom of Fig. 7.8. It is also important to note that all the chromatographic calculations and simulations are only accurate for a sample diluted in a sample diluent composition strictly equivalent to the mobile phase itself.

In RPLC, the injected volume should represent between 0.5% and 5% of the column volume to achieve a good compromise between sensitivity and peak broadening. Therefore, the user can directly see the impact of a too large injected volume on the chromatogram (severe band broadening). However, it is important to keep in mind that this behavior also largely depends on the retention of the compounds. With strongly retained compounds (high log P values), the impact of injected volume on band broadening will be limited, while it will be detrimental for poorly retained substances. Therefore, a compromise needs to be found to achieve a sufficient sensitivity and reasonable band broadening.

7.10 The Impact of Tubing Geometry in RPLC Conditions

In HPLC, the volumes between injector and detector also contribute to band broadening, similar to what was previously reported for injected volume. Therefore, the tubing located not only between the injector and column inlet, but also the tubing between the column outlet and the detector inlet, should be optimized in terms of dimensions to limit band broadening and efficiency loss under isocratic conditions.

The dispersion related to the tubing (σ^2_{tube}) can be expressed as a function of the tubing radius, r_{tube} and its length, L_{tube} [21]:

$$\sigma^2_{\text{tube}} = \frac{r^4_{\text{tube}} \cdot L_{\text{tube}} \cdot F}{7.6 \cdot D_m} \tag{32}$$

In this Excel spreadsheet, the dispersions related to injection and detector were neglected, for the sake of simplicity. Therefore, tubing volume impacts the plate count, according to the following equation:

$$N_{\text{obs}} = N_{\text{col}} \cdot \frac{1}{1 + \frac{\sigma^2_{\text{tube}}}{\sigma^2_{\text{col}} + \sigma^2_{\text{tube}}}} \tag{33}$$

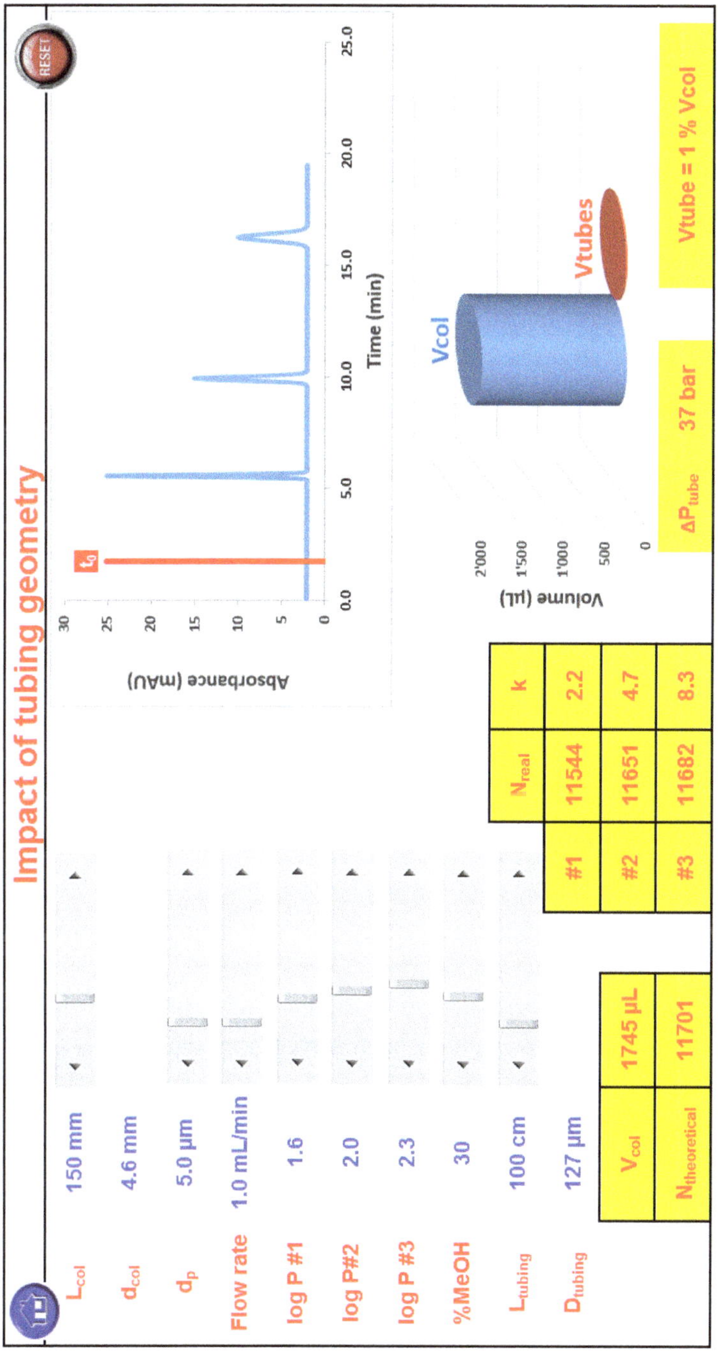

Figure 7.9: Excel spreadsheet highlighting the impact of tubing geometry.

Besides the impact of tubing dimensions on band broadening, the pressure generated by the tubing itself is also modified when changing tubing diameter (d_{tube}) and/or length (L_{tube}), according to the Hagen–Poiseuille equation:

$$\Delta P = 128 \frac{\eta \times L_{\text{tube}} \times F}{\pi \times d_{\text{tube}}^4} \tag{34}$$

In this Excel spreadsheet (see Fig. 7.9), the impact of tubing geometry (d_{tube} and L_{tube}) on chromatographic performance (chromatographic efficiency and generated pressure) was assessed. The user has the possibility to set the column dimensions (L_{col}, d_{col}, d_p), mobile phase flow rate, log P values of three model compounds, % MeOH in the mobile phase and the tubing geometry (L_{tube} and d_{tube}). Various conventional tubing diameters widely used in HPLC can be selected, namely 65, 127, 250 and 500 μm. A chromatogram is then simulated under the suggested conditions, and a graph representing the column volume vs. tubing volume is also provided at the bottom of Fig. 7.9. The pressure generated by the tubing is also provided, to select the best tubing dimensions. It is worth mentioning here that the impact of tubing becomes particularly critical when using column of low volume (short and thin). Therefore, the tubing volume has to be decreased in line with the reduction of column dimensions, to limit as much as possible the loss in plate count. Finally, a compromise has to be found between plate count and pressure, when selecting the optimal tubing geometry for plumbing an HPLC system. It is also important to remember that a given chromatographic system cannot accommodate a wide range of column internal diameters. Currently, even the best HPLC/UHPLC systems on the market are not adapted to columns of 1 mm I.D. in isocratic mode, and it would be relevant to have a technical solution for limiting the contribution of tubing.

7.11 The Impact of Compound Molecular Weight in RPLC Mode

As shown in Eq. (21), the diffusion coefficient (D_m) of a compound is directly proportional to the analyte molecular weight (MW). In addition, it is clear that there is a direct relationship between the optimal linear velocity (u_{opt}) and D_m, according to Eq. (20). Therefore, the kinetic performance (and van Deemter curve shape) could be strongly modified when analyzing compounds of different MW. In other words, the efficiency obtained at a given flow rate might be different, depending on the compound MW. A decrease of performance (broader peaks) is generally observed in RPLC when increasing the MW of the analyzed substances, since experiments are often conducted at a flow rate much higher than the optimum linear velocity of the van Deemter curve.

Besides kinetic performance, the size of the analyzed compound impacts the slope of the LSS curve (log k vs. % organic solvent), which corresponds to the S parameter in Eq. (12). It has been empirically demonstrated that the relationship between S and MW can be expressed by the following empirical relationship [4]:

$$S = 0.5 \times MW^{1/2} \quad (35)$$

Based on the LSS theory (Eq. (12)), the retention factor may vary strongly even for a minor mobile phase composition modification (% MeOH) at high S value (large molecules). In some cases (for example with large proteins), it will be even impossible to find out an isocratic composition allowing to analyze several large molecules, and gradient elution becomes mandatory.

In this Excel spreadsheet (see Fig. 7.10), the user can set the column dimensions (L_{col}, d_{col}, d_p), mobile phase flow rate, % ACN in the mobile phase and the MW of three model compounds. By taking into account these values, some kinetic curves for the three different model compounds are drawn, describing the dependence of efficiency vs. mobile phase flow rate. In addition, the LSS curves (log k vs. % ACN) obtained from Eq. (12) are also provided to illustrate the variation in curve slopes (S values), and the way by which retention factors are varying with % ACN. Finally, a chromatogram was also simulated for the three model compounds. Here, a color code (blue, purple or green) was used to distinguish the three analytes. In this Excel spreadsheet, the three selected compounds have log P values of 2.2, 2.3 and 3.5, while the S values have been estimated using Eq. (35), based on the compound MW. Equation (12) was employed to calculate the log k of each substance and their corresponding retention times.

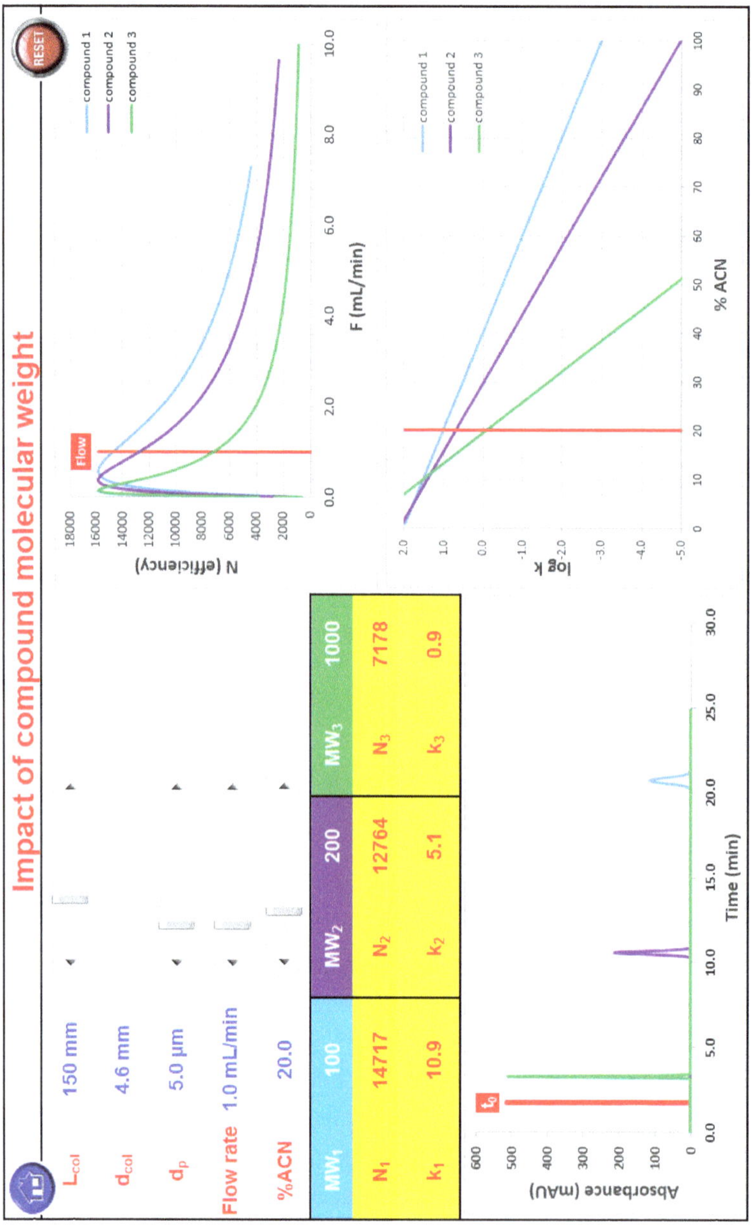

Figure 7.10: Excel spreadsheet highlighting the impact of compound molecular weight.

7.12 Conclusions

In conclusions, our free Excel software "HPLC teaching assistant" allows for learning for and teaching LC in an innovative and efficient way, using virtual (simulated) chromatograms obtained under numerous analytical conditions. This tool can be used by academic teachers as well as company training instructors who are interested in using innovative technology to better convey message during their courses.

References

[1] I. Molnar, Computerized design of separation strategies by reversed-phase liquid chromatography: development of DryLab software, *J. Chromatogr. A* 965 (2002) 175–194.
[2] E.F. Hewitt, P. Lukulay, S. Galushko, Implementation of a rapid and automated high performance liquid chromatography method development strategy for pharmaceutical drug candidates, *J. Chromatogr. A* 1107 (2006) 79–87.
[3] S. Heinisch, J.L. Rocca, A computer routine for the selection and optimization of multisolvent mobile phase systems in reversed-phase liquid chromatography, *Chromatographia* 32 (1991) 559–565.
[4] R.C. Rittenhouse, A computer simulation of high-performance liquid chromatography, *J. Chem. Educ.* 72 (1995) 1086–1086.
[5] I.C. Bowater, I.G. McWilliam, Using computers to replace some HPLC laboratory work, *J. Chem. Educ.* 71 (1994) 674–678.
[6] R.A. Shalliker, S. Kayillo, G.R. Dennis, Optimizing chromatographic separation: an experiment using an HPLC simulator, *J. Chem. Educ.* 85 (2008) 1265–1268.
[7] J.C. Reijenga, Training software for high-performance liquid chromatography, *J. Chromatogr. A* 903 (2000) 41–48.
[8] P.G. Boswell, D.R. Stoll, P.W. Carr, M.L. Nagel, M.F. Vitha, G.A. Mabbott, An advanced, interactive, high-performance liquid chromatography simulator and instructor resources, *J. Chem. Educ.* 90 (2013) 198–202.
[9] J. Li, P.W. Carr, Estimating diffusion coefficients for alkylbenzenes and alkylphenones in aqueous mixtures with acetonitrile and methanol, *Anal. Chem.* 69 (1997) 2550–2553.
[10] Y. Henchoz, D. Guillarme, S. Martel, S. Rudaz, J.L. Veuthey, P.A. Carrupt, Fast log P determination by ultra-high-pressure liquid chromatography coupled with UV and mass spectrometry detections, *Anal. Bioanal. Chem.* 394 (2009) 1919–1930.
[11] D. Benjaim, E. Grushka, Characterization of Ascentis RP-Amide column: Lipophilicity measurement and linear solvation energy relationships, *J. Chromatogr. A* 1217 (2010) 65–74.
[12] S. Martel, D. Guillarme, Y. Henchoz, A. Galland, J.L. Veuthey, S. Rudaz, P.A. Carrupt, Chromatographic approaches for measuring log *P*. In: *Drug Properties — Measurement and Computation*, Mannhold, R. (Ed), Wiley-VCH, 2008 pp. 331–355.

[13] R.A. Scherrer, S.M. Howard, Use of distribution coefficients in quantitative structure-activity relations, *J. Med. Chem.* 20 (1977) 53–58.
[14] H.N. Po, N.M. Senozan, The Henderson-Hasselbalch equation: Its history and limitations, *J. Chem. Educ.* 78 (2001) 1499–1503.
[15] R. Sitaraman, S.H. Ibrahim, N.R. Kuloor, A generalized equation for diffusion in liquids, *J. CHem. Eng. Data* 8 (1963) 198–201.
[16] D. Guillarme, S. Heinisch, J.L. Rocca, Effect of temperature in reversed phase liquid chromatography, *J. Chromatogr. A* 1052 (2004) 39–51.
[17] L.R. Snyder, J.W. Dolan, *High Performance Gradient Elution: The Practical Application of the Linear-Solvent-Strength Model*, Wiley, 2007.
[18] U.D. Neue, Peak capacity in unidimensional chromatography, *J. Chromatogr. A* 1184 (2008) 107–130.
[19] D. Guillarme, E. Grata, G. Glauser, J.L. Wolfender, J.L. Veuthey, S. Rudaz, Some solutions to obtain very efficient separations in isocratic and gradient modes using small particles size and ultra-high pressure, *J. Chromatogr. A* 1216 (2009) 3232–3243.
[20] L.R. Snyder, J.J. Kirkland, J.W. Dolan, *Introduction to Modern Liquid Chromatography*, 3rd Edition, Wiley, 2010.
[21] D. Guillarme, D. Nguyen, S. Rudaz, J.L. Veuthey, Method transfer for fast liquid chromatography in pharmaceutical analysis: application to short columns packed with small particle. Part I: Isocratic separation, *Eur. J. Pharm. Biopharm.* 66 (2007) 475–482.

Chapter 8

Examples on Small Molecule Pharmaceuticals (From the Beginning to the Validation)

Róbert Kormány* and Norbert Rácz

Egis Pharmaceuticals PLC, Keresztúri út 30–38, H-1106 Budapest, Hungary

**kormany.robert@egis.hu*

8.1 Introduction

Method development is the bottleneck in liquid chromatography (LC) even today, when more and more fast chromatographic systems are available and routinely used. Expert or intelligent programs can be applied to reduce the time spent on method development and offer extra information about the robustness of the separation. In spite of the fact that the exact separation (retention) mechanism is often poorly understood, some practical approaches are often used to predict a separation under any conditions based only on some preliminary experimental runs (4–12 runs). The solvophobic theory of reversed-phase liquid chromatography (RPLC) generally gives us guidance for planning (designing) the experiments for method development or optimization [1]. The DryLab chromatographic modeling software is based on this concept [2] and allows to perform a multifactorial optimization procedure [3].

LC is still the main analytical tool in pharmaceutical industry as about the 80% of analysis are carried out with high-performance liquid chromatography (HPLC). According to new instructions and regulatory needs, quality by design (QbD) approach must be used during the production and for the quality control of products which are analyzed mostly by HPLC.

Therefore, the analytical procedure should meet the same criteria. Not going into detail, this means that the space of method variables (design space, DS) should be well planned and all variables should be well controlled in the selected range [4].

Since a LC separation depends on several parameters and variables, any tool which helps reducing the number of experiments and provides additional information about method robustness and retention properties clearly improve the quality of the applied procedure. The concept of validation of an analytical procedure has been defined by the International Council on Harmonization (ICH) Q8 guideline [5]. The key steps for QbD-related method development are the following [6]:

- Clear definition of the method goals, where the QbD term is the analytical target profile (ATP). In HPLC method development, one important method goal is almost the same, namely to get sufficient — mostly baseline — resolution between the critical peaks of the given sample.
- Performing a risk-assessment means to evaluate which variables may have a negative influence on the defined method goals, i.e. which method parameters affect the resolution of peaks in a negative way.
- Experimental evaluation of how the critical variables affect the method goals. This should be done in a systematic, multifactorial way using a scientifically based design of experiments (DoE). As a result of the experimentation, a DS can be created, which describes the ranges of parameters in which the method goals are fulfilled.
- If a final method has been found in the way described above, a robustness study should be performed to evaluate how the method goals (for example the critical resolution) are influenced by small unintentional deviations from the defined method parameters.
- The results of the robustness study will help to set up a control strategy for the method or even for any of the critical separation variables (the next important step in QbD-related method development).
- The advantage of going through this — despite that at the first sight this approach seems to be complex and extensive — is the possibility to apply a Continual Improvement as the analytical task. Altering the variables of a method (the set point) within the calculated DS is not

considered as a "change" by the regulatory authorities and therefore can be done without a revalidation of the method. This allows, in many laboratories, for a flexibility in the use of the method with a freedom that was unknown before.

The examples reviewed in this section were planned according to QbD principles using a simple DoE with only three measured variables: the gradient time (*tG*), mobile phase temperature (*T*) and mobile phase pH or ternary composition (*tC*). In this section, the following topics are discussed and explained:

- LC method development and validation become more scientific and shorter if the principle of QbD approach is used with ultra-high-pressure liquid chromatographic (UHPLC) technology [7].
- A good alternative way to identify a certain peak (peak tracking) in different chromatograms is the molecular mass of the compound, due to its high specificity [8].
- Searching for alternative columns, while keeping the quality of a given separation is always one of the key purposes of method robustness testing [9].
- How a 3D multifactorial retention model can be built up and carried out [10]?

8.2 Case Study 1: Method Optimization and Robustness Testing

This case study describes a new and fast UHPLC separation of amlodipine and bisoprolol and all their closely related compounds, for impurity profiling purposes. Computer-assisted method development was applied, and the achievable selectivity and resolution was investigated. The work was performed according to QbD principles using a DoE with three experimental factors; namely the gradient time (*tG*), temperature (*T*) and mobile phase pH.

Thanks to modeling software, it was proved that the separation of all compounds was feasible by proper adjustments of variables. It was also demonstrated that the reliability of predictions was good, as the predicted retention times and resolutions were in good agreement with the experimental ones. The final, optimized method separated 16 peaks related to amlodipine and bisoprolol within 7 min, ensuring baseline resolution between all peak pairs.

8.2.1 *Chromatographic conditions*

UHPLC experiments were performed on an Acquity UPLC system equipped with binary solvent delivery pump, auto-sampler, photodiode array detector and Empower software. This UHPLC system had a 5 μL injection loop and 500 nL flow cell. The dwell volume of the system was measured as 125 μL. For the initial model runs, the mobile phase flow rate was set to 0.5 mL/min and gradients were run from 10% to 90% B. The injection volume was set to 1 μL. The column used this study was a 50 $\times$ 2.1 mm, 1.7 μm Acquity CSH C18. Mobile phase "A" was 30 mM phosphate buffer (pH was set at three levels: 2.0, 2.6 and 3.2) and mobile phase "B" was acetonitrile. Sample solvent was a mixture of acetonitrile/water = 10/90 v/v%.

Representative real-life sample of Amlodipine, Bisoprolol and their European Pharmacopoeia (Ph. Eur.) impurities contained 1 mg/mL Amlodipine besilate and Bisoprolol fumarate and their impurities at 0.1% level was prepared by spiking all the impurities to the API solution.

8.2.2 *Design of experiments (DoE)*

The selected example describes a fast and efficient method development for the determination of impurities and degradation products of combined active pharmaceutical ingredients, utilizing the separation power of a very efficient state-of-the-art column. A general methodology of optimization is to simultaneously model the effect of temperature and gradient steepness on selectivity with a given RPLC column. Thanks to the current developments in chromatographic modeling software products, it is now possible to model the effect of three variables simultaneously for a given separation. In our case, gradient steepness (tG), temperature (T) and mobile phase pH were selected as model variables to create a cube resolution map, showing the critical resolution of the peaks to be separated against the three factors. Probably, these selected variables have the most significant effect on the selectivity and resolution for such analytes. In RPLC, the retention can be described as a function of gradient steepness, with the linear solvent strength (LSS) theory — in most cases — and its temperature dependence follows a van't Hoff type relationship. Both relationships can be transformed to linear functions. When separating ionizable compounds, strong pH-related changes in retention occur for pH values within ± 1.5 units of the pK_a value. Outside this range, the compound is considered as mostly ionized or non-ionized, and its retention is not significantly altered with pH. In a relatively small pH range — within the ± 1.5 units of the pK_a value — the dependence of retention on the mobile phase pH can generally be described using quadratic polynomials.

Therefore, in the proposed final model, two variables (tG and T) were set at two levels ($tG_1 = 3$ min, $tG_2 = 9$ min and $T_1 = 20°C$ and $T_2 = 50°C$), while the third factor (pH) was set at three levels ($pH_1 = 2.0$, $pH_2 = 2.6$ and $pH_3 = 3.2$). This full factorial experimental design required 12 experiments ($2 \times 2 \times 3$) on a given column.

8.2.3 *Finding the optimal conditions*

First, the criteria for the minimum required resolution were set. The impurities have to be separated from (a) each other, (b) the APIs and (c) other possible disturbing compounds such as the fumaric acid and benzenesulfonic acid. For the baseline separation of the critical peak pairs, the value of $R_{s,\mathrm{crit}}$ should be higher than 1.5. But considering that impurities are present in small concentrations (at ~0.1 %), and have to be separated from the APIs which are present in high concentration, the $R_{s,\mathrm{crit}} > 1.5$ might not be enough. In this case, it is better to select $R_{s,\mathrm{crit}} > 2.5$ as criteria — if feasible. Figure 8.1 shows the obtained 3D resolution map. Red color represents the regions inside the DS where the resolution criteria is fulfilled, while blue colors indicate co-elutions ($R_s = 0$). There are four robust spaces that meet the criteria (Fig. 8.1(b)). At low pH ($pH > 2.5$), and at low temperature (below 30°C) or at high temperature (above 40°C), the resolution between fumaric acid and B-Impurity A was the lowest one, while at higher mobile phase pH ($pH > 2.5$) and at low temperature (<30°C), Bisoprolol and B-Impurity G were considered as the

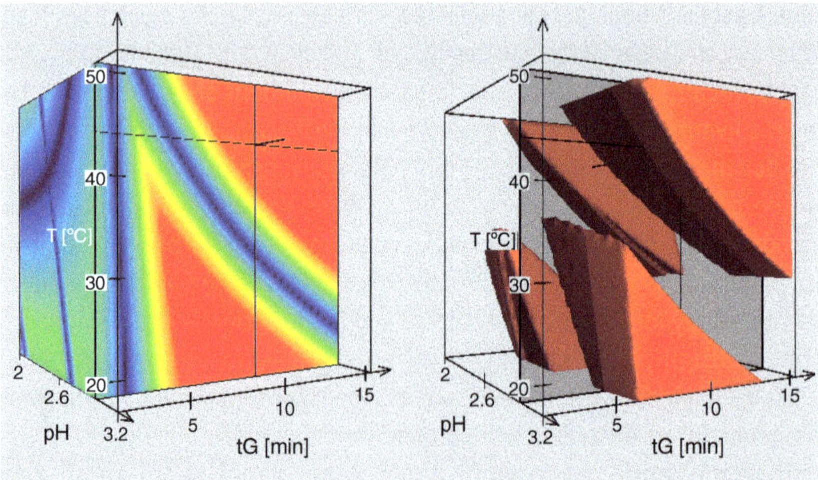

(a) (b)

Figure 8.1: Three-dimensional resolution map based on *tG*-*T*-pH model. The map shows the critical resolution as function of three variables. The warm colors (red) correspond to high resolution while the cold colors (blue) correspond to low resolution. Figure (a) shows the evolution of critical resolution in the entire design space while (b) shows only the robust zones where the pre-defined resolution criteria is fulfilled.

critical peak pair. Furthermore, a steeper gradient decreases the resolution between Bisoprolol and its B-Impurity G. Taking all these observations into account, the best working point (WP) is located into the robust space at high pH (pH > 2.5) and at high temperature ($T > 40°$C). The final conditions were set as $tG = 10$ min starting from 10% B up to 90% B (slope $=$ 8.0% B/min), column temperature $T = 45°$C and mobile phase pH $= 3.0$. Please note that the selected 10 min long gradient is outside the 3 and 9 min calibrated model, but the accuracy of the extrapolation is still valid in this range. Moreover the reliability of the model was verified (see later).

8.2.4 *Simulated robustness testing*

The reliability of the software's new simulated robustness testing feature has already been proven [11]. Similar to this previous work, the robustness of the optimized method was also assessed by the built-in robustness module. Beside the three model variables (*tG*, *T*, pH), the flow rate as well as initial and final composition of the mobile phase have been included as factors in the robustness model. The effect of these six factors was evaluated at three levels. The modeled deviations from the nominal values were the following: The gradient time was set to 9.9, 10 and 10.1 min, temperature was set to 44°C, 45°C and 46°C, mobile phase pH was set to 2.9, 3.0 and 3.1, flow rate was set to 0.495, 0.500 and 0.505 mL/min, initial mobile phase composition was set to 9.5%, 10% and 10.5% B, and its final composition was set to 89.5%, 90% and 90.5% B. Then, the 729 experiments (3^6) were simulated. A criterion of $R_{s,\mathrm{crit}} > 1.5$ was considered. Figure 8.2(a) shows the results of the experiments expressed in

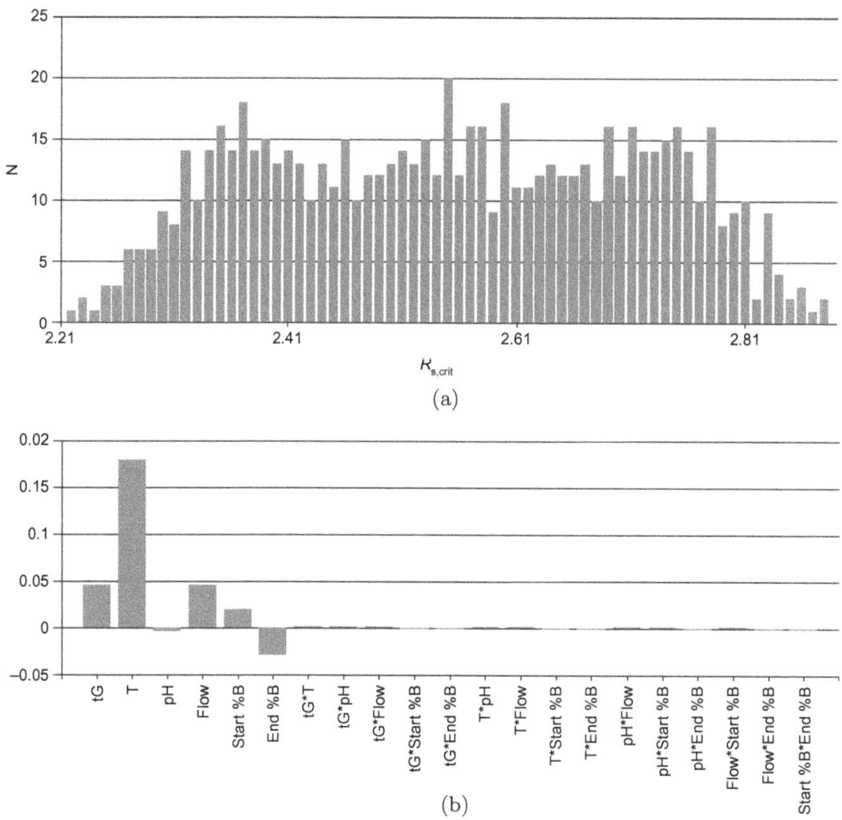

Figure 8.2: Results of simulated robustness testing. Frequency of critical resolution (a) and the relative effects of the chromatographic parameters on separation (b).

frequency as a function of critical R_s. As shown, the most frequent resolution value was $R_{s,\text{crit}} = 2.55$ (20 conditions provided this R_s value), while the lowest predicted resolution was $R_{s,\text{crit}} = 2.21$. Therefore, the method can be considered as robust, since the failure rate was 0% in the studied DS. Another feature of the modeling software used in this study is the calculation of individual and interaction parameter effects. Figure 8.2(b) describes the importance of each parameter, related to the selected deviation from the nominal value, for the critical resolution. This figure indicates that the column temperature has the most important influence on the critical resolution while the mobile phase pH plays a less important role.

8.2.5 *Reliability of the modeled results*

As a final step, the accuracy of the predicted results was evaluated. Experimental verifications of predicted chromatograms were performed. First, the optimal method was verified. Figures 8.3 and 8.4 show the predicted and experimentally observed chromatograms when operating the column at the optimal WP. The predicted retention times were in good agreement with the experimental ones, since the average retention time relative errors were $<1.0\%$ (see Table 8.1), which can be considered as an excellent prediction for such a fast gradient. The accuracy of critical resolution prediction was also assessed. As illustrated in Table 8.1, the predicted critical resolution was also in good agreement with the experimental one (2.55 vs. 2.52).

To estimate the reliability of the modeled robustness testing, 3 of the 729 experiments were selected and experimentally performed. In the first case, the conditions that provided the lowest critical resolution were set ($tG = 9.9$ min, $T = 44°$C, pH $= 3.1$, flow rate $= 0.495$ mL/min, start %B $= 9.5$ and end %B $= 90.5$). Next, the case where all parameters were set at their lowest levels was evaluated ($tG = 9.9$ min, $T = 44°$C, pH $= 2.9$, flow rate $= 0.495$ mL/min, start %B $= 9.5$ and end %B $= 89.5$). Finally, the third case corresponds to all parameters set at their highest levels ($tG = 10.1$ min, $T = 46°$C, pH $= 3.1$, flow rate $= 0.505$ mL/min, start %B $= 10.5$ and end %B $= 90.5$). In all of these three cases, the predicted retention times and $R_{s,\text{crit}}$ values were in good agreement with

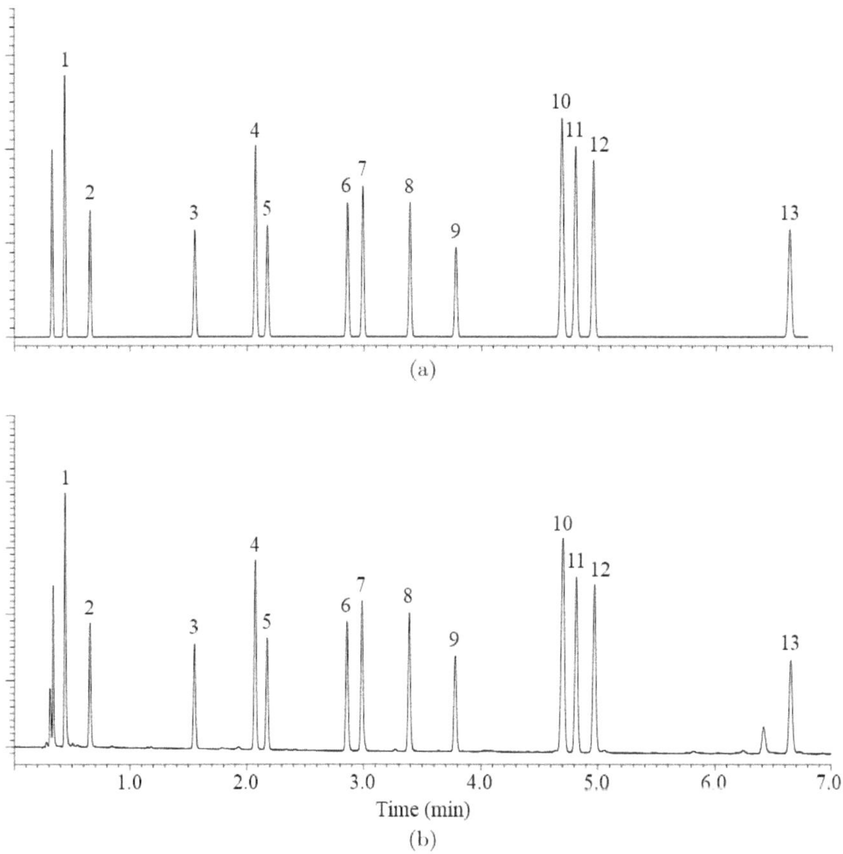

Figure 8.3: Predicted (a) and experimental (b) chromatograms of the model reference solution. B-Impurity A (1), B-Impurity L (2), B-Impurity R (3), Bisoprolol (4), B-Impurity G (5), A-Impurity D (6), A-Impurity F (7), Amlodipine (8), A-Impurity E (9), A-Impurity G (10), A-Impurity B (11), A-Impurity H (12), A-Impurity A (13).

the experimental ones, the errors in retention times were less than 0.05 min and errors in $R_{s,\text{crit}}$ values were less than 0.03 (see Table 8.1).

8.3 Case Study 2: Mass Spectrometry Supported Peak Tracking and High pH Separation

A previously developed method for Terazosin was revised in order to reduce the analysis time from 90 min (2 $\times$ 45 min) down to below 5 min. The

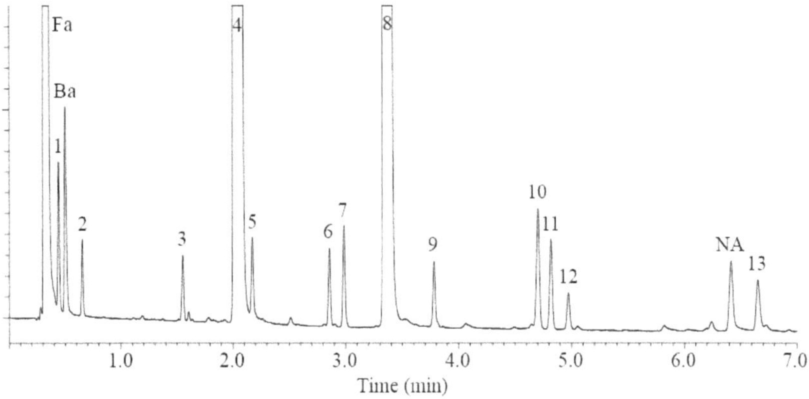

Figure 8.4: Real sample spiked with all impurities at 0.1%. Fumaric acid (Fa), B-Impurity A (1), Benzenesulfonic acid (Ba), B-Impurity L (2), B-Impurity R (3), Bisoprolol (4), B-Impurity G (5), A-Impurity D (6), A-Impurity F (7), Amlodipine (8), A-Impurity E (9), A-Impurity G (10), A-Impurity B (11), A-Impurity H (12), Unknown (NA), A-Impurity A (13).

method in Ph. Eur. investigates the specified impurities separately. The reason for the different methods is that the retention of two impurities is not adequate in RPLC, not even with 100% water as mobile phase. Therefore ion-pair chromatography has to be applied, and since the two impurities absorb at low UV-wavelength they had to be analyzed by a different method than the other specified impurities. It was assumed that it is possible to increase the retention of such compounds by increasing the pH. This could be done by using new type of stationary phases, which are nowadays available for higher pH applications. Also, a detection wavelength could be selected that is appropriate for the detection and quantification of all impurities.

During the method optimization, the identification of the peaks in the different runs is often challenging due to changes in elution order. Generally peak tracking is based on UV peak area, but it has some limitations and uncertainity especially when huge number of peaks possessing similar peak areas have to be tracked. Taking the molecular weights of the solutes into account for the peak tracking obviously facilitates the alignment of the peaks between the different runs. LC-MS is now a routine technique and available in most laboratories. In the latest versions of the modeling

Table 8.1: Experimental verification of retention time and resolution predictions.

		Original method		Worst method		Low parameters		High parameters	
Peak number	Name	Predicted t_R (min)	Experimental t_R (min)	Predicted t_R (min)	Experimental t_R (min)	Predicted t_R (min)	Experimental t_R (min)	Predicted t_R (min)	Experimental t_R (min)
1	B-Impurity A	0.44	0.44	0.48	0.48	0.47	0.46	0.41	0.43
2	B-Impurity L	0.65	0.65	0.72	0.71	0.70	0.68	0.60	0.63
3	B-Impurity R	1.55	1.55	1.63	1.63	1.59	1.60	1.51	1.52
4	Bisoprolol	2.07	2.07	2.15	2.14	2.11	2.11	2.02	2.04
5	B-Impurity G	2.17	2.18	2.25	2.24	2.21	2.21	2.12	2.14
6	A-Impurity D	2.86	2.86	2.93	2.92	2.89	2.89	2.81	2.82
7	A-Impurity F	2.99	2.99	3.05	3.04	3.03	3.02	2.93	2.94
8	Amlodipine	3.39	3.39	3.45	3.44	3.42	3.42	3.33	3.35
9	A-Impurity E	3.78	3.78	3.84	3.82	3.81	3.81	3.72	3.75
10	A-Impurity G	4.69	4.70	4.72	4.72	4.74	4.76	4.61	4.64
11	A-Impurity B	4.80	4.82	4.83	4.82	4.85	4.86	4.73	4.77
12	A-Impurity H	4.95	4.97	4.97	4.96	5.01	5.03	4.86	4.91
13	A-Impurity A	6.63	6.65	6.65	6.63	6.67	6.69	6.56	6.61
$R_{s,crit}$	A-ImpG - A-ImpB	2.55	2.52	2.22	2.19	2.29	2.29	2.81	2.84

Notes: The "original method" corresponds to the optimal method, the "worst method" corresponds to the conditions where the lowest resolution can be achieved, while "Low" and "High parameters" corresponds to conditions where all the variables were set at their lower and higher levels.

software, the m/z-MS-data can be combined with the UV-peak-area-based tracking technology.

8.3.1 *Chromatographic conditions*

UHPLC measurement was performed using an Acquity UPLC system equipped with binary solvent delivery pump, an auto-sampler, a photo diode array (PDA) detector and QDa mass detector. Detection was performed at 220 nm, and the cone voltage of QDa mass detector was set as 15 V and was used in positive mode. The system injector contained 1 μL loop and the volume of detector flow cell was 500 nL. The dwell volume of the system was measured to be 0.125 mL. A 50 $\times$ 2.1 mm, 1.7 μm Acquity CSH C18 column was used in this work.

8.3.2 *Preliminary experiments, stationary phase*

One of the most important steps during HPLC method screening is the selection of an appropriate stationary phase. The studied components are basic molecules mostly possessing high pK_a values (see Table 8.2). Therefore, a pH-resistant stationary phase is required. Acquity CSH C18 column is a good choice for the separation of basic compounds, due to its special manufacturing technology. This stationary phase is pH resistant between

Table 8.2: Calculated exact masses and pK_a values.

Name	Exact mass[a]	pK_a[a]
Terazosin	387.19	7.24
Impurity A	239.05	2.94
Impurity B	388.17	4.90
Impurity C	289.15	8.72
Impurity E	492.22	7.54
Impurity J	389.45	7.24
Impurity K	383.16	7.24
Impurity L	180.09	7.82
Impurity M	274.10	—
Impurity N	184.12	7.82
Impurity O	282.16	—

Note: [a]Determined by MarvinSketch software.

$1 < pH < 11$ and below 45°C. The dimension of the chosen column was 50×2.1 mm with 1.7 μm particle size for the UPLC system. The maximum pressure during the measurements was ca. 800 bar, due to the higher viscosity of the water-methanol eluent system used for this study.

8.3.3 *Design of experiments (DoE)*

In this case study, the DoE factors included gradient time (tG), temperature (T) and ternary eluent composition (tC). For the better handling, the experiments were numbered from 1 to 12 (see Chapter 2). The mobile phase A consisted of 0.1 v/v% ammonium hydroxide in water, while mobile phase B1 was acetonitrile, B2 was methanol and their 50/50-mixture with 0.1 v/v% ammonium hydroxide. Two linear gradients, tG_1 with 3 min and tG_2 with 9 min, (corresponding to a factor 3 of difference) and a gradient range from 10% to 90% B were carried out at two different mobile phase temperatures as $T_1 = 20$ and $T_2 = 40$°C. The flow rate was set to 0.5 mL/min. The injection volume was 0.2 μL.

8.3.4 *Sample preparation*

To obtain an appropriate signal-to-noise ratio for impurities, the concentration of the API was relatively high (1 mg/mL). Using the UPLC system with columns of 50×2.1 mm size, a 1 μL sample injection is commonly for used suitable efficiency. This injection volume cannot be applied in every situation. In our case, the solubility of the sample at the starting composition water/methanol = 80/20 v/v% was not adequate. Furthermore, the components 6 and 11 precipitated with time (Figs. 8.5(a) vs. 8.5(b)). Methanol is a good solvent for Terazosin, but in this case the peak of component 1 split, as shown in Fig. 8.5(b).

This problem can be solved by injecting a smaller volume. Using a loop with 1 μL volume in the UPLC system allows to inject 0.2 μL with the "partial loop with needle overfill" (PLNO) injection mode. This solution requires higher sample concentration to maintain the quantity of the injected sample. 1 mL of methanol can dissolve 5 mg of Terazosin, so the amount of the injected sample is the same as before. In addition, the previous problems were solved (Fig. 8.6).

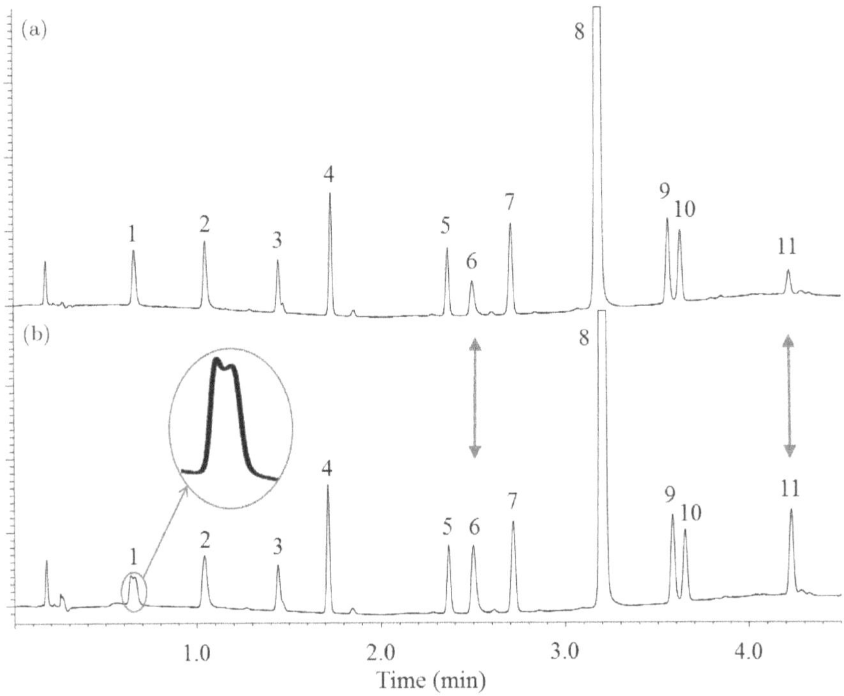

Figure 8.5: 1 μL sample injection dissolved in the weak solvent A (a) and in pure methanol (b).

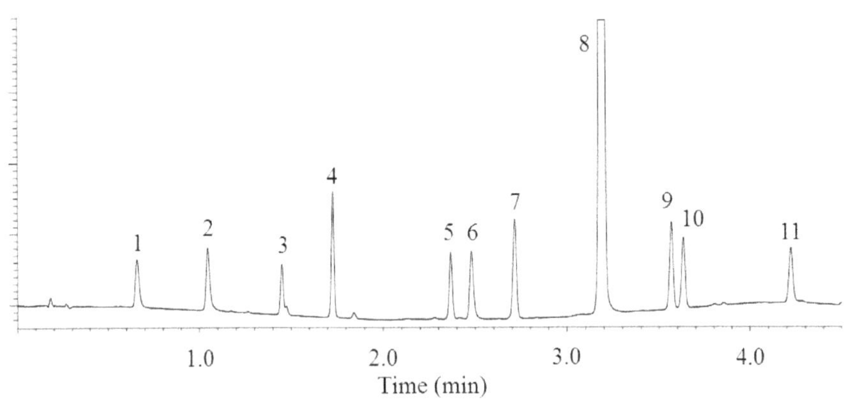

Figure 8.6: Obtained chromatogram using a sample injection of 0.2 μL. Impurity N (1), Impurity L (2), Impurity O (3), Impurity B (4), Impurity M (5), Impurity C (6), Impurity A (7), Terazosin (8), Impurity K (9), Impurity J (10) and Impurity E (11).

Flow-through-needle (FTN) injection mode is also feasible. This is important because nowadays many UHPLC instruments apply FTN injection mode and not off-line loop samplers.

8.3.5 *Effect of mobile phase pH*

During the investigation of the pH-dependence of the method, it was observed that there was no retention time change between pH = 10 and 11 (Figs. 8.7(a)–8.7(c)). However, the retention time of Impurity B, which contains a phenolic OH-group, increased with decreasing pH. At pH = 9, the retention time of Impurity B increased so much that it eluted after Impurity M and Impurity C and co-eluted with Impurity A (Fig. 8.7(d)). If

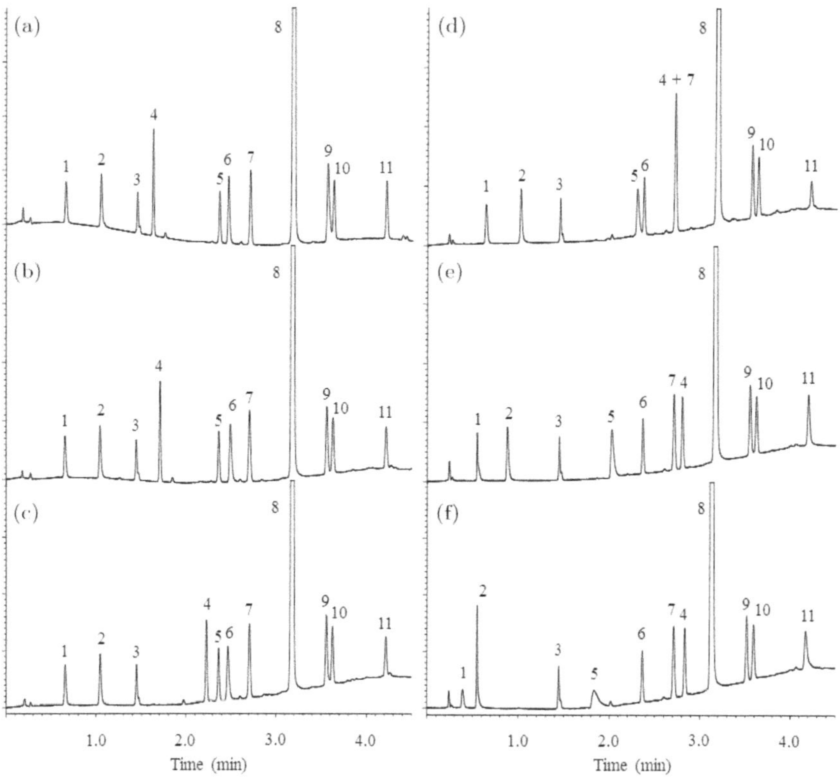

Figure 8.7: Effect of mobile phase pH. (a) pH = 11.0; (b) pH = 10.7 (0.1 v/v% ammonium hydroxide solution); (c) pH = 10.0; (d) pH = 9.0; (e) pH = 8.0 and (f) pH = 7.0.

pH $<$ 9, the form of Impurity M continuously deteriorated, and the retention of Impurity N and Impurity L decreased (Figs. 8.7(e) and 8.7(f)). For pH adjustment to pH $<$ 10, ortho-phosphoric acid was used. The 0.1 v/v% ammonium hydroxide solution was titrated until the desired pH value was obtained. To prepare the pH $=$ 11 solution, the addition of 0.4 v/v% ammonium hydroxide solution was required.

To conclude, the use of 0.1 v/v% ammonium hydroxide solution was optimal, because it ensured that the pH value was above 10, and it could be maintained in water/organic mixtures without any precipitation. The preparation of the eluent is extremely simple: Add 500 μL of 25 v/v% ammonium hydroxide in 500 mL eluent. The pH value of this solution is 10.7.

8.3.6 *Peak tracking*

Chromatographic modeling requires a decent peak tracking. It means that all the peaks should be identified during all initial runs, which are created and further used to calculate a separation model. This can be done by comparing the peak areas, because peak areas are expected to remain constant in a *tG–T*-model, so long as the sample, the injected sample amount, the flow-rate and detection wavelength remain constant (and degradation does not occur at high temperature).

This procedure reaches its limitations if the peaks possess similar areas, or the integration leads to stronger variation of peak areas, which often occurs with very small peaks (typically with impurities which are present at low concentration in the sample). In case of similar peak area, UV-spectra might support the correct peak assignments.

A good alternative to find a certain peak in different chromatographic runs is the property of the molecular mass of the compound, due to its high specificity. HPLC-MS is now a proven technique and is increasingly available in HPLC labs. The main disadvantage of using mass spectra (MS) for peak tracking may be that not all compounds will give a suitable MS signal under different elution conditions (see Fig. 8.8).

The ideal case to use MS data for peak tracking would be if all sample components and their masses are known. If there are unknown components, but retention times are known from UV detection, it is possible to

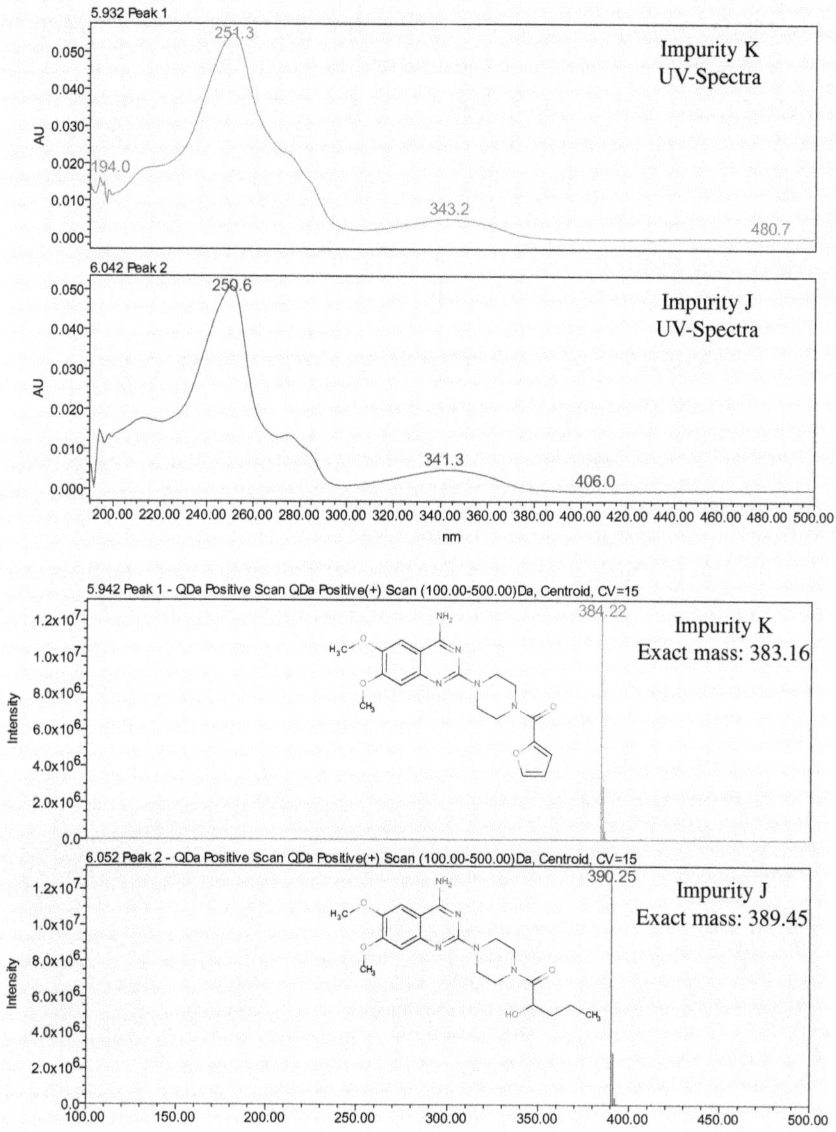

Figure 8.8: UV- and mass-spectra of Impurity K and Impurity J. In this case, the Impurity K and Impurity J peak areas very similar and also their UV spectra are comparable, so there is not much chance to differentiate them. Using m/z values allow for a unique identification of these impurities.

look for the molecular masses under each peak which follows the typical peak shape — increasing at the peak start, going through a maximum and decreasing to baseline at the tail of the peak. This mass then would belong to that certain peak, and this mass can be found in the other chromatograms obtained during the DoE. It is therefore advisable to allow entering the molecular mass into the peak tracking data table. The latest DryLab4 version offers both, UV peak area and molecular mass values, for tracking peak movements. In this way, the automation of the complex task "Peak Tracking" becomes much easier.

8.3.7 *Calculation of a 3D critical resolution space (CRS) called also method operable design region (MODR)*

In Fig. 8.9(c) we can see that at 100% methanol, the DS has a range where $R_{s,\text{crit}} > 1.5$, but this is not suitable for the system with acetonitrile/methanol mixture and with acetonitrile-rich eluents. With 100% acetonitrile, the resolution between Impurity M, Impurity C and Impurity A is not sufficient ($R_{s,\text{crit}} < 1.5$) and the peak elution order is altered. In case of acetonitrile/methanol mixture, the peak pair Impurity J — Impurity K is exhibiting a partial overlap and the retention of peak pair Impurity B — Impurity O is changed. Using 100% methanol as eluent B, a much better

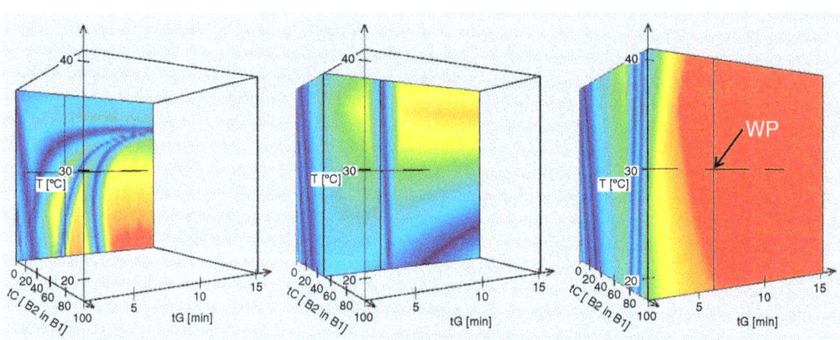

(a) (b) (c)

Figure 8.9: Three-dimensional resolution maps obtained by using 100% acetonitrile (a), 50/50 v/v% acetonitrile/methanol (b) and 100% methanol as organic modifier (mobile phase B) (c). Red colors mean regions above $R_{s,\text{crit}} > 1.5$ (baseline resolution of the critical peak pair) and blue colors indicate co-elution ($R_{s,\text{crit}} = 0$) of the closest ("critical") peak pair.

Table 8.3: Predicted and experimental retention times and measured masses.

Peak number	Name	Predicted t_R (min)	Experimental t_R (min)	Measured mass
1	Impurity N	0.65	0.66	185.09
2	Impurity L	1.00	1.05	181.03
3	Impurity O	1.44	1.45	283.12
4	Impurity B	1.71	1.73	389.21
5	Impurity M	2.36	2.37	275.07
6	Impurity C	2.47	2.48	290.13
7	Impurity A	2.70	2.72	240.00
8	Terazosin	3.18	3.19	388.20
9	Impurity K	3.56	3.57	384.22
10	Impurity J	3.62	3.63	390.25
11	Impurity E	4.21	4.22	493.29

separation can be achieved. In this case, we can see that methanol is a better solvent compared with acetonitrile or ternary mixtures of acetonitrile with methanol, as suggested in the Ph. Eur. method.

The best WP was: $tG = 6$ min (10–90% B), $T = 30$°C, $tC = 100$% methanol + 0.1 v/v% cc. ammonium hydroxide solution in water. The gradient steepness is 13.33% B/min. The last component is eluted at 4.2 min (Table 8.3), so the analysis time can be reduced to 4.5 min, which means 70% B final eluent composition at 13.33% B/min gradient slope.

Figure 8.10 shows the predicted and the measured chromatograms. Table 8.3 shows that correlation between modeled and measured retention times was excellent. The average deviation of the retention times is 0.5 s, as long as the $R_{s,crit}$ in both cases is 1.82. This precision is revolutionary in separation modeling.

8.4 Case Study 3: Simulated Column Interchangeability

Nowadays, thousands of LC columns are available on the market. If only octadecyl (C18) phases are taken into account, then we still have the possibility to choose from more than 500 products. On one hand, this can make the method development easier since the chromatographer can select

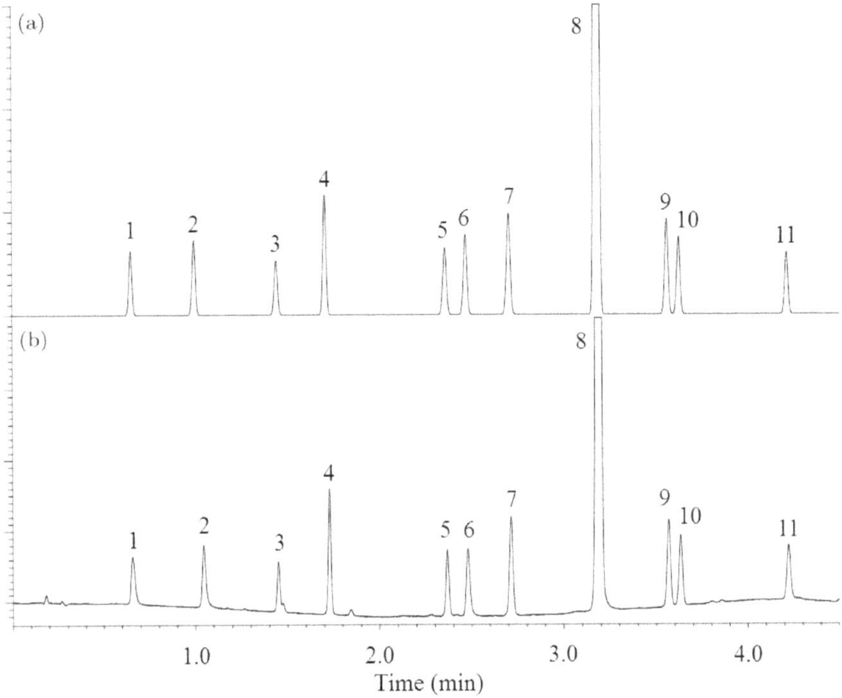

Figure 8.10: Predicted (a) and experimental (b) chromatograms. $R_{s,\text{crit}} = 1.82$ between Impurity K (9) and Impurity J (10) in predicted chromatogram, and $R_{s,\text{crit}} = 1.82$ between Impurity K and Impurity J in experimental chromatogram.

the most suitable stationary phase for a given separation. On the other hand, it can be a heavy task to find an appropriate replacement (alternative) column, which provides a very similar separation as our original column. Today, it is indeed required to suggest an alternative column in pharmaceutical analytical laboratories and to prove its equivalency during the method validation process. In fact, the pharmaceutical regulatory guidelines mention that method robustness has to be checked on columns from different batches and also on other manufacturer's column providing similar separation quality.

In previous studies, the simulated robustness testing was systematically studied and compared to experimental measurements and DoE-based predictions [7, 12]. The reliability of this "early-stage" simulated robustness approach was critically evaluated for real-life separations applying short narrow-bore columns (50 × 2.1 mm) and fast separations. Moreover, as a continuation of robustness study, the column interchangeability was further investigated, using four different C18 columns packed with sub-2 μm particles. By properly varying the method variables, the separation was feasible on all columns within the same timescale (less than 4 min). This work demonstrates that nearly the same quality of separation can be achieved on different stationary phases.

The novelty of the present work is the practical use of the recently introduced Column Comparison module in DryLab 4.3 modeling software. In this module, various 3D resolution maps can be compared (overlapped), which can help studying the measured points — in a DS — of the different phases and find a common zone where the sample components are all separated with sufficient selectivity and resolution.

8.4.1 *Chromatographic conditions*

UHPLC experiments were performed on a Waters Acquity UPLC I-Class system equipped with binary solvent delivery pump, auto-sampler and photodiode array detector. This system had flow-through-needle (FTN) sample injector and 500 nL flow cell. The dwell volume of the system was measured as 0.1 mL.

The mobile phase used in this work was a mixture of acetonitrile and water buffered with 10 mM ammonium acetate buffer. The sample was prepared from Amlodipine (0.5 mg/mL) and spiked with all the impurities at 0.5% level; sample solvent was acetonitrile/water = 30/70 v/v%.

The columns used in this study were selected on the basis of the following criteria: all of them should be based on porous silica gel (to neglect differences in morphology), with similar particle size (to have comparable specific surface area and efficiency). We focused on differences and effects of accessible free silanols (see Table 8.4).

8.4.2 *Preliminary experiments*

As previously mentioned, the goal of this study was to introduce a strategy where — beside method optimization — a substitution (alternative) column can be offered as part of the robustness testing.

Table 8.4: Characteristics of the columns packed with sub-2 μm particles tested in this study.

	HSS C18	HSS C18 SB	Hypersil GOLD C18	Titan C18
Pore size (Å)	130	130	175	80
Surface area (m^2/g)	230	230	220	410
Surface coverage ($\mu mol/m^2$)	3.2	1.6	NA	NA
Carbon load (%)	15	8	11	13
Endcapping	Yes	No	Yes	Yes

Based on former experiments, the amlodipine and its impurities were found to be relatively lipophilic, so the starting mobile phase composition was set as 30% acetonitrile. However, the Impurity A compound was highly lipophilic, so high acetonitrile content (90%) was required at the end of the gradient to elute this substance. In addition, it is also important to mention that there is structural similarity between Amlodipine, Impurity D, Impurity E and Impurity F, and all of them contain a primary amino group ($pK_a > 10$). Therefore, all these substances will be ionized under common RP conditions. Impurity H has an acidic character due to the carboxylic acid group attached to an aromatic structure ($pK_a \sim 4$), so depending on the RPLC conditions it can be either fully ionized or neutral.

During the preliminary experiments, four C18 columns belonging to the USP L1 group were chosen. The reference column was the Acquity HSS C18, and our goal was to find the appropriate replacement column. During the initial experiments at pH $= 4.5$, it was observed that Acquity HSS SB C18 column showed high silanol activity under these conditions, since the peak of the basic substances were broad and tailed, with a significant increase in retention (Fig. 8.11(b)). For all these reasons, this column was excluded.

In the case of Titan C18 column, which has medium surface coverage and endcapping, the peak shape of basic compounds were more asymmetrical than the peaks of acidic or neutral compounds, but they could be evaluated during method optimization (Fig. 8.11(d)).

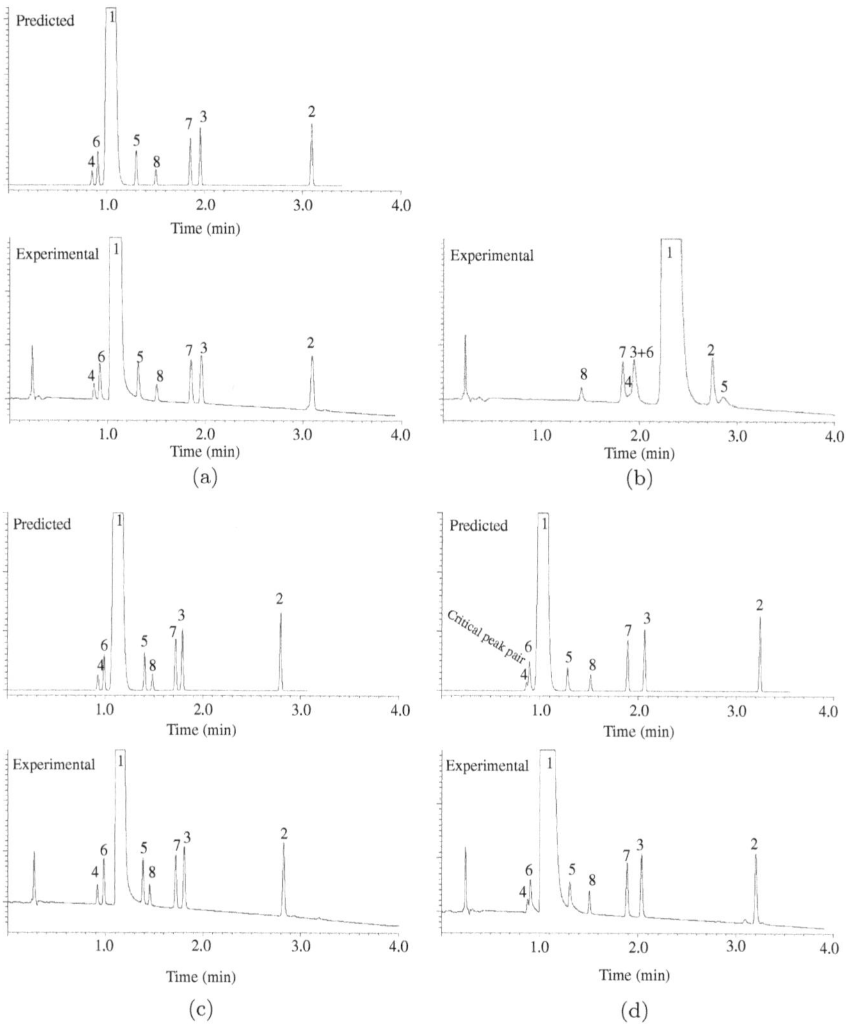

Figure 8.11: Predicted (top) and experimental (bottom) chromatograms of the four tested 50 × 2.1 mm C18 columns packed with sub-2 μm particles. Acquity HSS C18 (a), Acquity HSS C18 SB (b), Hypersil GOLD C18 (c) and Titan C18 (d). Amlodipine (1), Impurity D (4), Impurity E (5) and Impurity F (6) contain free amino groups. Impurity H (8) contains free carboxylic group. There is a movement of Impurity H with increasing pH to shorter retention times, which has a strong influence on the elution order. Impurity A (2), Impurity B (3) and Impurity G (7) are neutral in the tested chromatographic conditions.

With the Acquity HSS C18 (Fig. 8.11(a)) and Hypersil GOLD C18 (Fig. 8.11(c)) columns, which have both high surface coverage and end-capping, the peak shape of all compounds was symmetrical.

8.4.3 *Design of experiments (DoE)*

For this optimization, gradient steepness (tG), temperature (T) and mobile phase pH were selected as model variables to create a cube resolution map, showing the critical resolution of the peaks to be separated against the three factors. Probably, these selected variables have the most significant effect on selectivity and resolution for such analytes. Therefore, in our proposed final model, two variables (t_G and T) were set at two levels ($tG_1 = 3$ min, $tG_2 = 9$ min and $T_1 = 20°$C and $T_2 = 50°$C), while the third factor (pH) was set at 3 levels ($pH_1 = 4.0$, $pH_2 = 4.5$ and $pH_3 = 5.0$). This full factorial experimental design required 12 initial experiments ($2 \times 2 \times 3$) on a given column. These experiments have been performed on the selected three columns.

8.4.4 *Calculation of a 3D-critical resolution space (CRS) also called method operable design region (MODR)*

As illustrated in Fig. 8.12(a), at low temperature and short gradient time (the left bottom side of the resolution cube), the DS has a range where the resolution ($R_{s,\text{crit}}$) is larger than 1.5. At intermediate temperature (and intermediate gradient time), the separation is not acceptable. However, at high temperature and long gradient time (the right top side of resolution cube), the $R_{s,\text{crit}} > 1.5$ criterion is also fulfilled, but probably column life time would be shorter at high temperature conditions. For these reasons, the best WP was selected as: $tG = 4$ min (30–90% B), $T = 25°$C, pH $= 4.2$. The WP is indicated as the intercept of horizontal and vertical black lines.

Figure 8.12(a) shows the predicted and measured chromatograms on Acquity HSS C18 column at the selected WP. The correlation between calculated and measured retention times was excellent. The average deviation of the retention times between model and measured data was 0.5 s.

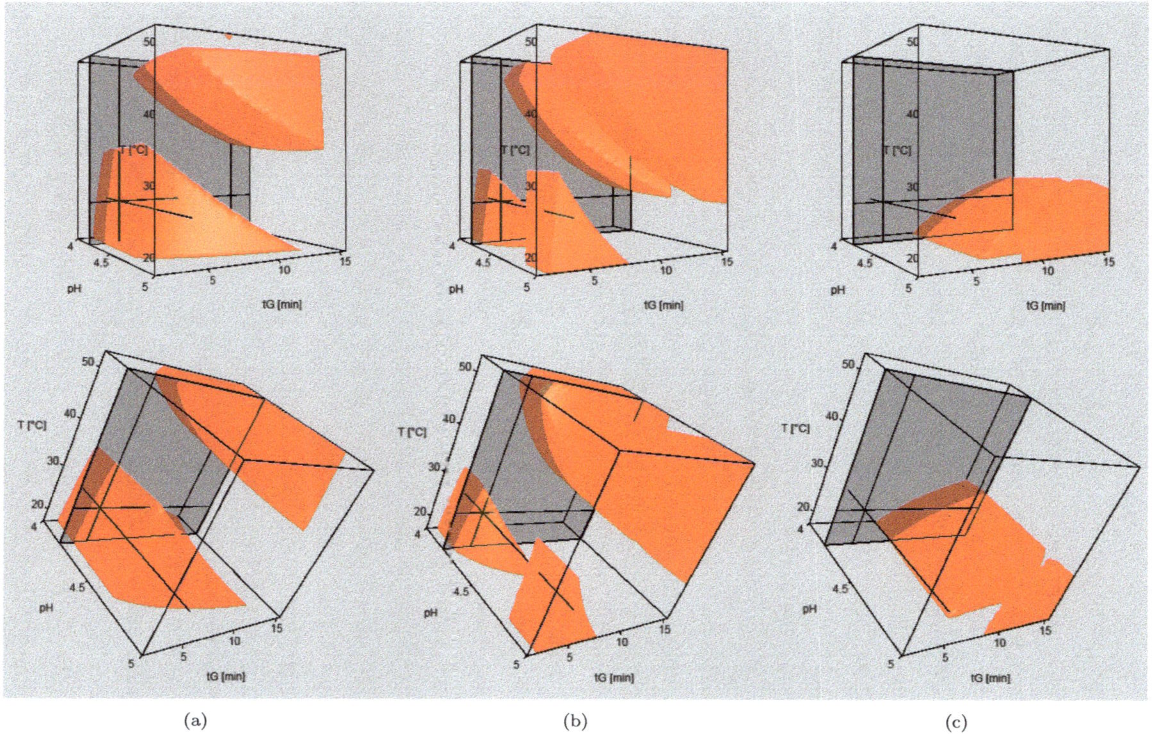

Figure 8.12: DryLab 3D models of different columns, Acquity HSS C18 (a), Hypersil GOLD C18 (b) and Titan C18 (c). Baseline resolution regions are shown in red. The different geometric bodies form a DS, which allows for altering the position of the set point (working point, WP) without the need for a new validation, as the alteration of the WP inside the DS is not considered as a "change", and so far no change management is necessary. The robustness of the individual WPs is different between the different red regions.

8.4.5 *Column interchangeability*

These 12 experiment-based approaches seem to be a reliable procedure when comparing the achievable analysis time, resolution and working point. By applying 50 × 2.1 mm columns, it takes approximately only 2–3 h of experimental work for one given column. The advantage of this column screening approach is that the suitability of a column — for a given application — can be evaluated at a very early stage of method development. In addition, the column interchangeability can also be estimated during the method development. Therefore, our column screening approach seems to be a promising method development strategy, as it consists of performing initial runs and building up 3D models using different columns at the early phase of method development.

The same procedure as described in Secs. 3.2 and 3.3 was then applied on Hypersil GOLD C18 and Titan C18 columns. In Fig. 8.12, the resolution cubes were compared for these two additional columns and the reference one. When comparing Figs. 8.12(a) and 8.12(c), it is clear that the Titan C18 column cannot be considered as a suitable replacement column under the conditions described in the section on Preliminary experiments, since it provides suitable separation in the opposite part of the DS, as the Acquity HSS C18 column does. Even if appropriate separation is feasible on the Titan C18 column, it is not comparable to the one obtained on the reference column. When comparing Figs. 8.12(a) and 8.12(b), some differences can be observed in the low temperature range in the resolution cube, due to the acidic Impurity H, which has variable ionic characteristics at pH between 4 and 5. Nevertheless, under the conditions described in the section of Preliminary experiments, the Hypersil GOLD C18 seems to be an appropriate replacement column, as it also provides $R_{s,\text{crit}} > 1.5$.

To help in selecting the most interesting alternative column, the new version of DryLab software allows the user to compare the parts of the resolution cubes where the $R_{s,\text{crit}} > 1.5$ criteria is fulfilled. It has to be mentioned that the retention order has to be checked in every case because the software does not indicate if there was an alteration (change in elution order).

By comparing the two resolution cubes (Fig. 8.13(b)), it can be established that the WPs obtained on the Acquity HSS C18 and Hypersil GOLD

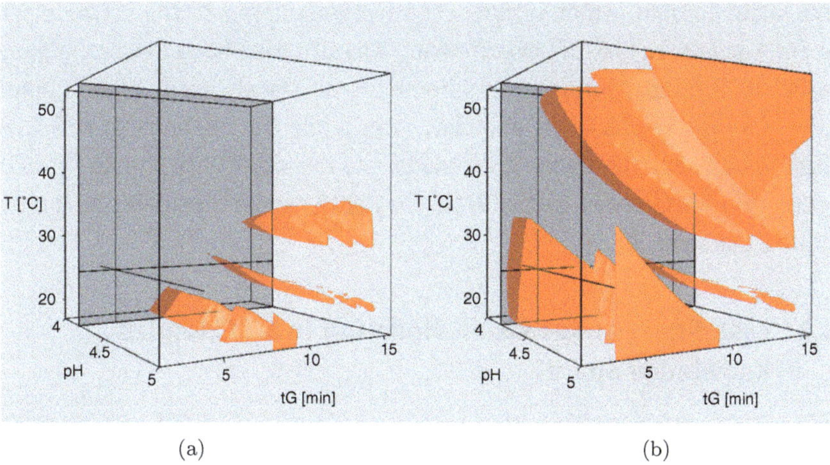

Figure 8.13: Comparison of the resolution maps of three columns (a) and two columns (b) in the same design space. The red colors correspond to the overlapping robust zones where the resolution criterion is fullfilled. Panel (a) compares the Acquity HSS C18, Hypersil GOLD C18 and Titan C18 columns while (b) compares the Acquity HSS C18 and Hypersil GOLD C18 columns.

columns at $tG = 4$ min (30–90% B), T = 25°C and pH = 4.2 are interchangeable for the measurement of amlodipine and its related impurities, as they share a relatively large zone in the DS around the selected WP, while the Titan C18 is clearly inappropriate (Fig. 8.13(a)).

Using the analytical strategy mentioned above, it is possible to quickly develop a robust method and easily find out an appropriate replacement column for the method.

8.4.6 *Robustness testing*

In this study, the robustness of the method around the WP was compared for the Acquity HSS C18 and Hypersil GOLD C18 columns. Nominal deviations from the WP were set as: $T = 25 \pm 1$°C, mobile phase pH $= 4.2 \pm 0.1$, gradient time $tG = 4 \pm 0.1$ min, initial mobile phase composition: $30 \pm 1\%$ B, final mobile phase composition: $90 \pm 1\%$ B and flow rate $F = 0.5 \pm 0.1$ mL/min. A required resolution of $R_{s,\text{crit}} > 1.5$ was considered. Performing the 729 virtual experiments resulted in 100% and 96.3% success rate using the Acquity HSS C18 and Hypersil GOLD C18 columns, respectively. The lowest resolution ($R_{s,\text{crit}}$) was equal to 1.4 with this latter column, which occurs when five parameters of the six were set on its + levels. In real-life experiments, this situation has a low probability to occur. The most influencing parameters on the Hypersil GOLD column were the mobile phase pH and flow rate, while on the Acquity HSS C18 these were t_G and flow rate. To conclude, the two stationary phases showed some minor differences, but, overall, they both can be considered as robust around the same WP.

8.5 Case Study 4: Retention Modeling in an Extended Knowledge Space

It was interesting to study the limits of modeling range on the gradient time (*tG*), the temperature (*T*) and the pH, respectively, in relation to the separation of small molecules.

During the selection of the molecules, different aspects have been considered. In addition to the appropriate retention on the column, the substances must have a proper peak shape in the whole studied pH range. This was important in estimating the resolution, and it was essential to read the retention times properly. In another aspect, they must have sufficient UV absorption in the selected wavelength range to avoid overloading of the column. Third, molecules should have p*Ka* values in the mapping pH range. It was exciting to estimate the prediction accuracy of the software as these molecules should have a relatively high retention change through the experimental space. A wide range of molecules have been studied (covering differences in molecular mass and retention properties) in order to obtain differences between the "S" parameters in LSS model as large as possible. Thus, the following compounds were selected: acetylsalicylic acid, amlodipine, cetirizine, diclofenac, ketoprofen, loratadine, nipagin M, phenacetin, rosuvastatin and trimethoprim.

Three factors were investigated by the software: gradient time (1), temperature (2) and mobile phase pH (3). Generally, two gradients should be performed having a factor three difference between their slopes. The *tG* settings were extended to five different values; such as 5 min (starting point), 10 min (factor two difference), 15 min (factor three difference), 20 min (factor four difference) and 25 min (factor five difference), respectively. Temperature setting was also extended as 20°C (starting point),

30°C, 40°C, 50°C and 60°C. Regarding mobile phase pH, the experiments were carried out in a large range from pH = 2.7 to pH = 6.9. In order to verify the accuracy of the established models, intermediate points were added. For the gradient time, two mid-points between 5 min and 25 min (9 min and 18 min), for the temperature two mid-points between 20°C and 60°C (35°C, 55°C), and for the pH five mid-points (3.5, 4.0, 5.0, 5.5 and 6.5) were used as approval experiments (Fig. 8.14 and Table 8.5). This includes $5 \times 5 \times 15 = 375$ and 20 additional approval experiments to obtain retention prediction information of the examined molecules on the entire pH range of the citrate buffer.

8.5.1 *Chromatographic conditions*

UHPLC separations were performed using an Acquity H-Class system with quaternary pump, flow-through-needle injector system, column thermostat and a PDA detector. The system has 400 μL dwell volume and 30 μL extra-column volume. An Acquity BEH C18 1.7 μm, 50×2.1 mm column was used for the experiments.

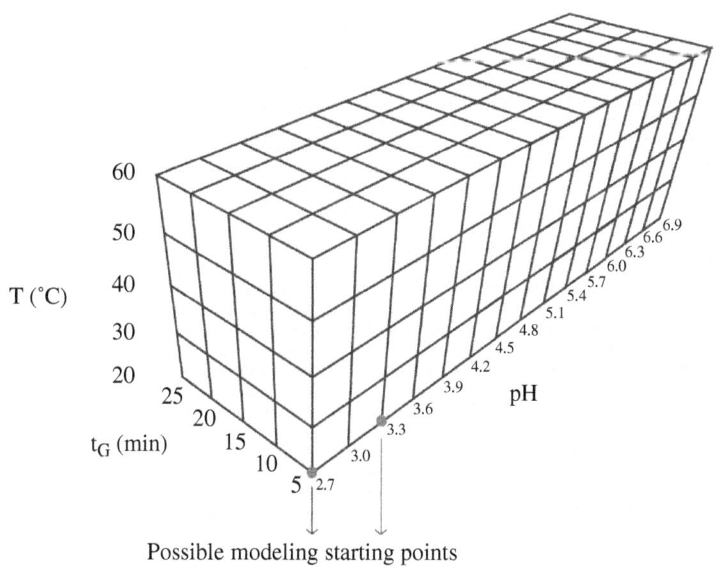

Figure 8.14: The conditions of the experiments in a huge, extended DS.

Table 8.5: The conditions of approval experiments to verify the prediction accuracy of the software.

No. of approval experiment	*tG* (min)	*T*(°C)	pH	No. of approval experiment	*tG* (min)	*T*(°C)	pH
1	9	35	3.5	11	18	35	5.0
2	9	55	3.5	12	18	55	5.0
3	18	35	3.5	13	9	35	5.5
4	18	55	3.5	14	9	55	5.5
5	9	35	4.0	15	18	35	5.5
6	9	55	4.0	16	18	55	5.5
7	18	35	4.0	17	9	35	6.5
8	18	55	4.0	18	9	55	6.5
9	9	35	5.0	19	18	35	6.5
10	9	55	5.0	20	18	55	6.5

8.5.2 *The change in prediction accuracy when extending the gradient time range*

The maximum permissible gradient time difference in modeling depends on an approximation. A relationship between apparent log*k* and gradient time can be approximated with a linear relationship [13, 14]. The suitability of the prediction highly depends on the behavior of the molecules. During the evaluation, the temperature difference was kept at the software's maximum suggested value (30°C), and the predictions were carried out from factor two difference (5–10 min) to factor five difference (5–25 min) through the entire pH range (see above for details) with a ±0.6 unit difference [15]. Table 8.6 contains the difference between the predicted and experimentally measured retention times for different (minimal and maximal) prediction ranges. The average accuracy of the nine peaks was somewhat lower when working in a larger DS, but it has not become considerably worse. When considering the gradient time as variable, it is permitted to work with two gradients possessing a factor five difference in slopes since the average accuracy of retention time prediction has not decreased below 96.45%. Longer gradient time is not suggested on short (5 cm long) columns; if we do not succeed separating all the compounds, it is preferable to choose a more efficient

Table 8.6: Retention prediction accuracies at the extension of DS in case of gradient time.

Prediction range			Experiment run condition			Retention prediction accuracy (%)
tG(min)	*T*(°C)	pH	*tG*(min)	*T*(°C)	pH	
5–10	20–50	2.7–3.3–3.9	9	35	3.5	98.27
5–25	20–50	2.7–3.3–3.9	9	35	3.5	98.33
			18	35	3.5	97.86
5–10	20–50	3.9–4.5–5.1	9	35	4.0	97.26
			9	35	5.0	98.26
5–25	20–50	3.9–4.5–5.1	9	35	4.0	97.19
			9	35	5.0	97.87
			18	35	4.0	96.45
			18	35	5.0	97.21
5–10	20–50	5.7–6.3–6.9	9	35	6.5	98.51
5–25	20–50	5.7–6.3–6.9	9	35	6.5	98.24
			18	35	6.5	97.93

column of the same type (with smaller particle diameter or core–shell particles) or a column with alternative selectivity [9]. Column length can also be increased to improve the separation efficiency through plate counts.

8.5.3 *The change in prediction accuracy when extending the temperature range*

The beneficial effect of temperature on selectivity has already been shown [16]. However, one should keep the advice of manufacturers on the temperature range in order to not shorten the column lifetime. The chosen column had a maximum of 60°C upper limit at high pH, and thus that was the limit of the model. The column thermostat had a cooling function; thus, the lowest experiment point was set at 20°C. A lower temperature (e.g. 10°C) has not been tried because of temperature fluctuations. In order to achieve low temperature measurements, a special pre-cooler and a liquid-based thermostat are required. The laboratory where the experiments were performed was not equipped with such systems. Results obtained in two pH ranges are shown in Table 8.7. For this evaluation, the gradient time was kept constant (difference of factor three), and the accuracy of

Table 8.7: Retention prediction accuracies at the extension of DS in case of temperature.

Prediction range			Experiment run condition			Retention prediction accuracy (%)
tG(min)	*T*(°C)	pH	*tG*(min)	*T*(°C)	pH	
5–15	20–40	2.7–3.3–3.9	9	35	3.5	98.91
			18	35	3.5	98.50
5–15	20–60	2.7–3.3–3.9	9	35	3.5	97.63
			9	55	3.5	98.91
			18	35	3.5	96.98
			18	55	3.5	99.01
5–15	20–40	4.5–5.1–5.7	9	35	5.5	98.62
			18	35	5.5	97.94
5–15	20–60	4.5–5.1–5.7	9	35	5.5	97.96
			9	55	5.5	98.36
			18	35	5.5	97.12
			18	55	5.5	98.39

temperature modeling was observed through the pH range (with ±0.6 difference). The average accuracy of retention time prediction was not lower than 95% in the larger DS, so the software can be used in extended ranges when modeling temperature.

8.5.4 *The change in prediction accuracy when extending the pH range*

When calculating the pH prediction accuracy, gradient time (difference of factor three) and temperature (30°C difference) were maintained constant. The accuracy in ±0.3, 0.6, 0.9 and 1.2 pH unit ranges were studied, respectively. Some of the results are shown in Table 8.8. The average accuracy decreased when modeling in wider ranges but did not fall below 95%, so the estimation of a range of 2.4 pH unit can be used. It is important to note that it was more difficult to perform the peak identification as most of the p*Ka* values fell within the modeling range. In this case, MS detection can be a great help (with the appropriate buffers). Working in a pH range larger than ±1.2 unit is not recommended due to the drastic reduction in buffer capacity.

Table 8.8: Retention prediction accuracies at the extension of DS in case of pH.

Prediction range			Experiment run condition			Retention prediction accuracy (%)
tG(min)	*T*(°C)	pH	*tG*(min)	*T*(°C)	pH	
5–15	20–50	3.3–3.6–3.9	9	35	3.5	98.80
			18	35	3.5	98.51
5–15	20–50	5.1–5.4–5.7	9	35	5.5	98.46
			18	35	5.5	97.90
5–15	20–50	6.3–6.6–6.9	9	35	6.5	98.45
			18	35	6.5	98.18
5–15	20–50	2.7–3.9–5.1	9	35	3.5	97.29
			18	35	3.5	96.43
			9	35	4.0	97.78
			18	35	4.0	96.90
			9	35	5.0	96.96
			18	35	5.0	96.27
5–15	20–50	4.5–5.7–6.9	9	35	5.0	96.59
			18	35	5.0	96.14
			9	35	5.5	98.55
			18	35	5.5	98.01
			9	35	6.5	97.64
			18	35	6.5	97.31

8.5.5 *The combined effect of the three factors on the reliability of prediction*

After examining the three factors individually, the only question that remains is how the three factors together affect the accuracy of the simulation. Thus, the resolution cubes were built up on the basis of the largest DS, with a difference in gradient times by a factor of three, temperature with $\Delta T = 40°\text{C}$ and pH in ± 1.2 unit range, respectively. Retention prediction accuracy is shown in Table 8.9. If the three factors are extended at the same time, the average accuracy decreases more than in individual cases, but the accuracy did not fall below 95% even in that case. So, it can be concluded that in an extended DS, it is also possible to estimate the retention time with a proper accuracy for each component.

Table 8.9: Retention prediction accuracies in extended DS by combining three factors (gradient time, temperature and pH).

Prediction range			Experiment run condition			Retention prediction accuracy (%)
tG(min)	*T*(°C)	pH	*tG*(min)	*T*(°C)	pH	
5–25	20–60	2.7–3.9–5.1	9	35	3.5	96.55
			9	55	3.5	97.73
			18	35	3.5	95.64
			18	55	3.5	97.45
			9	35	4.0	96.84
			9	55	4.0	99.35
			18	35	4.0	95.98
			18	55	4.0	99.23
			9	35	5.0	97.22
			9	55	5.0	98.30
			18	35	5.0	96.39
			18	55	5.0	98.22
5–25	20–60	4.5–5.7–6.9	9	35	5.0	96.21
			9	55	5.0	96.75
			18	35	5.0	95.81
			18	55	5.0	96.60
			9	35	5.5	98.22
			9	55	5.5	98.87
			18	35	5.5	97.70
			18	55	5.5	98.79
			9	35	6.5	96.75
			9	55	6.5	98.38
			18	35	6.5	96.30
			18	55	6.5	98.21

8.5.6 *Visual inspection of the extended variables*

In Fig. 8.15, chromatograms (both experimental and modeled ones) are shown. Table 8.10 represents the conditions of the different runs shown in Fig. 8.15. The main difference of the predictions comes from peaks A, B and G. As we can see, the non-extended prediction ranges (2, 6 and 8) show better correlation than the extended ones. For the runs at recommended conditions, the co-elution of ketoprofen and rosuvastatin, and the "distance" between trimethoprim and acetylsalicylic acid is modeled with good accuracy. Because of the choice of model compounds — having

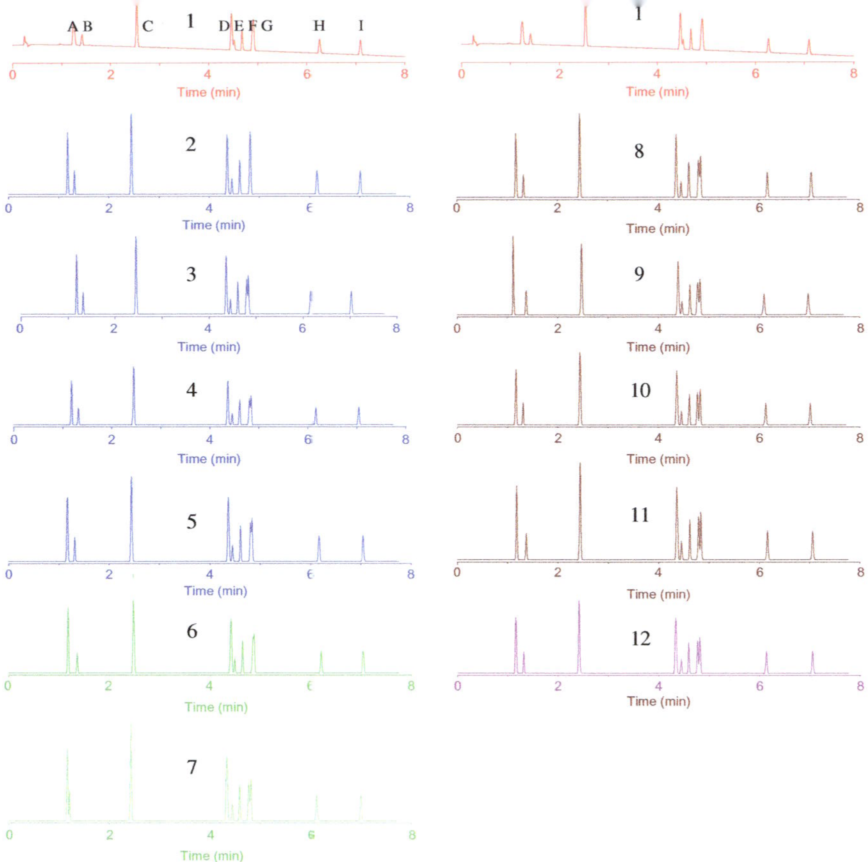

Figure 8.15: Chromatograms in a selected condition (t_G = 9 min, T = 35°C, pH = 4.0). Chromatograms marked with 1 indicate the experimental run (marked with red), 2, 3, 4 and 5 are from studying the limit of gradient time (shown in blue), 6 and 7 are from studying temperature (shown in green), 8, 9, 10 and 11 are from studying pH (shown in brown) and 12 is from studying the combined effect of the three factors (shown in purple). For further details, see Table 8.10. The retention order is trimethoprim (A), acetylsalicylic acid (B), phenacetin (C), nipagin M (D), cetirizine (E), amlodipine (F), rosuvastatin and ketoprofen (G), diclofenac (H) and loratadine (I).

Table 8.10: The levels of selected variables used for chromatogram prediction.

	Design space		
No. of chromatogram	tG(min)	T(°C)	pH
2	5–10	20–50	3.9–4.5–5.1
3	5–15	20–50	3.9–4.5–5.1
4	5–20	20–50	3.9–4.5–5.1
5	5–25	20–50	3.9–4.5–5.1
6	5–15	20–40	3.9–4.5–5.1
7	5–15	20–60	3.9–4.5–5.1
8	5–15	20–50	3.9–4.2–4.5
9	5–15	20–50	2.7–3.6–4.5
10	5–15	20–50	3.9–4.8–5.7
11	5–15	20–50	2.7–3.9–5.1
12	5–25	20–60	2.7–3.9–5.1

p*Ka* in the examined range — there is a small deviation between the other chromatograms, but the retention times are estimated with great certainty even in extended conditions.

8.6 Conclusions

The aim of this case study was to examine the applicable range of method variables. The recommended ranges for modeling retention are: $\Delta tG = 3 \times tG_1$, $\Delta T = 30$°C and ΔpH $= \pm 0.6$ unit.

The effect of the three factors (tG, T and pH) on the retention prediction accuracy was studied individually. Each factor could be extended to map a larger DS. In case of gradient time, a five time extension was proven to be accurate with a minimum of 96.45% accuracy. For the temperature, a 40°C difference could be modeled without any significant loss of accuracy. For the mobile phase pH, establishing the models proved to be harder than for other variables when the studied pH range included the solute p*Ka* values. If the molecules are unknown, then peak tracking is hardly doable. A mass detection can aid to extend the pH range (even to a ± 1.2 unit). The three factors can be extended at the same time as well, and the accuracy of retention time prediction has not decreased significantly.

References

[1] C. Horváth, W. Melander, I. Molnár, Solvophobic interactions in liquid chromatography with nonpolar stationary phases *J. Chromatogr.* 125 (1976) 129–156.
[2] I. Molnár, Computerized design of separation strategies by reversed-phase liquid chromatography: development of DryLab software, *J. Chromatogr. A* 965 (2002) 175–194.
[3] I. Molnár, H.-J. Rieger, R. Kormány, Chromatography modelling in high performance liquid chromatography method development, *Chromatography Today* 6 (2013) 3–8.
[4] M. Pohl, K. Smith, M. Schweitzer, M. Hanna-Brown, J. Larew, G. Hansen, P. Borman, P. Nethercote, Implications and opportunities of applying QbD principles to analytical measurements, *Pharm. Technol. Eur.* 22 (2010) 29–34.
[5] ICH Q8 (R2) — Guidance for Industry, Pharmaceutical Development, 2009.
[6] I. Molnár, H.-J. Rieger, A. Schmidt, J. Fekete, R. Kormány, UHPLC method development and modeling in the framework of Quality by Design (QbD), *The Column* 10/6 (2014) 16–21.
[7] R. Kormány, I. Molnár, J. Fekete, D. Guillarme, S. Fekete, Robust UHPLC Separation Method Development for Multi-API product containing amlodipine and bisoprolol: the impact of column selection, *Chromatographia* 77 (2014) 1119–1127.
[8] R. Kormány, I. Molnár, J. Fekete, Renewal of an old European Pharmacopoeia method for Terazosin using modeling with mass spectrometric peak tracking, *J. Pharm. Biomed. Anal.* 135 (2017) 8–15.
[9] R. Kormány, K. Tamás, D. Guillarme, S. Fekete, A workflow for column interchangeability in liquid chromatography using modeling software and quality-by-design principles, *J. Pharm. Biomed. Anal.* 146 (2017) 220–225.
[10] N. Rácz. R. Kormány, Retention modeling of DryLab software in an extended design space, *Chromatographia* (2018) https://doi.org/10.1007/s10337-017-3466-0.
[11] R. Kormány, J. Fekete, D. Guillarme, S. Fekete, Reliability of simulated robustness testing in fast liquid chromatography, using state-of-the-art column technology, instrumentation and modelling software, *J. Pharm. Biomed. Anal.* 89 (20147) 67–75.
[12] A.H. Schmidt, M. Stanic, I. Molnár, In silico robustness testing of a compendial HPLC purity method by using of a multidimensional design space build by chromatography modeling — case study pramipexole, *J. Pharm. Biomed. Anal.* 91 (2014) 97–107.
[13] L.R. Snyder, J.W. Dolan, D.C. Lommen, DryLab computer simulation for high-performance liquid chromatographic method development, I. isocratic elution, *J. Chromatogr.* 485 (1989) 65–89.
[14] J.W. Dolan, D.C. Lommen, L.R. Snyder, DryLab computer simulation for high-performance liquid chromatographic method development, II. gradient elution, *J. Chromatogr.* 485 (1989) 91–112.
[15] DryLab® 4 User's Manual, 2012.
[16] J.W. Dolan, Temperature selectivity in reversed-phase high performance liquid chromatography, *J. Chromatogr. A* 965 (2002) 195–205.

Chapter 9

Computer-assisted Method Development in Characterization of Therapeutic Proteins by Reversed-phase Chromatography

Szabolcs Fekete

School of Pharmaceutical Sciences,
University of Geneva, University of Lausanne,
CMU — Rue Michel Servet 1, 1211 Geneva 4, Switzerland
szabolcs.fekete@unige.ch

9.1 Introduction

In contrast to small molecules, large molecules such as proteins show different retention mechanisms in several modes of chromatography, such as (1) an on/off mechanism retaining the macromolecules at the column inlet until at some point in the gradient they are desorbed and then move through the column without any further interaction; (2) precipitation–re-dissolution, i.e. separation based on solubility instead of interaction with the stationary phase and (3) multi-point attachment to the surface of the stationary phase [1].

While these mechanisms are fundamentally different from those observed with small molecules, the gradient separation of macromolecules, in most cases, can still be predicted from the linear solvent strength (LSS) theory or from slightly modified models. The reason is that, in most cases, a relatively limited range of the method variables has to be studied because sufficient retention, recovery and peak shape can only be obtained in a limited design space (DS). As an example, for monoclonal antibodies (mAbs) in the reversed-phase (RP) mode, the temperature has to be kept between

70°C and 90°C to obtain acceptable recovery and peak shape. Therefore, performing measurements at a lower temperature makes no sense. It is known that mAbs show deviations from the common linear van't Hoff type behavior (temperature dependency of the retention) in a wide temperature range, due to possible conformational changes. But it was also shown that within a narrow temperature range (e.g. $\Delta T = 20°C$), a linear retention model provides excellent prediction accuracy [2]. The situation is similar to that seen in organic modifier, ion-pairing reagent and pH since only a limited range has to be studied due to the on–off retention mechanism of proteins. In such a narrow range of method variables, simple linear models (or polynomial ones) can be used in most of the cases.

The other advantage with large molecules is that generic conditions can be applied for different protein classes (e.g. cytokines, mAbs, antibody-drug conjugates (ADCs)). Indeed, the structures of the different proteins within a class are very similar: the amino acid sequence is very close and the global conformation is similar. It is also clear that the variants, which have to be separated from the native protein and from each other, possess relatively small differences compared to the native protein (such as the oxidation of some amino acids, deamidation, reduction of a disulfide bonds, etc.). In the whole protein structure, those modifications are minor compared to the native amino acid sequence (e.g. modifications of 2–5 amino acids from the total few hundred or thousand amino acids in the protein backbone) [3].

To conclude on protein HPLC method development, generic conditions can be used for the optimization in most cases and the impact of method variables on the separation has to be studied only in a narrow range.

This chapter presents some specific examples, but the concept can be applied for most of the protein samples.

9.2 Protein Analysis at Different Levels

The comprehensive characterization of protein biopharmaceuticals, e.g. mAbs, is typically performed at different levels, such as the protein, sub-units, peptide, and glycan and amino acid levels [4, 5]. Due to the limited resolving power of different separation modes on large intact proteins, partial enzymatic digestion and/or reduction of disulfide bonds are frequently used to ease the separation of smaller protein fragments. Pepsin, papain, or the immunoglobulin-degrading enzyme of *Streptococcus pyogenes* (IdeS) are commonly employed to obtain relatively large fragments and simplify the investigation of their micro-heterogeneity [6]. Papain is used to generate Fc and Fab fragments of ∼50 kDa each, while pepsin and IdeS generate F(ab′)2 and Fc/2 fragments of ∼100 and ∼25 kDa, respectively. The reduction of disulfide bonds can easily be performed by the addition of strong reductive agents (e.g. dithiotreitol — DTT or tris(2-carboxyethyl)phosphine — TCEP) to produce the light chain (Lc) and heavy chain (Hc) fragments of 25 and 50 kDa, respectively. Following IdeS digestion, further reduction generates three fragments of 25 kDa each, namely, Lc, Fc/2 and Fd. A next level of detail is obtained upon analyzing peptides that can be generated from the protein following their proteolytic digestion using enzymes like trypsin (cleavage next to arginine and lysine), chymotrypsin (preferably cleaves C-terminal of aromatic amino acids) AspN (cleavage of N-terminal of aspartic acid), GluC (cleaves C-terminal of glutamic acid and aspartic acid), LysC (cleavage next to lysine), etc. In case information on S–S bridges is mandatory, digestion can be performed under non-reducing conditions; otherwise, digestion is preceded by a reduction and alkylation step, e.g. using iodoacetamide, to prevent the reformation of S–S bridges. A detailed characterization of glycans requires their removal from the protein backbone. N-glycans can be enzymatically liberated using 'universal' endoglycosidases like PNGase F, PNGase A, Endo S or Endo H. Amino acid compositional analysis requires the quantitative liberation of amino acids typically through acid hydrolysis at 110°C for 24 h using 6M HCl.

9.2.1 *Peptide mapping*

The smaller the protein fragment (e.g. 0.5–2 kDa peptides obtained after tryptic digestion) the more similar the retention behavior to common small molecules. Therefore, similar approaches can be applied as for small pharmaceutical compounds (e.g. impurity profiling). Figure 9.1 shows an example on peptide mapping of a 20 kDa therapeutic protein. A *tG–T*, 2D retention model was built by using a 150 × 4.6 mm column operated at

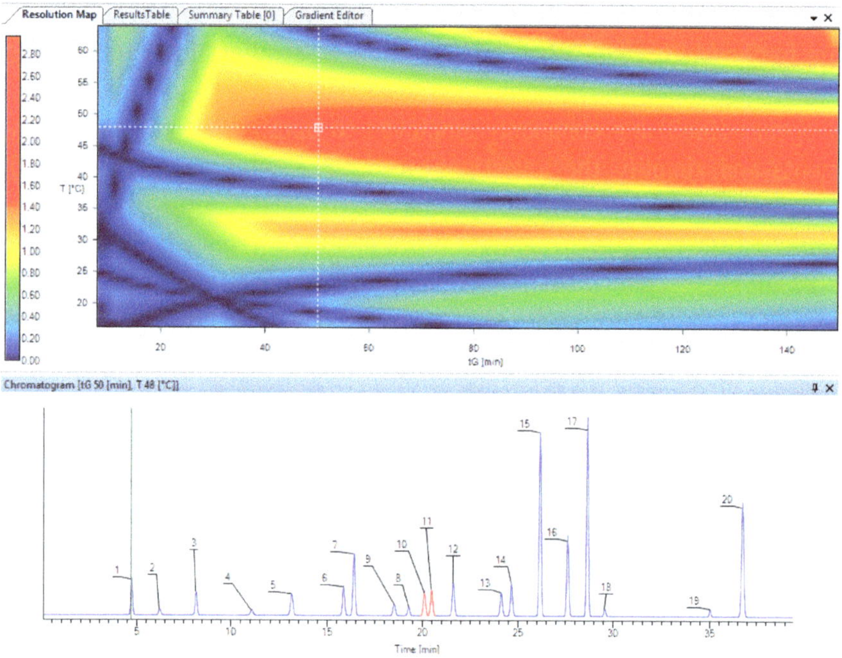

Figure 9.1: Optimization of a peptide mapping of a 20 kDa therapeutic protein followed by tryptic digestion.

1 mL/min flow rate. The studied levels of the factors were as follows: $tG_1 =$ 30 min, $tG_2 =$ 120 min, $T_1 =$ 20°C and $T_2 =$ 60°C. The linear gradient run from 5% to 60% B acetonitrile and the mobile phase contained 0.1% TFA.

Please note that relatively long gradient time has been set for the input run. It is due to the fact that those tryptic samples are often complex, including several closely eluted peaks. The long gradient time allows better separation of closely eluted peaks and thus helps the peak-tracking procedure. The resolution map shows that co-elution may occur by changing the temperature (blue horizontal lines on the resolution map) and the elution order of the peak can be changed. It is often the case for peptide mapping. In most cases, this 2D retention model gives a fast and efficient way for the optimization.

As shown in Fig. 9.1, a $tG =$ 50 min long gradient at $T =$ 48°C provided an appropriate separation. Moreover, the last peak eluted at $tr =$ 36.7 min,

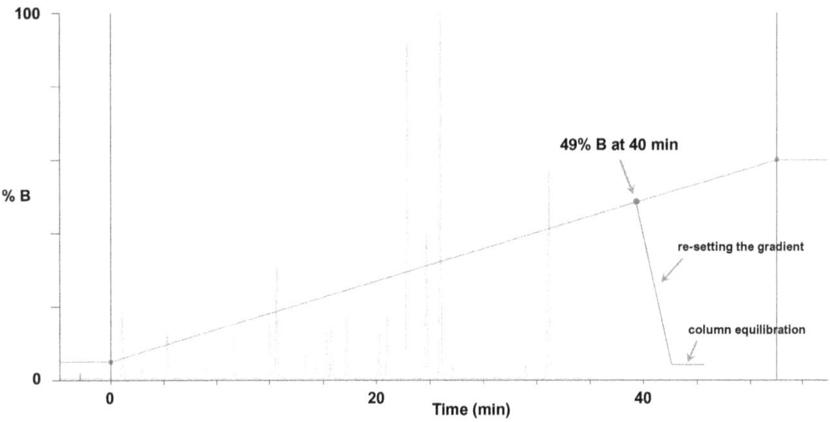

Figure 9.2: Optimization of the gradient program and the final mobile phase composition through the "Gradient Editor".

leading to further decrease in the analysis time. By clicking on the "Gradient Editor", the mobile phase composition can be obtained at any time during the gradient (Fig. 9.2). At 40 min run time, the mobile phase contains 49% B eluent. It suggests that there is no need to go up to 60% B as it was done during the input runs. The gradient can be stopped at 40 min (49% B eluent), then resetting and equilibration steps can be added. In total, there is no need for longer than 43–44 min separation.

9.2.2 *Analysis of mAb sub-units*

The number of approved mAbs has been growing continuously in the pharmaceutical field. Antibodies are large tetrameric glycoproteins of approximately 150 kDa, composed of four polypeptide chains: two identical heavy chains (≈50 kDa) and two identical light chains (≈25 kDa) that are connected through several inter- and intra-chain disulfide bonds at the hinge region. The resulting tetramer has two similar halves that form a Y-like shape [7]. Functionally, mAbs consist of two regions: the crystallizable fraction (Fc) and the antigen-binding fraction (Fab) [8]. Because this structure is made of four polypeptide chains, monoclonal antibodies can display considerable micro-heterogeneity. There are several common modifications that produce charge variants (or isoforms) (e.g.

deamidation, C-terminal lysine truncation, N-terminal pyroglutamation, methionine oxidation, and glycosylation variants) and size variants of the peptide chains (e.g. aggregation or incomplete formation of disulfide bridges). Due to the increasing importance of this class of therapeutic compounds, the development of analytical methods for their detailed characterization is an active area of study. Complete proteolytic digestion of mAbs (peptide mapping) followed by gradient RPLC-MS analysis is the method of choice for the identification and quantification of chemical modifications of mAbs [9, 10]. However, peptide mapping is time-consuming and can induce putative modifications during the lengthy and complex sample preparation [10]. Alternatively, the analysis of large mAb fragments, such as Fab, Fc, F(ab′)2, Hc and Lc, requires very little sample preparation and can provide a high-throughput alternative to peptide mapping (Fig. 9.3). For these reasons and due to advances in RPLC columns and instrumentation, the second approach is currently preferred to the traditional peptide mapping.

Recent studies have showed that mAb fragments (IgG1 and IgG2) generally elute using a 25–40% acetonitrile (containing 0.1% TFA) gradient at elevated temperatures [11, 12]. In ultra high-pressure liquid chromatography (UHPLC), narrow bore columns (2.1 mm ID) are generally used to increase the sensitivity, reduce frictional heating effects and decrease the

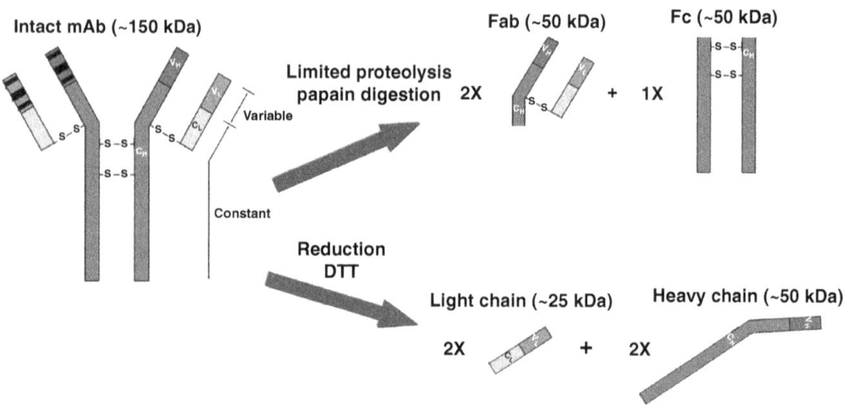

Figure 9.3: Schematic view of the limited proteolytic digestion and reduction of monoclonal antibodies (adapted from Ref. [2] with permission).

solvent and sample consumption. By taking into account the fact that (i) only a 15% change in B produces an adequate gradient for eluting all the different mAb fragment variants and (ii) that 2.1-mm columns are used, then applying the rules of geometrical method transfer, and considering the fact that larger molecules elute in broader peaks, the following conclusions can be drawn. For 150×2.1 mm columns, gradient times in the range of $tG_1 = 4$ min to $tG_2 = 12$ min (at a flow rate of 0.3–0.4 mL/min, starting from 25% to 40% B) should provide appropriate initial data for constructing resolution maps and predicting retention times.

It was recently demonstrated [12] that the use of elevated temperatures (up to 80–90°C) is necessary for the RPLC separation of mAb fragments due to the adsorption phenomena on both silica-based and hybrid stationary phases. At elevated temperatures, thermal degradation is however possible and becomes relevant for gradient times longer than 20 min [12]. A compromise must be found between the residence time and separation temperature. Therefore, the use of $tG_1 = 4$ min and $tG_2 = 12$ min gradients can be employed to avoid issues with stability. Finally, the effect of temperature on selectivity and resolution should be investigated only in a limited temperature range (e.g. $\Delta T = 20$–30°C). The mobile phase temperature should thus be set, e.g. $T_1 = 70$°C and $T_2 = 90$°C (or $T_1 = 60$°C and $T_2 = 90$°C, dependinq on the thermal stability of the stationary phase).

Since linear retention models are not always applicable for large molecules, if the DS is large, quadratic models can be used for method optimization. A 3^2 factorial design can be used in an extended DS, but 2^2 designs work well when working a limited — practically useful — DS. Figure 9.4 demonstrates and suggests the use of 3^2 and 2^2 2D designs ($tG - T$) depending on the set levels of the factors.

The optimization software packages generally employ a linear model for the simultaneous optimization of tG and T. The polynomial relationship of two variables can be written as

$$y = b_0 + b_1x_1 + b_2x_2 \tag{1}$$

where y is the response (retention time or its transformation), x_1 and x_2 are the model variables, e.g. tG and T, whereas b_0, b_1, b_2 are the model

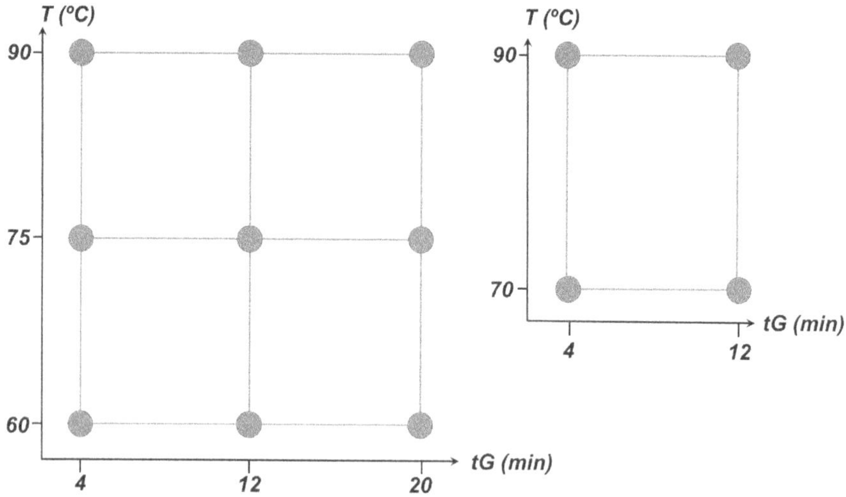

Figure 9.4: Suggested experimental designs for mAb fragment separation in extended (a) and limited (b) DS.

coefficients. As observed with antibody fragments, in a large DS, it is preferred to use a quadratic model to achieve maximum accuracy in the prediction of retention times. A general quadratic model for two variables can be written as

$$y = b_0 + b_1x_1 + b_2x_2 + b_{11}x_1^2 + b_{22}x_2^2 + b_{12}x_1x_2 \quad (2)$$

9.3 Optimization of the Separation of Fab and Fc Fragments

This example describes a fast and efficient method for the determination of variants and degradation products of a recombinant mAb (bevacizumab) from a commercial solution, using the separation power of a new wide-pore core–shell type column (150 mm long). The native mAb was digested with papain, and the aim of the method development was to separate as many variants of the Fab and Fc fragments as possible within the shortest achievable analysis time. Three initial gradients with different slopes were carried out at three column temperatures. Figure 9.5 shows the chromatograms obtained during the nine initial runs. Note that relatively large deviations

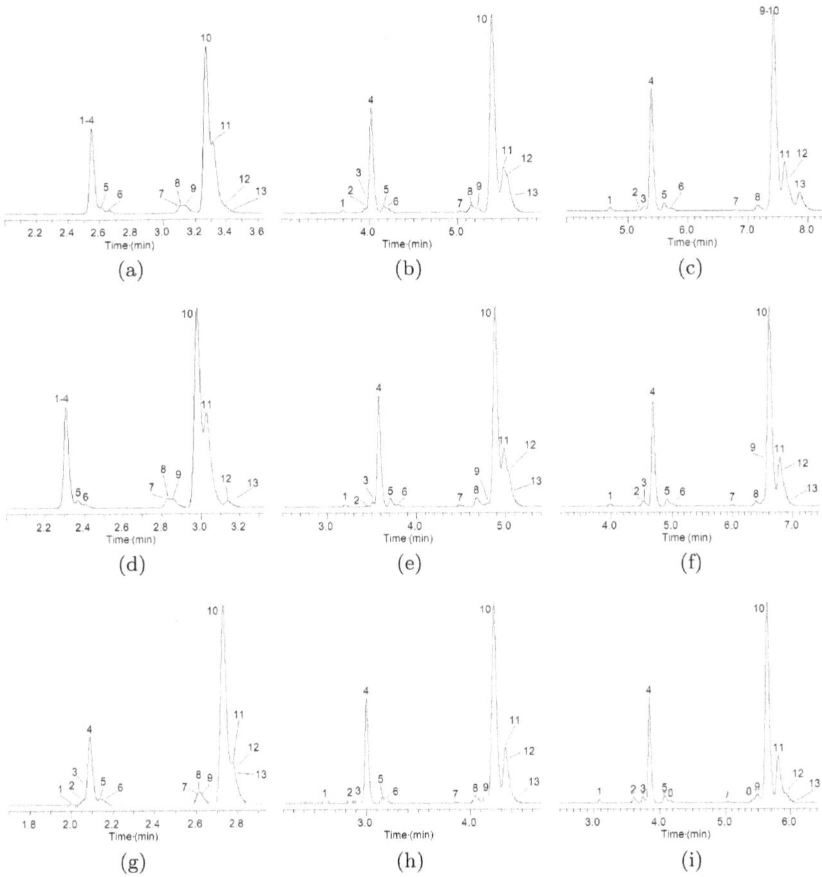

Figure 9.5: Experimental chromatograms of the nine initial runs (Bevacizumab Fc and Fab fragments). Column: Aeris WP C18 (150 mm × 2.1 mm), injected volume: 0.5 μL, detection: fluorescence (excitation at 280 nm, emission at 360 nm). Mobile phase A: 0.1% TFA in water, mobile phase B: 0.1% TFA in acetonitrile. Gradient: from 30% to 40% B, flow rate: 0.35 mL/min. Gradient time and temperature were set as 4 min, 70°C (a), 8 min, 70°C (b), 12 min, 70°C (c), 4 min, 80°C (d), 8 min, 80°C (e), 12 min, 80°C (f), 4 min, 90°C (g), 8 min, 90°C (h) and 12 min, 90°C (i). Peaks: 1–3: pre-Fc peaks, 4: Fc, 5,6: post-Fc peaks, 7–9: pre-Fab peaks, 10: Fab, 11–13: post-Fab peaks (adapted from Ref. [2] with permission).

in the peak areas (and sum of peak areas) are expected when tracking the peaks because of recovery issues with large antibody fragments at low temperatures. Moreover, the recovery of these fragments depends on their molecular weight (size). In contrast, the reproducibility of retention times,

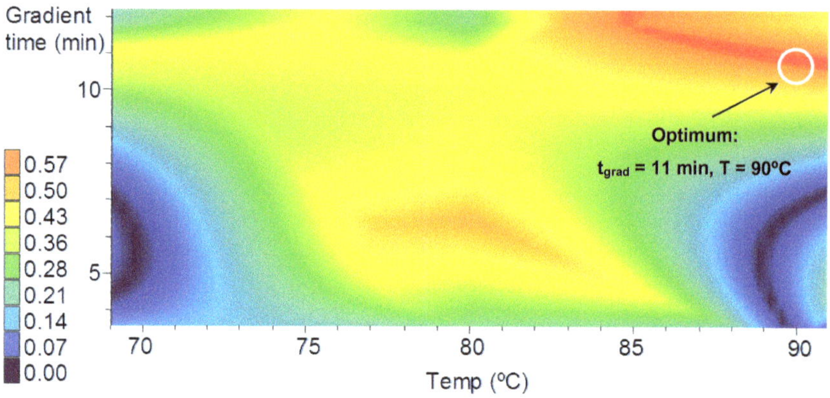

Figure 9.6: Two-dimensional resolution map of the column temperature (°C) against gradient time (t_G, min) for the separation of Bevacizumab Fc and Fab fragments (adapted from Ref. [2] with permission).

derived from consecutive runs at a constant temperature, was excellent. The result is presented in Fig. 9.6 as a resolution map. As shown, the 11-min gradient was found to provide the highest resolution when the column temperature was kept at 90°C. RPLC analysis was then performed using the optimum predicted conditions, and the resulting experimental chromatograms are provided in Fig. 9.7, along with the predicted data.

The accuracy of the quadratic approach — working with a 3^2 design — was evaluated using the 150 × 2.1 mm column. The predicted and experimentally derived chromatograms (retention times and resolution) are compared in Table 9.1, which reveals good agreement between the simulation and experimental results. The average relative error in the retention times was ~1.0%, which is considered an excellent prediction using such rapid gradient profiles. The mean error in the predicted resolution (R_s) was 16.1%. The error in the resolution values contains the retention time error as well as the uncertainty of peak width and peak symmetry prediction. Thus, this prediction is considered reliable and the suggested fast gradient runs can be applied in routine work, resulting in significant time savings. In this case, the time spent for method development was approximately 8 h (3 gradient times × 3 temperatures × 3 samples). The predicted method was then experimentally verified and the final separation required only an 11-min linear gradient, whereas a separation of similar quality using

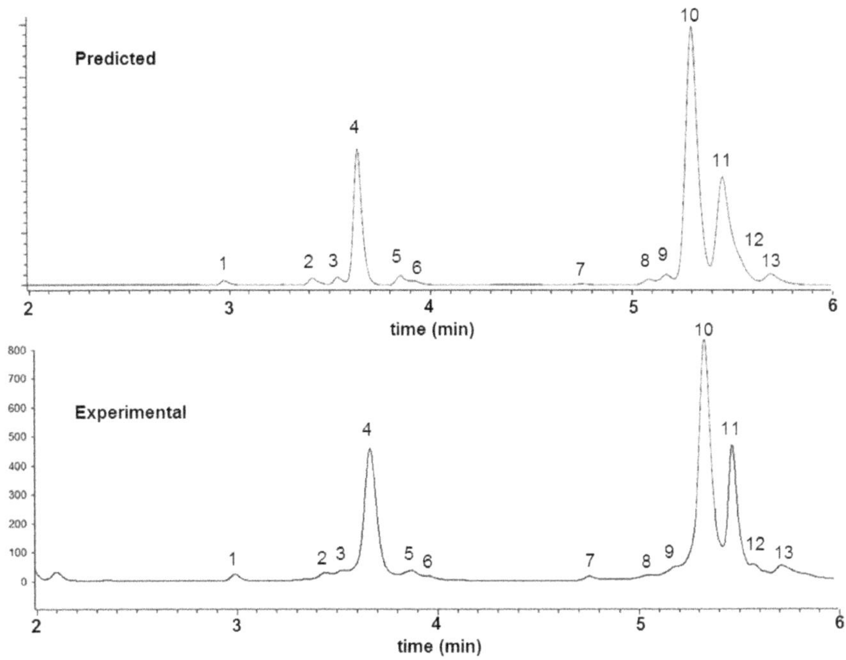

Figure 9.7: Predicted and experimental chromatograms of Bevacizumab Fc and Fab fragments optimized by quadratic model. Column: Aeris WP C18 (150 mm × 2.1 mm), injected volume: 0.5 μL, detection: fluorescence (excitation at 280 nm, emission at 360 nm). Mobile phase A: 0.1% TFA in water, mobile phase B: 0.1% TFA in acetonitrile. Gradient: from 30% to 40% B, flow rate: 0.35 mL/min. Gradient time: 11 min, $T = 90°$C. Peaks: 1–3: pre-Fc peaks, 4: Fc, 5,6: post-Fc peaks, 7–9: pre-Fab peaks, 10: Fab, 11–13: post-Fab peaks (adapted from Ref. [2] with permission).

conventional columns would require at least 60 min. By using the most advanced, highly efficient 150 mm long narrow bore columns, it is possible to well resolve both the Fc and Fab variants.

9.4 Optimization of the Separation of Antibody Drug Conjugate Species by Using 3D Model

This example presents the use of modeling software for the successful method development of an IgG1 cysteine conjugated antibody drug conjugate (ADC) in RPLC. The goal of such a method is to be able to calculate the average drug to antibody ratio (DAR) of an ADC product. A generic method

Table 9.1: Experimental retention times and resolutions vs. those predicted from the 2D gradient time–temperature quadratic model of bevacizumab fragments (Fc and Fab) (adapted from Ref. [2] with permission).

	Retention time				Resolution			
Peaks	Experimental	Predicted	Difference[a]	Abs error %[b]	Experimental	Predicted	Difference[a]	Abs error %[b]
Pre-Fc 1	2.99	2.97	0.02	0.67	5.34	4.93	0.41	7.68
Pre-Fc 2	3.45	3.41	0.04	1.16	1.11	1.27	−0.16	14.41
Pre-Fc 3	3.60	3.54	0.06	1.61	0.84	0.96	−0.12	14.29
Fc	3.68	3.64	0.04	1.11	2.02	2.46	−0.44	21.78
Post-Fc 1	3.89	3.85	0.04	0.95	0.58	0.65	−0.07	12.07
Post-Fc 2	3.96	3.92	0.04	1.09	7.22	6.84	0.38	5.26
Pre-Fab 1	4.78	4.75	0.03	0.54	2.89	2.42	0.47	16.26
Pre-Fab 2	5.15	5.08	0.07	1.30	0.5	0.6	−0.10	20.00
Pre-Fab 3	5.24	5.17	0.07	1.39	0.67	0.82	−0.15	22.39
Fab	5.35	5.3	0.05	0.92	1.36	1.25	0.11	8.09
Post-Fab 1	5.49	5.45	0.04	0.75	1.01	0.57	0.44	43.56
Post-Fab 2	5.60	5.54	0.06	1.00	1.02	0.95	0.07	6.86
Post-Fab 3	5.70	5.69	0.01	0.18				
			Average	0.97			Average	16.05

Notes: [a]Difference = experimental − predicted.
[b]% error = [(experimental − predicted)/predicted] × 100.

development strategy is proposed including the optimization of mobile phase temperature, gradient profile and mobile phase ternary composition (3D model). Based on a limited number of preliminary experiments, a fast and efficient separation of the DAR species is feasible. The prediction offered by the retention model is highly reliable, with an average error of retention time prediction always lower than 0.5% using a 2D or 3D retention models. For routine purpose, four to six initial experiments are required to build the 2D retention models, while 12 experiments are recommended to create the 3D model for a large DS. At the end, RPLC can therefore be considered as a good method for estimating the average DAR of an ADC, based on the observed peak area ratios of RPLC chromatogram of the reduced ADC sample.

Based on the previous works [13], a 3D design ($tG \times T \times tC$ model) is suggested. The levels (and values) of such an experimental design are illustrated in Fig. 9.8 for a 150 × 2.1 mm column operating at 0.3 mL/min. For most solutes (including proteins), a factor of three is used between the two set levels of tG to provide accurate retention modeling (e.g. $tG_1 =$ 6 min and $tG_2 =$ 18 min). However, with the ADC sub-units, combining any gradients shorter than 10 min with a longer one ($tG > 15$ min) resulted

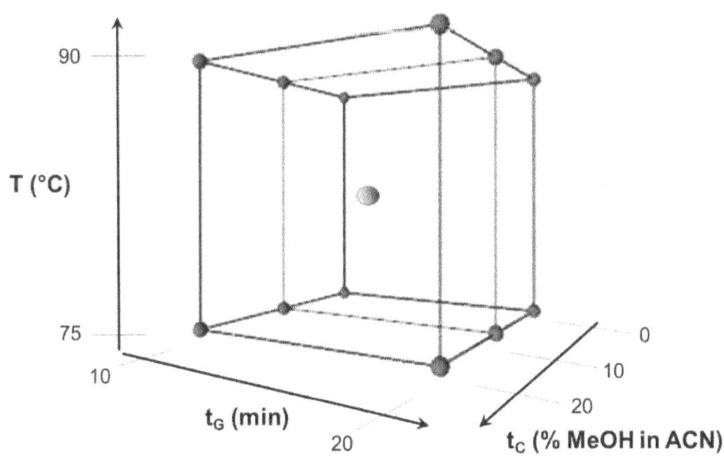

Figure 9.8: Suggested experimental design for 3D retention model (column: 150 × 2.1 mm, gradient: 25–50% B at 0.3 mL/min) to separate cysteine-linked ADC DAR species (adapted from Ref. [13] with permission).

in inaccurate retention model. This is probably due to the very high slope of the LSS model (S) for these large proteins. Finally, it was found that performing $tG_1 = 10$ min and $tG_2 = 20$ min gradients (difference of a factor two) resulted in accurate retention modeling and enables the precise prediction of retention times for any gradient program (linear and multi-linear too, and for extrapolated tG such as $tG < 10$ min).

After processing and checking the data accuracy, the retention times of 14 peaks of reduced ADC were matched in each of the chromatograms by using the PeakMatch module of the DryLab software. The peak tracking process was based on peak areas. All the data were automatically transferred into the modeling software, but small adjustments for the peak widths were required to get realistic peak capacity in the simulated chromatograms. Please note that peak tracking based on peak area was not obvious due to the fact that the sum of the peak areas was expectedly lower at 75°C vs. 90°C because of the significant on-column adsorption at lower temperature. Some ADC sub-units adsorb more intensively onto the stationary phase, while for other peaks (e.g. heavy chain including three drugs or naked light chain), the adsorption was less critical. Therefore, peak movements have to be followed and understood before matching the peak areas. Manual adjustment may have to be performed.

After building up the retention model, its accuracy was experimentally verified. The reduced ADC sample was run in the center point of the experimental design. Retention times and chromatograms were also predicted for this condition. Figure 9.9 shows the predicted and measured chromatograms, and the identification of the 14 peaks included in the model. As shown in Fig. 9.9, the experimentally observed and predicted chromatograms were in very good agreement. Table 9.2 present the difference and % error of measured and calculated retention times for the reduced ADC. There was no more than 0.5% error, and the average error of retention time prediction was below 0.3%.

The verification of the model was assessed by creating resolution map (Fig. 9.10). The color code in these resolution maps represents the value of the critical resolution ($R_{s,\mathrm{crit}}$), with warm "red" colors corresponding to high resolution values ($R_{s,\mathrm{crit}} > 1.0$) and cold "blue" colors corresponding to low resolution values ($R_{s,\mathrm{crit}} < 0.3$). The visual inspection of the cubes

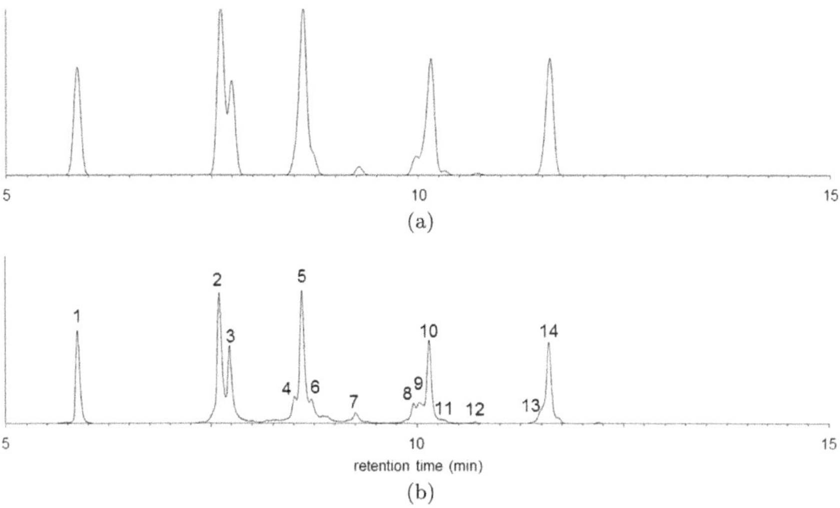

Figure 9.9: Model verification in the center point for reduced ADC. Column: Advance BioMAb RP C4. Mobile phase "A": 0.1% TFA in water, "B": 0.1% TFA in 90% acetonitrile +10% MeOH. Flow rate: 0.3 mL/min, gradient: 25–50% B in 15 min, temperature: 82.5°C, injected volume: 1 μL, detection at 280 nm. (a) corresponds to predicted, while (b) corresponds to experimental chromatograms (adapted from Ref. [13] with permission).

show the largest red region, where the method is probably robust and the resolutions of all peaks in the chromatogram are the best that can be achieved (when using the initial linear gradient). Based on the resolution cubes, the starting point of the optimization can easily be selected. Further optimization can be done by changing the B% of initial and final mobile phase composition. After changing the B%, the effects of temperature and ternary composition are worth re-studying. After further optimization, the optimal conditions were found as a gradient of 31–48% B, $tG = 18$ min, $T = 90$°C and $tC = 20\%$ MeOH (Fig. 9.11). The predicted and experimental chromatograms were in good agreement (lower than 0.5% error).

As illustrated by this example, this generic 3D retention model and optimization for cysteine-linked ADCs seems to be interesting. It can also be useful for laboratories working under regulated conditions, since all the possible combinations of method variables can quickly be checked.

The time required for this 12 runs-based design and its verification is about 7–8 h for one sample, assuming duplicate injections (2 × (6 ×

Table 9.2: Experimental retention times vs. predicted from the 3D gradient time–temperature-ternary composition model of cysteine-linked ADC (adapted from Ref. [13] with permission).

	t_r experimental (min)	t_r predicted (min)	Difference[a]	% error[b]
Peak 1	5.868	5.850	0.02	0.31
Peak 2	7.599	7.610	−0.01	−0.14
Peak 3	7.729	7.740	−0.01	−0.14
Peak 4	8.514	8.520	−0.01	−0.07
Peak 5	8.601	8.600	0.00	0.01
Peak 6	8.718	8.720	0.00	−0.02
Peak 7	9.258	9.280	−0.02	−0.24
Peak 8	9.961	9.960	0.00	0.01
Peak 9	10.031	10.070	−0.04	−0.39
Peak 10	10.148	10.150	0.00	−0.02
Peak 11	10.297	10.320	−0.02	−0.22
Peak 12	10.711	10.710	0.00	0.01
Peak 13	11.543	11.530	0.01	0.11
Peak 14	11.592	11.590	0.00	0.02
		Average	−0.01	−0.06

Notes: [a] Difference = experimental − predicted.
[b] % error = [(experimental − predicted)/predicted] × 100.

10 min + 6×20 min + 1×15 min) + system equilibration). Then the understanding of peak movements, peak-tracking, importing chromatograms and creating the model takes around 5–6 h. Finally, the optimization and then the experimental verification of the selected working point take an additional 2–3 h of work. In total, this optimization approach of ADC species separations in RPLC mode requires 2–3 working days.

9.5 Optimization of the Separation of ADC Species by Using 2D Model

Obviously, the 3D retention model can be simplified to 2D models if required (e.g. to gain in time or if 3D retention modeling software is not available). One possibility is to select a $tG \times T$ model which requires four initial runs, while the other choice is to perform a $tG \times tC$ model which necessitates six experiments (Fig. 9.12).

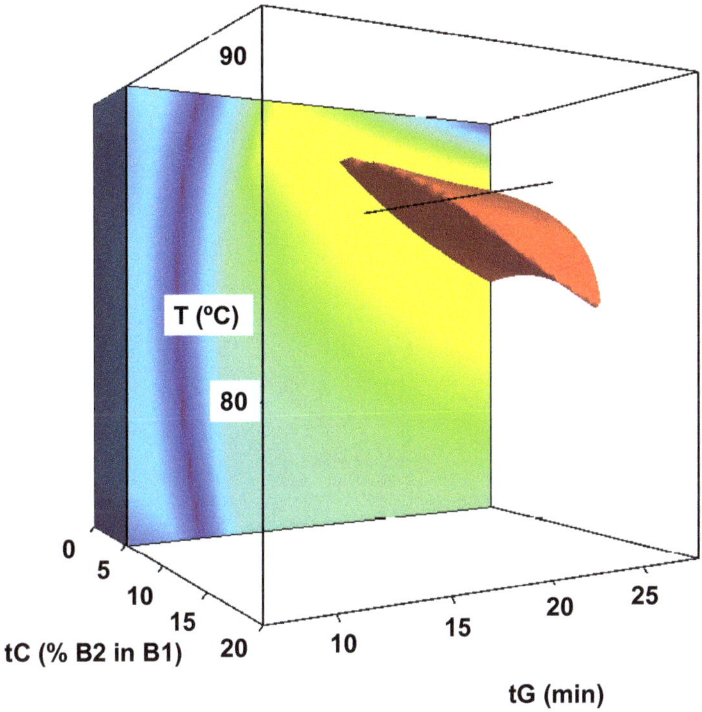

Figure 9.10: 3D resolution maps for reduced ADC, based on the initial experiments ($R_{s,\text{crit}} = 1.0$). Set conditions: $tG = 27$ min, $T = 87$°C and $tC = 5$% MeOH (adapted from Ref. [13] with permission).

To perform a $tG \times T$ model, ternary mobile phase composition is not suggested. Since the best recovery is mostly obtained with aprotic solvent (acetonitrile), the mobile phase B should preferably be 0.1% TFA in acetonitrile as a first choice. The time required for the experiments is only around 2–3 h (2 × (2 × 10 min + 2 × 20 min) + system equilibration) for one sample (with duplicate injections). Figure 9.13(a) shows the 2D resolution maps for the ADC sample. The blue lines indicate co-elutions (and therefore elution order changes). Since the blue lines oriented in both vertical and horizontal directions on the map, the DS indeed seems to be well selected since both method variables play an important role in the overall quality of the separation.

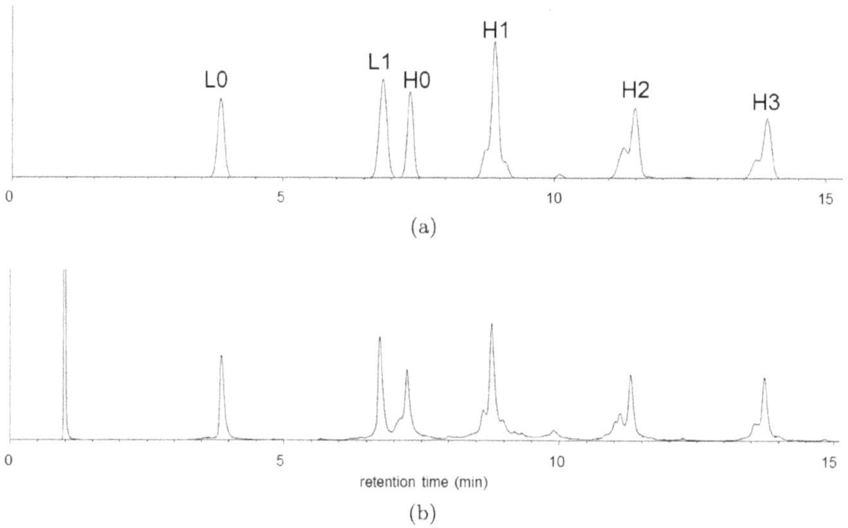

Figure 9.11: Predicted (a) and experimentally verified (b) chromatograms of reduced cysteine-linked IgG1 ADC under optimal conditions to optimize resolution between L1 and H0 species. Gradient: 31–48% B, $tG = 18$ min, $T = 90°$C and $tC = 20\%$ methanol (80% acetonitrile) (Column: Agilent Advance BioMAb RP C4, flow rate: 0.3 mL/min). L0, L1 correspond to light chain including 0 and 1 drug while H0, H1, H2 and H3 correspond to heavy chain species with 0, 1, 2 and 3 drugs (adapted from Ref. [13] with permission).

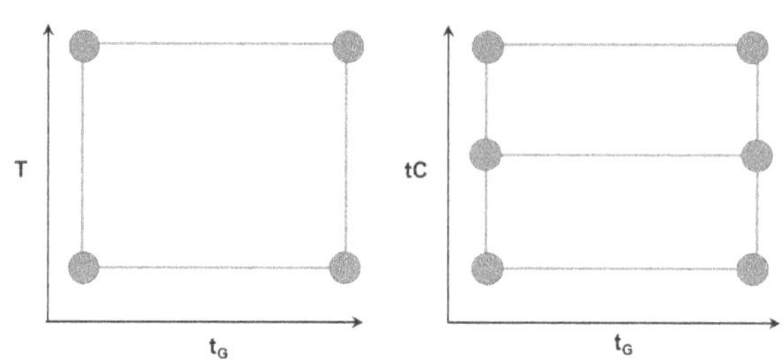

Figure 9.12: Simplified 2D experimental designs as $tG \times T$ model (4 runs) and $tG \times tC$ (6 runs) model for the optimization of ADC species separation (adapted from Ref. [13] with permission).

Figure 9.13(b) shows the obtained 2D resolution maps for a $tG \times tC$ model. In this case, the temperature should be set as high as possible to avoid recovery issues (e.g. $T = 90°$C). This experimental design takes around 3–4 h of work ($2 \times (3 \times 10$ min$+ 3 \times 20$ min$)+$system equilibration).

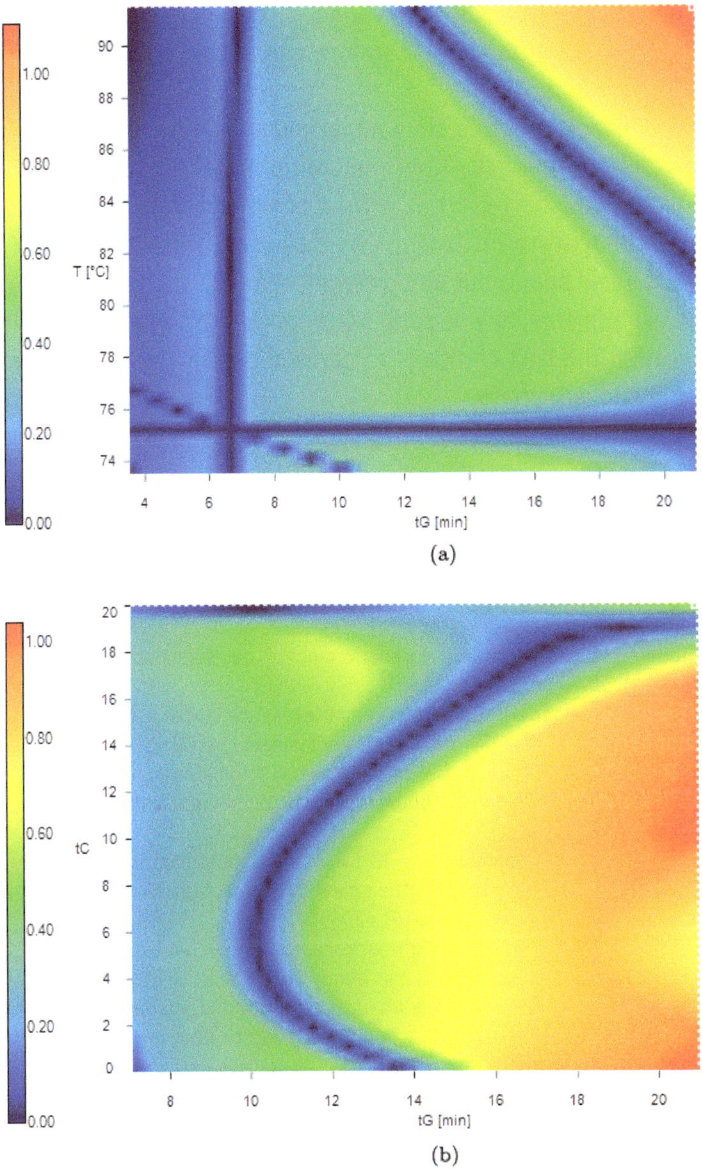

Figure 9.13: Simplified 2D resolution maps based on (a) four initial experiments ($tG \times T$ model). Gradient: 25–50% B, $tG_1 = 10$ min, $tG_2 = 20$ min, $T_1 = 75°$C, $T_2 = 90°$C and $tC = 0\%$ methanol (100% acetonitrile) and (b) six initial experiments ($tG \times tC$ model). Gradient: 25–50% B, $tG_1 = 10$ min, $tG_2 = 20$ min, $tC_1 = 0\%$ methanol, $tC_2 = 10\%$ methanol and $tC_3 = 20\%$ methanol, $T = 90°$C (adapted from Ref. [13] with permission).

The maps again suggest that both variables (tG and tC) have a huge impact on the critical resolution and therefore makes this model interesting for routine applications.

Both 2D models provided similar maximum resolution and analysis time as for an optimal method. If further optimization is required, then the $tG \times T$ model can be repeated with a ternary mobile phase (e.g. 20% methanol +80% acetonitrile as organic solvent) while the $tG \times tC$ model can be performed again, but at a different temperature (e.g. at 80°C). This repeated experiments may perform better quality of separation. If it is not the case, then the best choice is to perform one of these 2D models on a different stationary phase.

References

[1] E. Tyteca, J.L. Veuthey, G. Desmet, D, Guillarme, S. Fekete, Computer assisted liquid chromatographic method development for the separation of therapeutic proteins, *Analyst* 141 (2016) 5488–5501.

[2] S. Fekete, S. Rudaz, J. Fekete, D. Guillarme, Analysis of recombinant monoclonal antibodies by RPLC: towards a generic method development approach, *J. Pharm. Biomed. Anal.* 70 (2012) 158–168.

[3] S. Fekete, R. Kormány, D. Guillarme, Computer assisted method development for small and large molecules, *LC-GC, HPLC 2017* supplement, 30 (2017) 14–21.

[4] K. Sandra, I. Vandenheede, P. Sandra, Modern chromatographic and mass spectrometric techniques for protein biopharmaceutical characterization, *J. Chromatogr. A* 1335 (2014) 81–103.

[5] S. Fekete, D. Guillarme, P. Sandra, K. Sandra, Chromatographic, electrophoretic and mass spectrometric methods for the analytical characterization of protein biopharmaceuticals, *Anal. Chem.* 88 (2016) 480–507.

[6] S. Fekete, D. Guillarme, Ultra-high-performance liquid chromatography for the characterization of therapeutic proteins, *Trends Anal. Chem.* 63 (2014) 76–84.

[7] D.R. Mould, K.R.D. Sweeney, The pharmacokinetics and pharmacodynamics of monoclonal antibodies–mechanistic modeling applied to drug development, *Curr. Opin. Drug. Discov. Devel.* 10 (2007) 84–96.

[8] G.M. Edelman, B.A. Cunningham, W.E. Gall, P.D. Gottlieb, U. Rutishauser, M.J. Waxdal, The covalent structure of an entire gamma G immunoglobulin molecule, *J. Immunol.* 173 (2004) 5335–5342.

[9] N. Lundell, T. Schreitmuller, Sample preparation for peptide mapping — A pharmaceutical quality-control perspective, *Anal. Biochem.* 266 (1999) 31–47.

[10] K.R. Williams, K.L. Stone, Identifying sites of posttranslational modifications in proteins via HPLC peptide mapping, *Methods Mol. Biol.* 40 (1995) 157–175.

[11] S. Fekete, R. Berky, J. Fekete, J.L. Veuthey, D. Guillarme, Evaluation of a new wide pore core-shell material (Aeris™WIDEPORE) and comparison with other existing

stationary phases for the analysis of intact proteins, *J. Chromatogr. A* 1236 (2012) 177–188.
[12] S. Fekete, S. Rudaz, J.L. Veuthey, D. Guillarme, Impact of mobile phase temperature on recovery and stability of monoclonal antibodies using recent reversed-phase stationary phases, *J. Sep. Sci*, (2012) accepted.
[13] S. Fekete, I. Molnar, D. Guillarme, Separation of antibody drug conjugate species by RPLC: a generic method development approach, *J. Pharm. Biomed. Anal.* 137 (2017) 60–69.

Chapter 10

Computer-assisted Method Development in Characterization of Therapeutic Proteins by Ion-Exchange Chromatography

Szabolcs Fekete

School of Pharmaceutical Sciences,
University of Geneva, University of Lausanne,
CMU — Rue Michel Servet 1, 1211 Geneva 4, Switzerland
szabolcs.fekete@unige.ch

10.1 Introduction

Ion-exchange (IEX) chromatography is a historical and non-denaturing technique widely used for the characterization of charge variants of therapeutic proteins and is considered as a reference technique for the qualitative and quantitative evaluation of charge heterogeneity of therapeutic proteins [1, 2]. Among the different IEX modes, cation-exchange (CEX) chromatography is the most widely used for protein purification and characterization [3]. CEX is considered as the gold standard for charge sensitive analysis, but method parameters, such as column type, mobile phase pH, and salt concentration gradient, often need to be optimized for each individual protein [4]. IEX separates charge variants by differential interactions on a charged support. The number of possible charge variants increases with the molecular weight of the analyzed sample. In addition, changes in charge may be additive or subtractive, depending on any modifications. Thus, IEX profiles become more complex, and the overall resolution of individual variants may be lost [1]. This property is particularly apparent for large biomolecules. Therefore, not only the intact but also the reduced or digested forms (limited proteolysis or peptide mapping) of therapeutic proteins are commonly characterized by IEX.

In the late 1970s, chromatofocusing (with internal pH gradient) was recognized as the chromatographic analogy to isoelectric focusing (IEF) [5–7]. Chromatofocusing has been demonstrated to be useful for separating protein isoforms due to its high resolving power and ability to retain the protein's native state [8, 9]. Alternatively, pH gradient can be conducted externally by pre-column mixing of two eluting buffers at different pH values consisting of common buffer species [10]. The externally induced pH gradient has recently been applied for separation of deamidated variants of a mAb, resolving C-terminal lysine isoforms of a mAb after treating with carboxypeptidase B and also for the analysis of charge variants of intact mAbs [8, 11].

According to the literature, ionic strength-based IEX separations (classical salt gradient mode) have excellent resolving power and robustness, and still are the most frequently applied mode of IEX separations. Typically, sodium chloride or potassium chloride concentration is increased at a given pH during the gradient.

10.2 Salt Gradient-based Separations

IEX separates proteins based on differences in the surface charge of the molecules, with separation being dictated by the protein interactions with the stationary phase [12]. As a classical mode of IEX, a linear salt gradient is regularly applied for the elution. Several models for chromatographic retention of ion-exchange adsorbents have been proposed in the past years [13]. The retention models can be divided into stoichiometric and non-stoichiometric models. Stoichiometric models describe the multi-faceted binding of the protein molecules to the stationary phase as a stoichiometric exchange of mobile phase protein and bound counter-ions [14]. This stoichiometric displacement model (SDM) predicts that the retention of a protein under isocratic, linear conditions is related to counter-ion concentration. This model was extended to describe protein retention under linear gradient elution (LGE) conditions [15], as well as under non-linear protein adsorption conditions (Steric Mass Action, SMA, model) [16, 17] for isocratic and gradient elution mode. Another extension of the stoichiometric model for the ion-exchange adsorption which accounts for charge regulation was developed recently [18, 19].

Even if stoichiometric models are capable of describing the behavior of ion-exchange chromatographic systems, they assume that the individual charges on the protein molecules interact with discrete charges on the ion exchange surface. In reality, retention through ion-exchange is more complex, and this is primarily due to the interaction of the electrical fields of the protein molecules and the chromatographic surface [14]. Therefore, several non-stoichiometric models for describing protein retention as a function of the salt concentration in the mobile phase have also been proposed [20–23]. Quantitative structure–property relationship (QSPR) models have been derived for protein retention modeling in IEX by means of different numerical approaches that attempt to correlate retention to functions of descriptors derived from the 3D structure of the proteins [24–26].

The work of Snyder and co-workers showed that IEX systems follow non-linear solvent strength (nLSS) type retention mechanism [27, 28]. Consequently, solute-specific correction factors are required to use LSS model for retention predictions, thereby limiting the applicability of the LSS model. The retention factor (k) can be written in the following way according to the SDM model:

$$\log k = \log K - z \log C \tag{1}$$

where K is the distribution constant, z is associated with the protein net charge or number of binding sites (effective charge) and C is the salt concentration (that determines the ionic strength). This model is probably the most accepted one and is useful from a practical point of view. The nonlinearity of Eq. (1) is most pronounced for small values of z [28]. If $z > 6$ (which is very often the case of therapeutic proteins), an LSS type model may provide reliable data for retention factor (retention time) [29].

Proteins are eluted in order of increasing binding charge (correlates more or less with the isoelectric point (pI)) and equilibrium constant. The retention of large proteins in salt-gradient mode is strongly dependent on the salt concentration (gradient steepness or gradient time) — due to the relatively high z value — and a small change could lead to significant shift in retention. Therefore, isocratic conditions are impractical, and gradient elution is preferred in real-life proteins separations.

It was currently shown that LSS approach can be applied for large proteins (mAbs) possessing an important number of charges in the practically useful and interesting design space (DS) [30].

10.3 pH Gradient-based Separations

Ion-exchange chromatofocusing represents a useful alternative to linear salt-gradient elution IEX, in particular for separating protein isoforms with minor differences in the isoelectric point (pI). Chromatofocusing is performed on an ion-exchange column employing a pH gradient that can be generated internally within the column or by external mixing of a high-pH and a low-pH buffer using a gradient pump system. Highly linear, controllable, and wide-range pH gradients can be generated [9, 10, 31, 32].

The number of applications reported at the analytical scale is large, but the number of publications dealing with the mathematical modeling of linear pH gradient elution in IEX is rather limited [12]. To describe the elution behavior of proteins in linear pH gradient IEX, a pH-dependence parameter has to be incorporated into the ion-exchange model.

In pH-gradient mode, the proteins net charge is modified during the pH gradient, due to protonation–deprotonation of the functional groups. In CEX, the protein is expected to elute at, or close to its pI.

The applied pH range clearly determines the proteins that can possibly be eluted. The effect of gradient steepness (gradient time) on the retention of large proteins (intact mAbs and their variants) was recently studied and showed an LSS-like linear behavior [33].

In pH-gradient IEX mode, the use of a mixture of amine buffering species in the high-pH range and a mixture of weak acids in the low-pH range is quite common [31,32,34]. In such a system, maintaining linearity of the pH gradient slope may be somewhat difficult. It was shown that an appropriate mixture of Tris base, piperazine and imidazole provides a linear pH gradient from pH 6 to 9.5 [4]. Triethylamine- and diethylamine-based buffer systems also offered linear pH gradient in the pH range of 7.5–10.0 [10]. For mass spectrometric (MS) detection, 5 mM ammonium hydroxide in 20% methanol yielded a reasonable pH gradient in a limited pH range (between 9.5 and 10.5) [10]. Zhang *et al.* applied a salt-mediated improved pH gradient that was used in a wide pH range (between 5 and 10.5) [35]. In their study, a 0.25 mM/min sodium-chloride gradient was performed together with the pH gradient. One of the benefits of pH-gradient-based IEX is that the salt concentration can be kept low, yielding less buffer interferences (e.g. online or offline 2D LC).

10.4 Method Optimization in IEX

Method development in IEX was mostly based on trial-and-error or one-factor-at-a-time (OFAT) approaches. However, there are some guidelines available from column providers, which explain the basic rules for method screening (e.g. column selection, buffer selection, etc.).

Bai *et al.* showed the dependence of retention and selectivity of IgG antibodies on mobile phase pH, stationary phase type and salt-gradient steepness in CEX mode [36]. They studied the effect of the three variables independently and found that mobile phase pH was the most important parameter in CEX separations of proteins. It had the biggest impact on the separation and therefore should be determined first. It was also found that (i) peak width of IgG-s mostly depends on the type of the stationary phase and (ii) resolution can be tuned by changing the gradient steepness.

The mobile phase linear velocity also has a strong influence on the separation quality of large proteins [37, 38]. Indeed, the longitudinal diffusion is negligible with large molecules, while band broadening is mostly determined by the mass transfer resistance. Therefore, low flow rate is always preferred for high resolution separations, but a compromise has to be found between resolution and analysis time.

The influence of salt type can also be important. Its effect on the retention of bovine serum albumin was reported by Al-Jibbouri [39].

Computer-assisted method development and optimization in RPLC protein separations is now quite common and was also recently applied in ion-exchange mode. Because of the system nonlinearity, finding the optimum for process optimization is challenging [40]. Thiemo *et al.* developed a software called ChromX for the estimation of parameters, chromatogram simulation and process optimization [40]. ChromX provides numerical tools for solving various types of chromatography models, including the model combination of transport dispersive model (TDM) and SMA. Similar to RPLC method development, a non-LSS and LSS type computer-assisted method development procedure was recently reported for both salt- and pH-gradient modes in agreement with quality by design (QbD) concept [30, 41].

10.4.1 *Optimization of IEX separations in salt gradient mode*

For the salt gradient-based protein separation, it was found that temperature was not a relevant parameter for tuning selectivity and should be kept at low value (e.g. at 30°C) to achieve high resolving power (elevated peak capacity) [30]. Because the relationship between apparent retention factors and gradient time (slope) can be described with a linear function — in the practically useful limited range — only two initial gradient runs of different slopes are required for optimizing the salt gradient program. When combining the experiments in a design of experiments (DoE), it appeared that method optimization can be performed rapidly, in an automated way thanks to a HPLC modeling software, using two gradient times and three mobile phase pH (e.g. 10 and 30 min gradient on a 100 mm long standard bore column at pH = 5.6, 6.0 and 6.4) in a $tG - \mathrm{pH}$ model requiring six initial experiments. Such a procedure can be applied routinely and the time spent for method development would be only around 9 h. The relative error in retention time prediction was lower than 1%, making this approach highly accurate [30]. Figure 10.1 shows a generic DoE for the method development of saltgradient-based CEX separation of mAbs (possessing a wide range of pI between 6.7 and 9.1) applied for conventional (4.6 mm) columns.

Separation of the Fc and Fab domains of an IgG has facilitated investigation of the micro-heterogeneity of human mAbs (confirmation of chemical and post-translational modifications such as N-terminal cyclization,

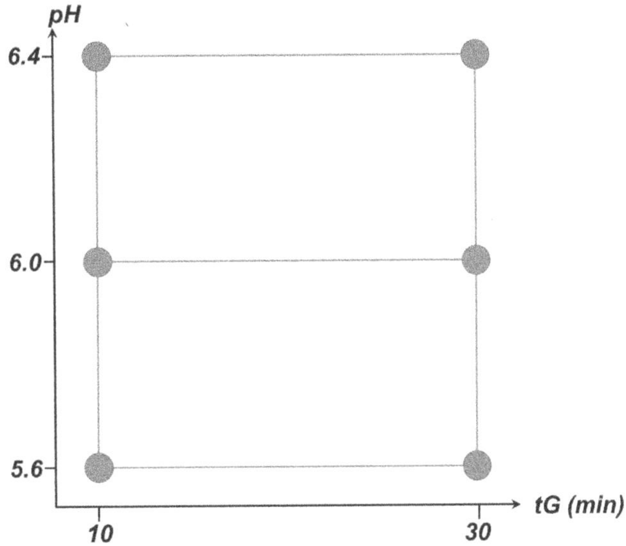

Figure 10.1: Suggested experimental designs for mAb fragment separation in salt gradient-based IEX, using 100 × 4.6 mm column dimension. The gradient time can be scaled in agreement with the column volume.

oxidation and deamidation, and C-terminal processed lysine residues). The present example describes a fast and efficient method development applied for the determination of charge variants of a recombinant mAb (cetuximab), using salt gradient approach in CEX mode. The native mAb was digested with papain, and the aim of the method development was to separate as many variants of the Fab and Fc fragments as possible, within the shortest achievable analysis time. The two initial gradients with different slopes were carried out at three pH values. Figure 10.2 shows the chromatograms of the six initial runs.

The corresponding resolution map is shown in Fig. 10.3. As shown, a 17 min gradient was found to provide the highest resolution when the mobile phase pH is ~5.6. The predicted optimum condition was set and experimental chromatograms recorded. Figure 10.4 shows the predicted and experimental chromatograms.

To evaluate the accuracy of this approach (with 10 and 30 min initial gradient runs) applied for 100 × 4.6 mm column, the predicted and

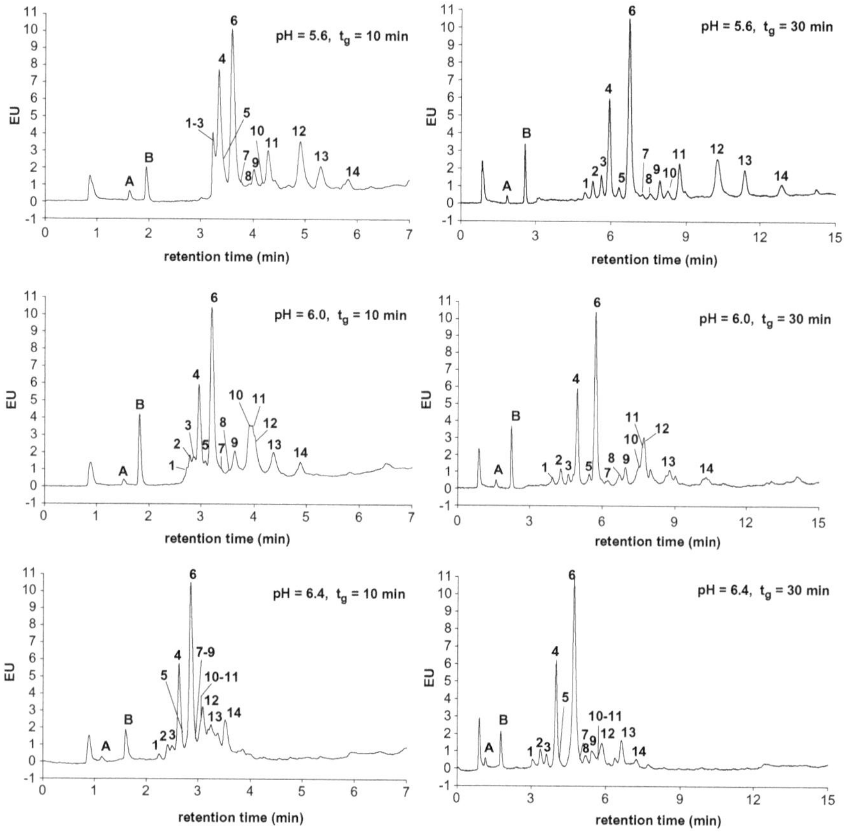

Figure 10.2: Cetuximab papain-digested sample (tG – pH model). Column: BioPro SP-F (100×4.6 mm). Mobile phase "A" 10 mM MES, "B" 10 mM MES +1 M NaCl. Flow rate: 0.6 mL/min, gradient: 0–20% B, temperature: 30°C, detection: FL (280–360 nm), injected volume: 2 μL. Gradient times: $tG_1 = 10$ min, $tG_2 = 30$ min, $pH_1 = 5.6$, $pH_2 = 6.0$, $pH_3 = 6.4$ (adapted from Ref. [30] with permission).

experimental chromatograms (retention times) were compared. The predicted retention times were in good agreement with the experimental ones; the average retention time relative errors was ~1.0%, which can be considered as excellent.

Please note that for more complex samples, the optimum conditions for high-resolution separations can be shifted to the lower pH and longer gradient time ranges. Therefore, for high resolution separations an extended model might be useful.

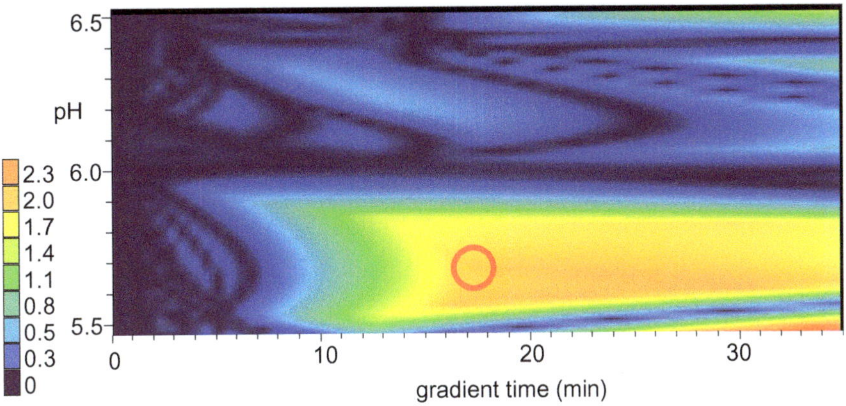

Figure 10.3: Resolution map of cetuximab papain-digested sample (*tG* – pH model). Conditions as defined in the caption of Fig. 10.2 (adapted from Ref. [30] with permission).

10.4.2 *Optimization of IEX separations in pH gradient mode*

An important thing in pH gradient mode is that a linear gradient of A to B buffers should provide a linear pH response, otherwise retention modeling becomes challenging. It was shown that an appropriate mixture of Tris base, piperazine and imidazole provides a linear pH gradient from pH 6 to 9.5 [4]. Triethylamine and diethylamine-based buffer systems also offered linear pH gradient in the pH range of 7.5–10.0 [10]. 5 mM ammonium hydroxide in 20% methanol yielded a reasonable pH gradient in a limited pH range [10]. Commercially available buffers, such as CX-1 pH gradient buffer A (pH = 5.6) and CX-1 pH gradient buffer B (pH = 10.2) from Thermo Fisher Scientific can also be used for routine work.

In the pH gradient mode, the two most important method variables were found as *tG* and *T* since they both have impact on selectivity and resolution [41]. As observed with mAbs, the dependence of retention time (or its transformation) on pH, gradient steepness and mobile phase temperature can be described by linear models. This observation suggests that method optimization with gradient steepness and mobile phase temperature as model variables requires the measurement of variable effects at two levels only.

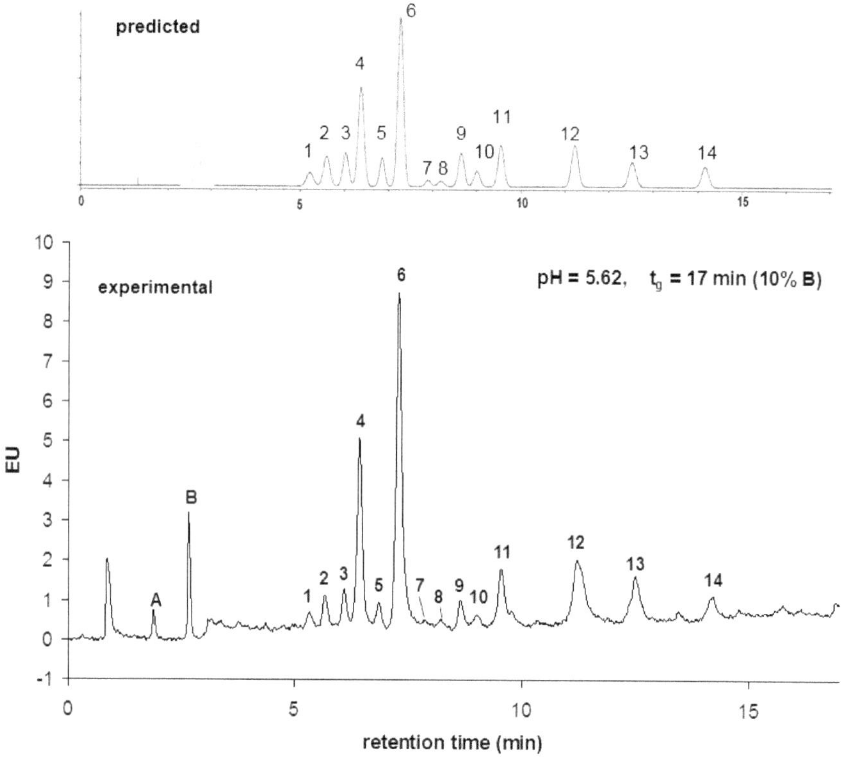

Figure 10.4: Comparison of predicted and experimental chromatograms. Column: BioPro SP-F (100 × 4.6 mm). Mobile phase "A" 10 mM MES, "B" 10 mM MES +1 M NaCl. Flow rate: 0.6 mL/min, gradient: 0–10% B, temperature: 30°C, detection: FL (280–360 nm), injected volume: 2 μL. Gradient times: $tG = 17$ min, pH $= 5.62$ (adapted from Ref. [30] with permission).

Gradient runs with two gradient times (again as $tG_1 = 10$ min and $tG_2 = 30$ min) at two temperatures ($T_1 = 25°C$ and $T_2 = 55°C$) on a 100 × 4.6 mm column can be performed to build up the model. The modeling software implements an interpretive approach, where the retention behavior is modeled on the basis of experimental runs, and then the retention times, peak widths, selectivity and resolution at other conditions are predicted in a selected experimental domain. This allows calculating the critical resolution and, accordingly, the optimal separation can be found.

For this purpose, retention times were transformed into retention factors, and linear models were chosen for both gradient time (steepness) and temperature. This modeling can be performed on a rectangular region in the $tG - T$ plane, determined by two gradient times (steepness) and two temperatures. Hence, this approach requires four initial experimental runs for creating the model. Following the execution of the input experimental runs, data (retention times, peak widths and peak tailing values) can be imported into DryLab and peak tracking can be done. Then, the optimization is carried out on the basis of the created resolution map, in which the smallest value of resolution ($R_{s,\text{crit}}$) of any two critical peaks in the chromatogram is plotted as a function of gradient time and mobile phase temperature.

An advantage of pH gradient-based separations using a CEX column is described as a multi-product charge sensitive separation method for various mAbs. For this illustration, we used our approach and developed a pH gradient for ten different mAbs (possessing *pI* between 6.7 and 9.1), by using commercially available buffers (pH 5.6–10.2).

The pH gradient steepness and mobile phase temperature were varied to find appropriate conditions for these ten mAbs and their variants. Figure 10.5 shows the obtained chromatograms of ten intact mAbs, and suggests that pH gradient CEX separation is indeed adequate for multi-product mAb separations. The optimal conditions on a strong cation exchanger resin were found to be 20 min long gradient (0–100% B) at 30°C. mAbs do not elute exactly in the order of their *pI*. The distribution of charges on the surface of proteins is generally considered as the reason for the minor deviations between the elution pH and *pI*. In our example, natalizumab clearly elutes earlier, while denosumab elutes at higher pH than expected. One possible explanation may be the differences in glycosylation profiles of these mAbs. Moreover, some supplementary interactions with the stationary phase can also occur that superposes to the charge-interaction-based elution mechanism. Based on these observations and the fact that retention times and *pI* are not perfectly correlated, care should be taken when evaluating the protein's *pI*, based on a pH-gradient CEX experiment.

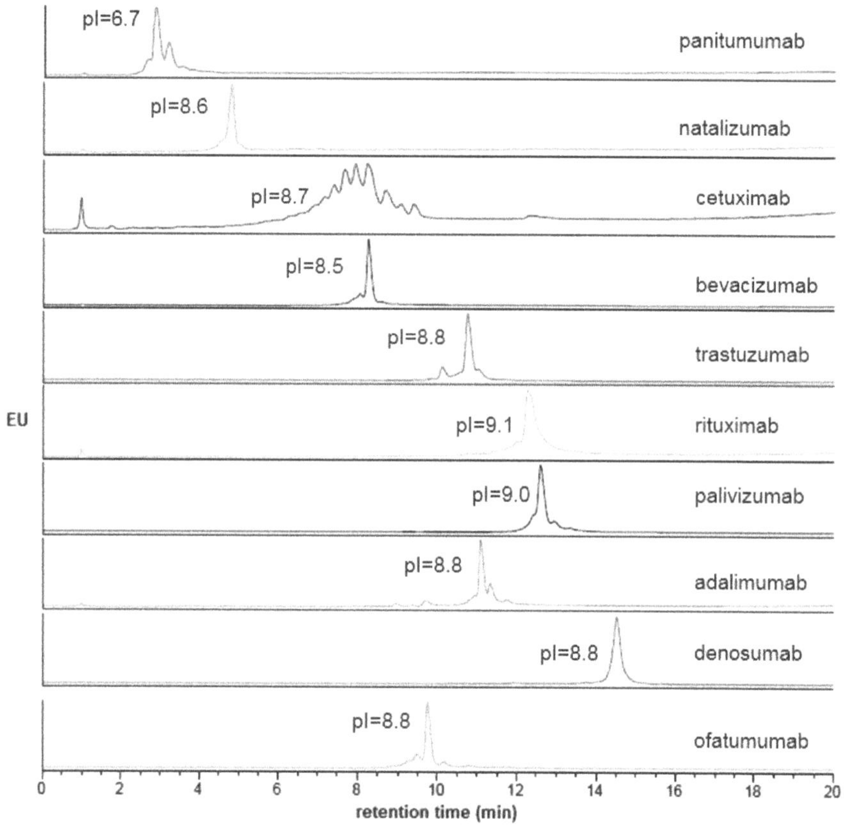

Figure 10.5: pH gradient for multi-mAb analysis. Column: BioPro SP-F (100 × 4.6 mm). Mobile phase "A" CX-1 Buffer A pH = 5.6, "B" CX-1 Buffer B pH = 10.2. Flow rate: 0.6 mL/min, gradient: 0–100% B in 20 min, temperature: 30°C, detection: FL (280–360 nm), injected volume: 2 μL.

References

[1] S. Fekete, A.L. Gassner, S. Rudaz, J. Schappler, D. Guillarme, Analytical strategies for the characterization of therapeutic monoclonal antibodies, *Trends Anal. Chem.* 42 (2013) 74–83.

[2] S. Fekete, A. Beck, J.L. Veuthey, D. Guillarme, Ion-exchange chromatography for the characterization of biopharmaceuticals, *J. Pharm. Biomed. Anal.* 113 (2015) 43–55.

[3] J. Svasti, C. Milstein, The disulphide bridges of a mouse immunoglobulin G1 protein, *J. Biochem.* 126 (1972) 837–850.

[4] J.C. Rea, G.T. Moreno, Y. Lou, D. Farnan, Validation of a pH gradient based ion-exchange chromatography method for high-resolution monoclonal antibody charge variant separations, *J. Pharm. Biomed. Anal.* 54 (2011) 317–323.

[5] L.A.Æ. Sluyterman, O. Elgersma, Chromatofocusing: Isoelectric focusing on ion exchange columns. I. General principles, *J. Chromatogr.* 150 (1978) 17–30.

[6] L.A.Æ. Sluyterman, J. Wijdenes, Chromatofocusing: Isoelectric focusing on ion exchange columns. II. Experimental verification. *J. Chromatogr.* 150 (1978) 31–44.

[7] L.A.Æ. Sluyterman, J. Wijdenes, Chromatofocusing: IV. Properties of an agarose polyethyleneimine ion exchanger and its suitability for protein separation, *J. Chromatogr.* 206 (1981) 441–447.

[8] A. Rozhkova, Quantitative analysis of monoclonal antibodies by cation-exchange chromatofocusing, *J. Chromatogr. A* 1216 (2009) 5989–5994.

[9] X. Kang, D. Frey, High-performance cation-exchange chromatofocusing of proteins, *J. Chromatogr. A* 991 (2003) 117–128.

[10] M. Talebi, A. Nordbog, A. Gaspar, N.A. Lacher, Q. Wang, X.Z. He. P.R. Haddad, E.F. Hilder, Charge heterogeneity profiling of monoclonal antibodies using low ionic strength ion-exchange chromatography and well-controlled pH gradients on monolithic columns, *J. Chromatogr. A*, 1317 (2013) 148–154.

[11] M. Perkins, R. Theiler, S. Lunte, M. Jeschke, Determination of the origin of charge heterogeneity in a murine monoclonal antibody, *Pharm. Res.* 17 (2000) 1110–1117.

[12] M. Schmidt, M. Hafner, C. Frech, Modeling of salt and pH gradient elution in ion-exchange chromatography, *J. Sep. Sci.* 37 (2014) 5–13.

[13] J. Ståhlberg, Retention models for ions in chromatography, *J. Chromatogr. A* 855 (1999) 3–55.

[14] T. Bruch, H. Graalfs, L. Jacob, C. Frech, Influence of surface modification on protein retention in ion-exchange chromatography — Evaluation using different retention models, *J. Chromatogr. A* 1216 (2009) 919–926.

[15] S. Yamamoto, K. Nakanishi, R. Matsuno, *Ion-Exchange Chromatography of Proteins*, Marcel Dekker, New York, 1988.

[16] S.R. Gallant, S. Vunnum, S.M. Cramer, Optimization of preparative ion-exchange chromatography of proteins: Linear gradient separations, *J. Chromatogr. A* 725 (1996) 295–314.

[17] C.A. Brooks, S.M. Cramer, Steric mass-action ion exchange: Displacement profiles and induced salt gradients, *AIChE J.* 38 (1992) 1969–1978.

[18] H. Shen, D.D. Frey, Effect of charge regulation on steric mass-action equilibrium for the ion-exchange adsorption of proteins, *J. Chromatogr. A* 1079 (2005) 92–104.

[19] H. Shen, D.D. Frey, Charge regulation in protein ion-exchange chromatography: Development and experimental evaluation of a theory based on hydrogen ion Donnan equilibrium, *J. Chromatogr. A* 1034 (2004) 55–68.

[20] G.S. Manning, Limiting laws and counterion condensation in polyelectrolyte solutions I. colligative properties, *J. Chem. Phys.* 51 (1969) 924–933.

[21] G.S. Manning, J. Chem, Limiting laws and counterion condensation in polyelectrolyte solutions III. An analysis based on the Mayer ionic solution theory, *Phys.* 51 (1969) 3249–3252.

[22] W.R. Melander, Z. ElRassie, Cs. Horvath, Interplay of hydrophobic and electrostatic interactions in biopolymer chromatography: Effect of salts on the retention of proteins *J. Chromatogr.* 469 (1989) 3–27.
[23] I. Mazsaroff, L. Varady, G.A. Mouchawar, F.E. Regnier, Thermodynamic model for electrostatic-interaction chromatography of proteins, *J. Chromatogr.* 499 (1990) 63–77.
[24] C.B. Mazza, N. Sukumar, C.M. Breneman, S.M. Cramer, Prediction of protein retention in ion-exchange systems using molecular descriptors obtained from crystal structure, *Anal. Chem.* 73 (2001) 5457–5461.
[25] G. Malmquist, U.H. Nilsson, M. Norrman, U. Skarp, M. Strömgren, E. Carredano, Electrostatic calculations and quantitative protein retention models for ion exchange chromatography, *J. Chromatogr. A* 1115 (2006) 164–186.
[26] W.K. Chung,Y. Hou,A. Freed, M. Holstein, G.I. Makhatadze, S.M. Cramer, Investigation of protein binding affinity and preferred orientations in ion exchange systems using a homologous protein library, *Biotechnol. Bioeng.* 102 (2009) 869–881.
[27] R.W. Stout, S.I. Sivakoff, R.D. Ricker, L.R. Snyder, Separation of proteins by gradient elution from ion-exchange columns: Optimizing experimental conditions, *J. Chromatogr.* 353 (1986) 439–463.
[28] M.A. Quarry, R.L. Grob, L.R. Snyder, Prediction of precise isocratic retention data from two or more gradient elution runs. Analysis of some associated errors, *Anal. Chem.* 58 (1986) 907–917.
[29] L.R. Snyder, J.J. Kirkland, J.L. Glajch, *Practical HPLC Method Development*, second ed., John Wiley & Sons Inc., 1997.
[30] S. Fekete, A. Beck, J. Fekete, D. Guillarme, Method development for the separation of monoclonal antibody charge variants in cation exchange chromatography, Part I: Salt gradient approach, *J. Pharm. Biomed. Anal.* 102 (2015) 33–44.
[31] L. Shan, D.J. Anderson, Effect of buffer concentration on gradient chromatofocusing performance separating proteins on a high-performance DEAE column, *J. Chromatogr. A* 909 (2001) 191–205.
[32] L. Shan, D.J. Anderson, Gradient chromatofocusing versatile pH gradient separation of proteins in ion-exchange HPLC: Characterization studies, *Anal. Chem.* 74 (2002) 5641–5649.
[33] S. Fekete, A. Beck, J. Fekete, D. Guillarme, Method development for the separation of monoclonal antibody charge variants in cation exchange chromatography, Part II: pH gradient approach, *J. Pharm. Biomed. Anal.* 102 (2015) 282–289.
[34] Y. Liu, D.J. Anderson, Gradient chromatofocusing high-performance liquid chromatography: I. Practical aspects, *J. Chromatogr. A* 762 (1997) 207–217.
[35] L. Zhang, T. Patapoff, D. Farnan, B. Zhang, Improving pH gradient cation-exchange chromatography of monoclonal antibodies by controlling ionic strength, *J. Chromatogr. A* 1272 (2013) 56–64.
[36] L. Bai, S. Burman, L. Gledhill, Development of ion exchange chromatography methods for monoclonal antibodies, *J. Chromatogr. A* 22 (2000) 605–611.
[37] T. Ishihara, S. Yamamoto, Optimization of monoclonal antibody purification by ion-exchange chromatography, application of simple methods with linear gradient elution experimental data, *J. Chromatogr. A* 1069 (2005) 99–106.

[38] S. Yamamoto, E. Miyagawa, Retention behaviour of very large biomolecules in ion-exchange chromatography, *J. Chromatogr. A* 852 (1999) 25–30.
[39] S. Al-Jibbouri, The influence of salt type on the retention of bovine serum albumin in ion-exchange chromatography, *J. Chromatogr. A* 1139 (2007) 57–62.
[40] R.R. Abzalimov, A. Frimpong, I.A. Kaltashov, Structural characterization of protein–polymer conjugates. I. Assessing heterogeneity of a small PEGylated protein and mapping conjugation sites using ion exchange chromatography and top-down tandem mass spectrometry, *Int. J. Mass Spec.* 312 (2012) 135–143.
[41] S. Fekete, A. Beck, J. Fekete, D. Guillarme, Method development for the separation of monoclonal antibody charge variants in cation exchange chromatography, Part II: pH gradient approach, *J. Pharm. Biomed. Anal.* 102 (2015) 282–289.

Chapter 11

Computer-assisted Method Development in Characterization of Therapeutic Proteins by Hydrophobic Interaction Chromatography

Balazs Bobaly* and Szabolcs Fekete

School of Pharmaceutical Sciences,
University of Geneva, University of Lausanne,
CMU — Rue Michel Servet 1, 1211 Geneva 4, Switzerland

**balazs.bobaly@unige.ch*

11.1 Introduction

Hydrophobic interaction chromatography (HIC) is a historical technique [1, 2] used for the purification [3–5] and analytical characterization [6, 7] of proteins. Similar to what can be done in reversed-phase liquid chromatography (RPLC) HIC can separate protein species based on their hydrophobicity, but using different conditions. Compared to RPLC, the main benefit of HIC is its ability to perform separations under non-denaturing conditions (i.e. physiological pH, ambient temperature and limited or no organic solvents). Native forms of the proteins are expected to be maintained, and the separated species can be collected for further activity measurements (e.g. cell-based potency, receptor binding, cell proliferation assay, enzyme assay, functional ELISA, etc.). In analytical HIC generally a buffered inverse salt gradient is used to elute proteins from a moderately apolar stationary phase. The sample is injected into the high salt concentration mobile phase to attain appropriate binding and retention. When the salt concentration is decreased, proteins elute from the stationary phase according to their increasing hydrophobicity. The main limitations of HIC are (1) the high mobile phase salt concentration, which — except in few applications — does not allow to directly hyphenate with mass spectrometry and (2) the slow mass transfer resulting in broad peaks and compromising kinetic efficiency.

The main applications of modern analytical HIC are the identity, heterogeneity, impurity and activity testing of monoclonal antibodies (mAbs) [8] and antibody–drug conjugates (ADCs) [9]. HIC is a complementary approach to RPLC in monitoring post-translational modifications (e.g. degradation, misfolding, oxidation, carboxy terminal heterogeneity, aspartic acid isomerization, unpaired cysteines, etc.) as well as mutations in the sequence [8]. HIC is an effective tool to separate different populations of ADC-loaded species that differ in their drug to antibody ratio (DAR). This enables the determination of ADCs' average DAR and drug load distribution [9]. Moreover, it can be used for the determination of heterodimerization efficiency of bispecific antibodies (bsAbs) [10], and HIC is the reference technique for determination of the relative hydrophobicity of mAbs [3, 7].

The goal of this chapter is to provide a general overview of theoretical and practical aspects of modern HIC method development applied for the characterization of therapeutic protein biopharmaceuticals. First, an overview of retention mechanisms is provided, and then mobile phase and stationary phase considerations are discussed. Finally, computer-assisted HIC method development is presented with real-life applications. Future perspectives and cutting-edge technologies implementing HIC separations will also be discussed.

11.2 Retention Theories in HIC

Many fundamental studies and retention models in HIC are available in the literature [4, 7]. Due to the complexity of retention mechanisms proposed for HIC they are often misunderstood and none of the established theories has received general acceptance. Various interpretations and approaches such as salting-out/salting-in effects, hydrophobic interaction and hydrophobic effects, solvophobic theory and dehydration of proteins or structural rearrangement of proteins are often confused. Here, we try to clarify the various concepts and briefly summarize the different variables affecting proteins retention in HIC.

11.2.1 *Salting-out and salting-in*

Tiselius described first the concept of protein chromatography based on hydrophobic interaction using the term "salting-out chromatography" [1]. In this seminal work salt solutions were applied as mobile phase. Salting-out effect is based on the interaction of electrolytes (mobile phase) and non-electrolytes (protein). In aqueous solutions, hydrophobic amino acid residues are usually folded into the inner, less solvent-exposed part of the protein, while hydrophilic species interact with the surrounding solvent molecules through H-bonding and polar interaction. In high salt concentration solutions non-electrolytes become less soluble, since water molecules will solvate predominantly salt ions. Under these conditions, the number of water molecules available to interact with the hydrophilic residues of the protein decrease. Protein–protein intermolecular or protein–surface interactions become more pronounced due to the limited accessibility of solvating water molecules. Finally, these conditions lead to the formation of protein associates through hydrophobic interactions (reversible aggregation) and/or adsorption of protein chains and aggregates to hydrophobic surfaces (stationary phase) (Fig. 11.1) [5].

The overall surface area of hydrophobic sites exposed to the polar solvent is decreased, which results in a less structured (higher entropy) condition, which is the favored thermodynamic state. This separation mode was later termed as "hydrophobic chromatography" or "hydrophobic affinity chromatography" [11]. Hjertén introduced the term "hydrophobic interaction chromatography" in 1973 [2]. HIC was also denoted as "salt-mediated separation of proteins" and "salt-promoted adsorption chromatography" [12]. In contrast to "salting-out", when adding salts having divalent cations and univalent anions such as $MgCl_2$ or $CaCl_2$, some specific interactions can occur with proteins [13]. These salts increase protein solubility (salting-in properties) [14], and the phenomenon was explained by the interaction of the salt with the protein surface. Salts exhibiting this behavior were called "chaotropic salts" [15]; their use is uncommon in HIC.

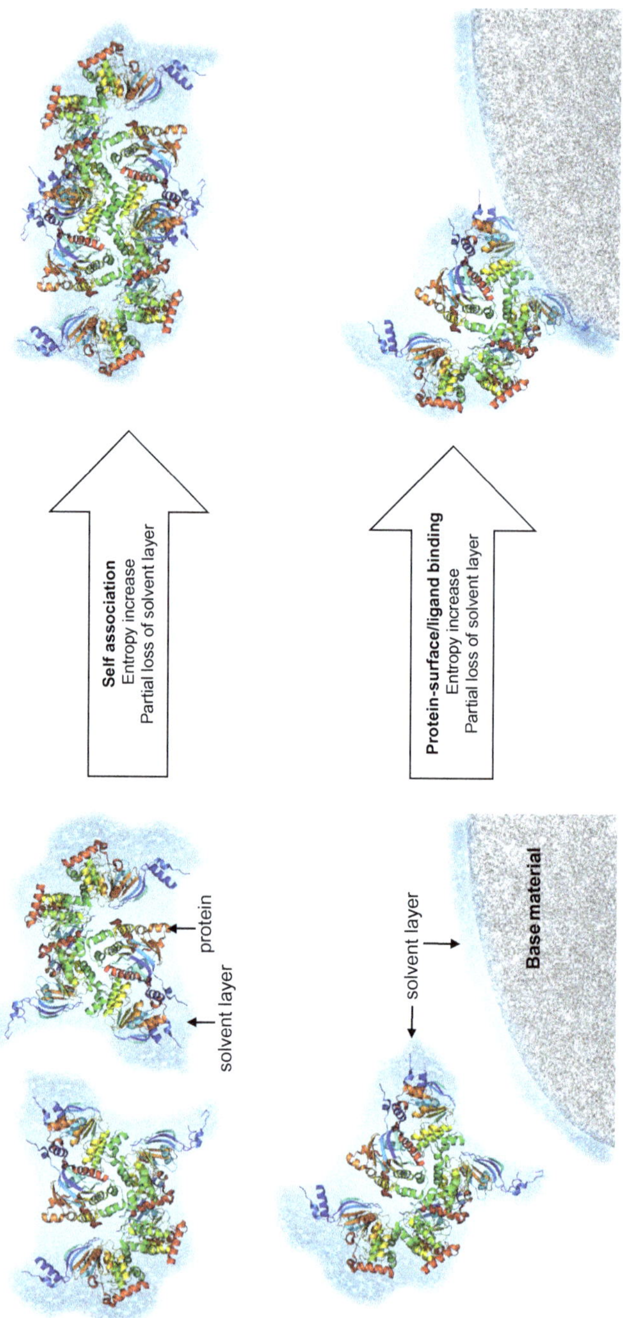

Figure 11.1: Schematic diagram showing hydrophobic interaction between proteins in an aqueous solution and between proteins and the hydrophobic surface of an HIC adsorbent.

11.2.2 *Hydrophobic effects*

Hydrophobic effect drives the retention process, which is generally defined as an interaction of apolar substances or moieties of molecules with water that is responsible for their low solubility [16]. In other words, hydrophobicity means the repulsion of the apolar protein moieties and the polar aqueous environment. The term "hydrophobic interactions" has also been used to describe driving forces resulting in the association of non-polar molecules or the binding of hydrophobic moieties in aqueous solutions [17]. Bulky water has an organized structure stabilized by H-bonds. Each oxygen atom has four hydrogens as neighbors in a tetrahedral configuration, and each hydrogen atom forms a bridge between two oxygen atoms (either covalent or H-bonds). Introducing hydrophobic moieties (such as hydrophobic protein residues) into this environment requires the separation of neighboring water molecules in order to form a cavity and to accommodate the protein [18]. This process requires a certain energy investment proportional to the surface of the cavity and to the surface tension of the solvent. If proteins are associated or adsorbed, their hydrophobic contact surface area is reduced and energy is released. Thus, the interaction of two hydrophobic entities in a polar medium takes place spontaneously and is mainly driven by entropy change. It was shown later that hydrophobic interactions are entropy driven at low temperatures, but enthalpy driven at elevated temperatures, when the heat capacity change remains constant in the range of experimental temperature [19]. The basis of a more detailed understanding of the influence of temperature on hydrophobic interaction was provided by such model experiments.

11.2.3 *Solvophobic theory*

The solvophobic theory generally explains the interactions between a polar solvent (aqueous mobile phase) and a less polar solute (protein). Due to strong cohesive forces existing between the solvent molecules and providing a strongly structured order for the solvent, less polar solutes tend to be less soluble. As the results of this strong solvent–solvent binding interaction, retention in RPLC can be explained by the solvophobic theory. According to this theory, solute molecules stick to the surface of the stationary phase due to their rejection form the solvent and their affinity to the hydrophobic stationary phase. Thus, retention is explained by a mixed effect of interactions between the solute and the stationary phase and by the rejection of solutes form the solvent. Horváth *et al.* described the basis for retention mechanisms in RPLC, employing the framework of the solvophobic theory [20]. More interested readers can find details in the review of Molnár on solvophobic theory [21]. A comprehensive treatment of the salting-out of proteins and the salt effect on HIC retention in the absence of specific salt binding is based on the adaptation made by Horváth and co-workers [17]. Using this adapted theory and accounting for the effect of salt concentration on the mobile phase surface tension, the magnitude of solute retention can be expressed as a function of the molar salt concentration in HIC [17]. The theory predicts that for sufficiently high salt concentrations — where the retention is controlled predominantly by hydrophobic interactions — the retention increases with both the molar salt concentration (in the mobile phase) and the size of the solute (protein) — or its hydrophobic moiety.

11.2.4 *Linear solvent strength theory for HIC applications*

The linear solvent strength (LSS) model is widely accepted and frequently applied in various modes of chromatography to describe the relationship between solute retention and experimental conditions (i.e. gradient steepness, mobile phase composition) [22–24]. Generally, LSS theory provides a good description for the retention behavior of various types of analytes. In some cases, slight deviations from the linear model can be observed for proteins. This is presumably due to conformational changes during elution affecting retention behavior. In HIC, the relationship between retention and mobile phase salt concentration (ionic strength) determines the applicability of the LSS model, and the following general equation can be given for isocratic elution:

$$\log k = \log k_0 + S \times c \tag{1}$$

where k is the retention factor in isocratic elution mode, k_0 corresponds to the retention factor observed in mobile phase containing no salt, c is the salt concentration of the mobile phase and S is the steepness of the linear function. In the case of large proteins, the use of isocratic conditions is impractical. The slope of the linear function is much higher compared to small molecules and the retention follows a so-called on-off mechanism, which is difficult to control under isocratic conditions. S and k_0 parameters can, however, be calculated from retention data obtained from two linear gradient runs possessing different gradient steepness. Then gradient retention times for any gradient can be derived from the LSS parameters [25, 26]. This approach was confirmed to be able to accurately predict HIC retention times for recombinant mAbs and ADC species [25–27] with a retention time error of less than 1–2%.

11.3 Method Development

In HIC, mobile phase and operating conditions are usually determined based on subjective experiences and historical references. For instance, most HIC applications are operated using butyl modified stationary phases and ammonium sulfate buffer. However, all of the salts possessing salting-out properties and appropriate solubility can be considered as potential mobile phase components. The detailed HIC characterization of novel therapeutic proteins may be challenging with historical HIC conditions. As an example, highly hydrophobic protein species may not completely elute from some stationary phases and the use of organic mobile phase additives may be necessary for acceptable recovery. The selection of the most appropriate conditions is essential and can be supported by modern method development approaches. Computer-assisted optimization of the separation based on initial experimental data provides straight knowledge on the method behavior and meets regulatory expectations.

The following sections are aimed to provide information regarding to the selection of mobile phase salt type and its concentration, pH, temperature and to the use of organic modifiers. Evaluation of modern HIC stationary phases for the analysis of therapeutic proteins is discussed. The mobile phase and the stationary phase together are considered as phase system. Current possibilities of computer-assisted phase system optimization are explained. At the end, a generic HIC method for the analysis of recombinant mAbs and ADC species is described.

11.3.1 *Mobile phase salt type and concentration*

Practical HIC gradient conditions should enable the elution of proteins possessing a wide range of hydrophobicity. Usually an inverse salt gradient is applied to elute the proteins from a moderately apolar (much less hydrophobic than conventional RPLC phases) stationary phase. Historically, the sample is injected into the mobile phase "A" containing 1.5–2 M aqueous ammonium sulfate buffered with 20–100 mM phosphate, in which appropriate binding (retention) is observed. Then elution occurs when increasing the volume fraction of the eluting mobile phase "B", which contains only the buffer component. It is worth keeping in mind that various salts can be applied. The effect of different salts on hydrophobic interactions, and thus on retention, follows the lyotropic (Hofmeister) series for the precipitation of proteins from aqueous solutions [28]. In this series, salt anions and cations are ranked based on their salting-out effect. Ions with high potency of salting-out (or precipitation) promote hydrophobic interactions. These ions are also characterized as being anti-chaotropic, such as phosphate, sulfate, acetate or chloride and ammonium, potassium or sodium. The combinations of the above anions and cations are the most frequently applied mobile phase components in HIC. Based on the hydrophobicity of the protein, salt nature can affect protein retention unexpectedly [29]. Thus, the effect of salt on retention cannot be predicted in advance but should always be determined experimentally.

Besides salt type, salt concentration is another important variable for tuning HIC retention. Based on the lyotropic strength of the salt, different concentrations are required to maintain the same retention. Stronger salts (such as ammonium sulfate) efficiently retain proteins at lower concentration, whereas weaker salts such as sodium chloride have to be used at higher concentrations (3–5 M) [26]. Peak widths also vary with salt concentration, since it impacts the gradient steepness and viscosity (mass transfer resistance) of the mobile phase [30]. It has been shown recently, that similar selectivity can be attained with various types of salts when their concentration is corrected on a given stationary phase (Fig. 11.2).

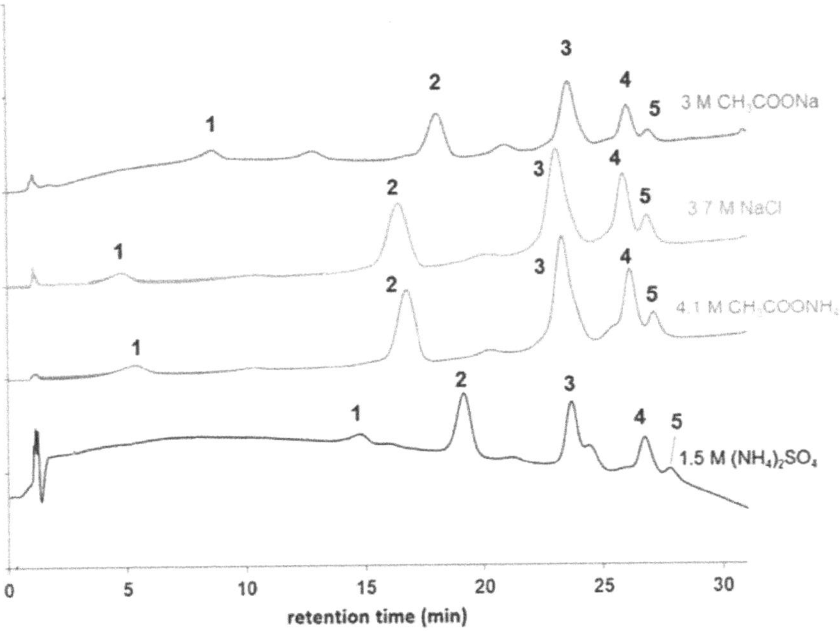

Figure 11.2: Representative chromatograms obtained on MabPac HIC 10 (100 × 4.6 mm, 5 μm) column by using different salt systems. Flow rate: 1 mL/min, gradient: 0–100% B in 30 min, temperature: 20°C, detection: fluorescence (ex: 280 nm, em: 360 nm), sample: brentuximab vedotin. Numbered peaks denote conjugated species from DAR0 to DAR8 (with permission from Ref. [26]).

Based on our experiences, we recommend the use of sodium chloride and sodium acetate. We commonly observed drifted baseline for ammonium sulfate, whereas ammonium acetate is hygroscopic, and therefore it may be challenging to maintain its constant quality once the container is opened.

11.3.2 *Modern HIC stationary phases for the separation of therapeutic proteins*

In HIC, proteins are retained using moderately hydrophobic stationary phases. Retention is not spontaneous such as in RPLC, and the solute must be salted-out to the stationary phase. In HIC, shorter alkyl chain ligands (e.g. ethyl, propyl, butyl, pentyl, hexyl, phenyl), ether- or amide-modified silica or polymeric materials are preferred [4, 5, 31]. The hydrophobicity of the stationary phase and the strength of the hydrophobic interaction are controlled by the length of the alkyl chain and by the ligand density. The longer the alkyl chain and the higher the ligand density, the stronger the interaction between the solute and the stationary phase is. The adsorption/partition mechanisms are complex and are controlled by mixed enthalpy- and entropy-driven processes, depending on the protein quality as well [32].

Recently introduced modern silica or polymeric materials possessing $2.5 - 10\,\mu$m particle size are able to withstand with 100–400 bar pressure drops [31]. Porous as well as non-porous particles are available on the market. Non-porous particles provide higher efficiency for proteins due to reduced mass transfer resistance. Hydrophobicity of the packing is a crucial point in method development. Minimum retention can be controlled by the salt type and concentration, but the ability to elute all the sample components (e.g. maximum retention) from the column is mainly determined by the hydrophobicity of the stationary phase. Complete elution/recovery of highly hydrophobic protein species may be challenging from some of the modern stationary phases with increased hydrophobicity (Fig. 11.3) [26].

11.3.3 *Optimization of the phase system*

Retention of a given protein is controlled by the salt type and concentration and by the stationary phase. A systematic study showed that various phase systems can be selected for tuning selectivity and retention in HIC [30]. Recently, phase systems have been optimized for the separation of mAbs and ADC species [26]. First, hydrophobicity indexes (c^*) can be calculated for different salt types. Hydrophobicity indeces can be derived from the previously discussed LSS parameters in Eq. (1). as follows:

$$c^* = \frac{\log k_0}{S} \tag{2}$$

where c^* corresponds to the salt concentration at which $\log k = 0$ $(k = 1)$ in isocratic elution mode. c^* value reflects well both the properties of the

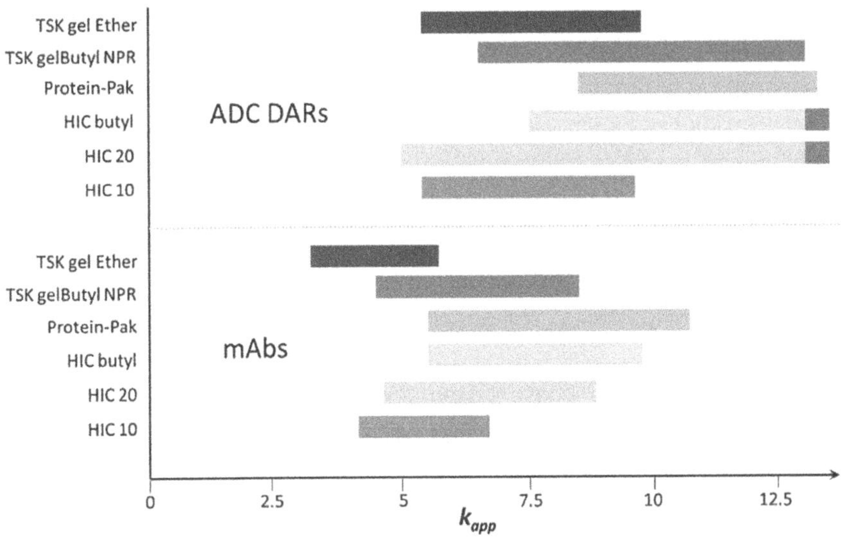

Figure 11.3: Comparison of elution windows (k_{app}) obtained on different columns by using ammonium sulfate buffer (1.5 M) (with permission from Ref. [26]).

salts system and the stationary phase and is useful in the characterization of phase systems. Various phase systems can be compared by using this approach, and then the optimal combination can be selected to set the elution window. Elution orders of therapeutic proteins remained the same, but retention and, thus, selectivity can clearly be tuned with the phase system approach [26].

11.3.4 *The use of organic modifiers in the mobile phase*

The use of water-miscible organic modifiers such as isopropanol, acetonitrile or methanol can help in modifying protein–ligand interactions to enhance recovery and tune selectivity. Organic modifiers are added in a limited concentration range (e.g. less than 20%) to avoid denaturation of the proteins and only to mobile phase "B" to avoid precipitation of the salt in mobile phase "A". The simultaneous use of organic and inverse salt gradients may be termed as a mixed mode HIC–RPLC separation. Depending on the protein and the salt system (presumably on protein conformation and hydrophobicity) organic modifiers may decrease or increase retention.

Generally, retention decrease can be observed for mAbs, whereas retention time usually increases for conjugated ADC species when increasing the concentration of organic mobile phase additives [25, 33]. Organic modifier concentration can be included as a variable into computer-assisted method development models [25] and play a crucial role in the complete elution of highly hydrophobic proteins [33]. At low organic modifier concentration (e.g. below 8–10%), these species could not be eluted completely, whereas high organic modifier concentration (e.g. above 12–15%) lead to denaturation of proteins, complicating the evaluation of the chromatographic profile [33]. The correct concentration of organic modifiers may vary depending on the phase system and the protein and cannot be predicted in advance but should always be determined experimentally in the early phase of method development.

11.3.5 *Effect of temperature and pH*

The effects of temperature on retention are often expressed by the Gibbs free energy (van't Hoff equation). In most chromatographic modes, solutes behave "regularly"; their retention decreases with the increase of temperature and plotting log (k) vs. $1/T$ gives a linear function. In HIC, the effect of temperature on retention is more complex, and irregular behavior is often reported. Retention of proteins in HIC is often increased with the temperature. This effect has been attributed to enhanced hydrophobic interactions resulting from temperature-induced conformational changes and concomitant increase of hydrophobic contact area upon binding to the stationary phase [34, 35]. Horváth and co-workers determined the individual relative contributions of enthalpy and entropy to the free energy change upon adsorption as the function of temperature [36], and they experimentally confirmed that enthalpy and entropy changes were large and positive at low temperatures, then decreased with increasing temperature, and finally became negative at high temperatures [37]. In practice, temperature in HIC can be used for tuning selectivity [26]. However, it has to be kept in mind that conformational changes are preferably avoided in HIC, and therefore working at moderate temperatures (e.g. below 40°C) is recommended. Moreover, if the mobile phase contains organic modifiers, temperature has to be kept even lower (e.g. 20–25°C) since, in this case, proteins can be more susceptible to denaturation [33].

In HIC, the effect of pH on retention is not straightforward [38, 39]. Charged residues are not directly involved in hydrophobic interactions, but changes in the overall protein charge affects protein hydrophobicity and local conformational changes (e.g. repulsion or attraction of charged residues) might affect the hydrophobic contact area. Generally, increasing pH reduces the hydrophobic interactions between the protein and the stationary phase due to the increased hydrophilicity promoted by the change in protein charge. On the contrary, a pH decrease may result in an apparent increase of hydrophobic interactions [39]. Changes in pH may result in different retention behavior depending on the physico-chemical properties of the particular protein. Thus, pH could be considered as an additional parameter for tuning selectivity and retention in HIC. On the other hand, it is recommended to use a pH close to physiological conditions to maintain non-denaturing chromatographic conditions. Close to physiological pH and in a narrow pH range (e.g. one pH unit), retention is expected

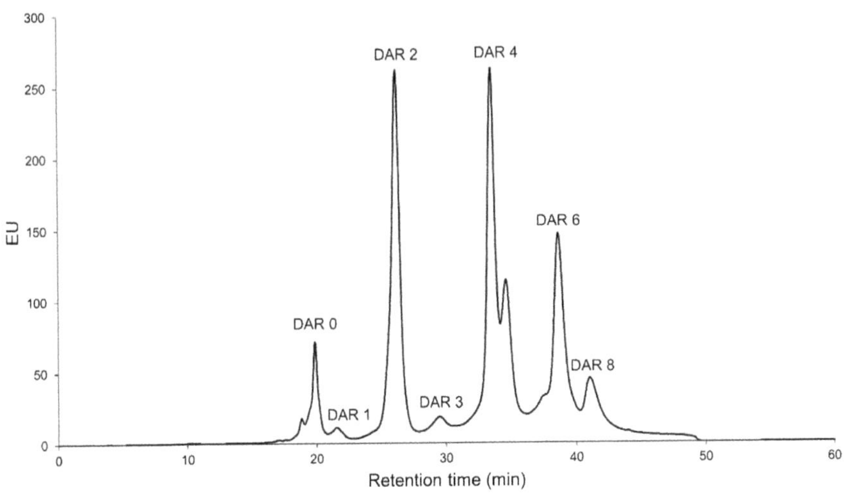

Figure 11.4: HIC chromatographic profile of the ADC sample, brentuximab vedotin using the generic conditions (with permission from Ref. [40]).

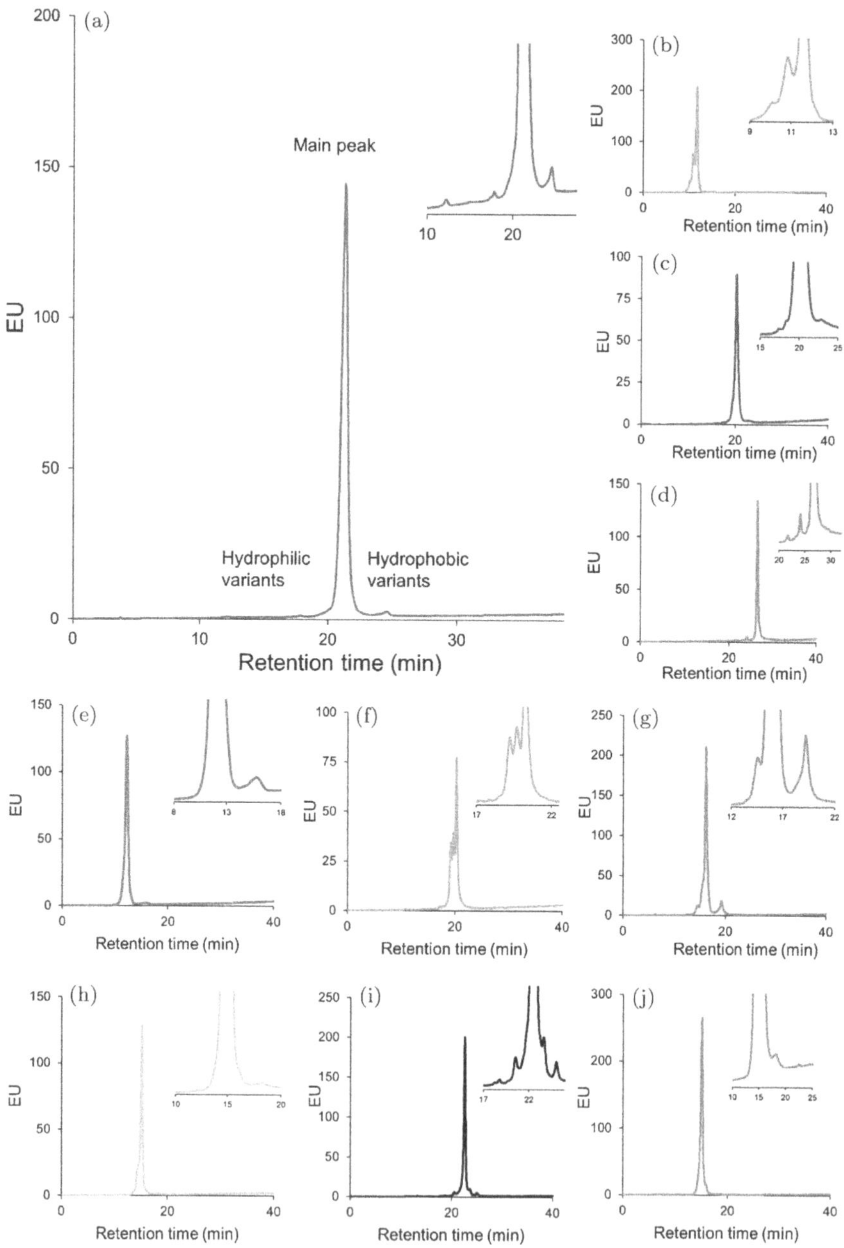

Figure 11.5: HIC chromatographic profiles of pertuzumab (a), adalimumab (b), belimumab (c), bevacizumab (d), denosumab (e), infliximab (f), ofatumumab (g), palivizumab (h), rituximab (i), trastuzumab (j) using the generic conditions (with permission from Ref. [40]).

to be unaffected, and so a method robust for slight pH variations can be developed [7].

11.3.6 *Generic HIC conditions*

Recently, Goyon *et al.* described generic HIC conditions that can be used for the characterization of various mAbs and for the reference cysteine-conjugated ADC, brentuximab-vedotin [40]. Thermo Fischer MabPac HIC-10 (5.0 μm, 250 × 4.6 mm, 1000 Å) column (obviously, a similar column such as TSKgel butyl-NPR can also be used) and an inverse salt gradient of 2 M ammonium sulfate buffered at pH 6.8 with 100 mM potassium phosphate were recommended. Gradient time was 40 min, flow rate was set to 0.8 mL/min and column oven was 25°C. 5 μL of samples at 1–5 mg/mL were injected. Gradient steepness can further be optimized depending on the separation of the mAb and their hydrophobic variants. Figures 11.4 and 11.5 show HIC chromatograms obtained with the proposed HIC conditions.

11.4 Computer-assisted Method Development in HIC

Method development can be assisted by specific software in several modes of LC. The possibility to use different retention models and the systematic optimization of method variables help to tune selectivity and retention with an excellent predictive power [41, 42]. The search for optimal chromatographic conditions can drastically be shortened by the simultaneous evaluation of chromatographic data obtained from initial chromatographic runs (multifactorial optimization). Several software also include structural information in retention models, but due to the complex and dynamic nature of protein conformation under chromatographic conditions, this feature is limited to small molecular applications. Once the best available conditions are found for the separation, method robustness can be evaluated by predicting the effects of slight variations in chromatographic parameters. Method development generally involves a scouting and an optimization phase. In the scouting phase, the possible variables and their range in which chromatographic profiles can advantageously be changed are monitored. Then in the optimization phase, the previously selected variables (and their practical ranges, based on the results of the scouting phase) can be built into an experimental design. After running and evaluating the experimental points of a design, the working point (optimum conditions) can be predicted and verified experimentally. The following section is aimed at providing insight into computer-assisted method development in HIC. Besides the phase system (which can be optimized by the hydrophobicity indexes, see earlier), gradient time (or steepness), organic modifier concentration (and type), mobile phase temperature and pH are the most relevant variables to be optimized in HIC. Multifactorial experimental designs exploring the effects of these variables are presented. It is worth keeping in mind that general linear gradients may provide limited resolution in certain cases. Application and optimization of nonlinear, or multi-step linear gradients in HIC is also discussed.

11.4.1 *Experimental designs in HIC*

It was recently shown, that the most important method variables for the HIC separation of mAbs are the gradient steepness and organic modifier concentration [25]. Authors also investigated pH (in the range of 6.3–7.0) and salt type (and molarity) in the scouting phase, but their effect on selectivity and resolution were found not to be significant. In the final experimental design, organic modifier concentration (0–10% isopropanol) and gradient steepness (3.33–10% B/min) have been included. The 2D model required only four experimental runs for the optimization, significantly decreasing the time spent for method development. The predicted method was verified experimentally, and results showed a good agreement between the predicted and experimental retention times (average relative retention time error was ∼1%). The optimized method was used for separation of mAbs possessing various hydrophobicity (Figs. 11.6 and 11.7).

Another study reported similar 2D designs including temperature (20–40°C) and gradient steepness (3.33–10% B/min) as method variables for the fast and automated optimization of the phase systems. At the end, gradient profiles were also edited in methods used for the separation of mAbs and ADC species [26]. A 3D model combining gradient steepness (1), temperature (2) and organic modifier (3) as variables was also proposed;

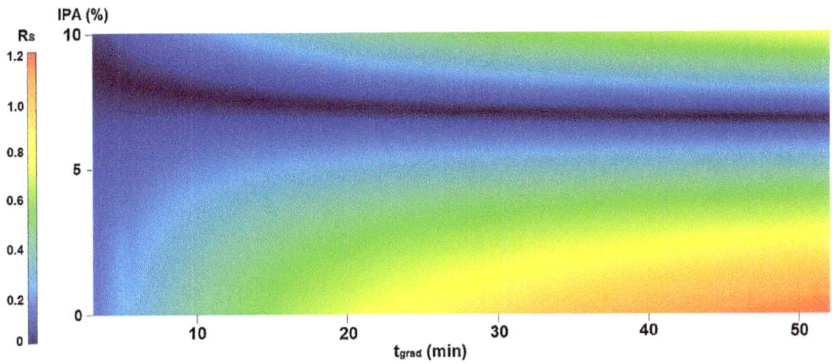

Figure 11.6: Two-dimensional resolution map for the optimization of mAb separation. Variables: gradient time (t_{grad}) and isopropanol % in mobile phase B (IPA%) (with permission from Ref. [25]).

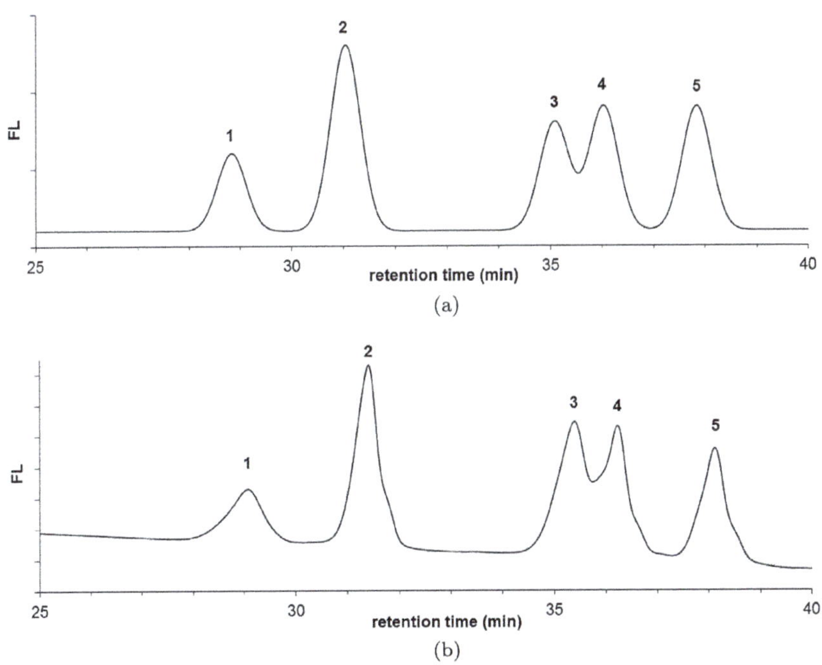

Figure 11.7: Comparison of predicted (a) and experimental (b) chromatograms for the separation of intact mAbs. Column: Thermo MAbPac HIC-10, 100 × 4.6 mm, mobile phase "A": 2 M ammonium-sulfate + 0.1 M phosphate (pH 7), "B": 0.1 M phosphate (pH 7). Flow rate: 1 mL/min, gradient: 0–100% B in 50 min, detection: FL (ex: 280 nm, em: 360 nm), mobile phase temperature: 25°C, peaks: denosumab (1), palivizumab (2), pertuzumab (3), rituximab (4), bevacizumab (5) (with permission from Ref. [25]).

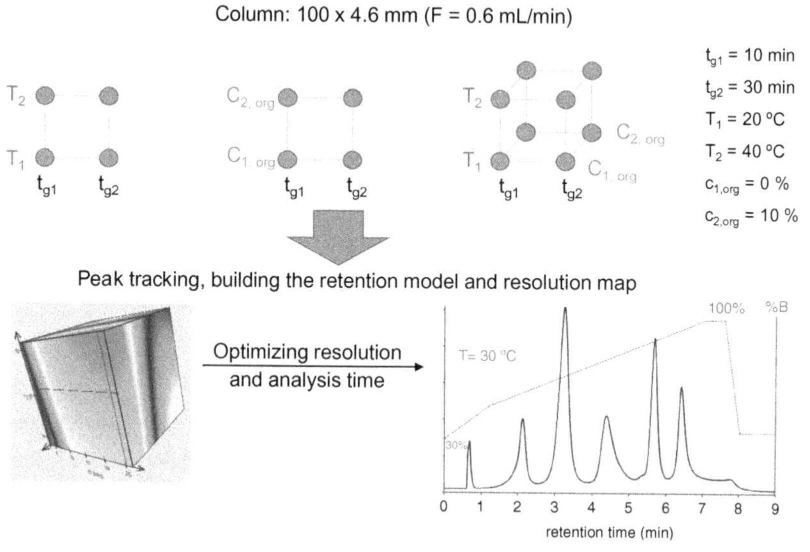

Figure 11.8: Proposed workflow of HIC method development for the separation of mAbs and related products (ADCs). Mobile phase "A" contains salt (high concentration) and buffer (low concentration), mobile phase "B" contains buffer (low concentration) (with permission from Ref. [7]).

however, experimental results have not been reported yet [7]. Figure 11.8 shows a possible setup and workflow for the 3D model.

11.4.2 *Optimization of gradient profiles*

Nonlinear, multi-step or segmented gradient profiles used for HIC separations have further been studied [26, 27]. ADC DARs represent homologous series of proteins. Unequidistant peak spacing is typical for such kind of solutes. In HIC, the cysteine-conjugated homologues of brentuximab vedotin follow a logarithmic-type elution profile. This results in unnecessarily large selectivity of low-DAR species, while the resolution of high-DAR species is compromised. It can be theoretically derived that logarithmic gradient profile provides a much better peak spacing across the whole chromatogram. It is currently not possible to perform a logarithmic gradient with any commercial LC systems; therefore, the logarithmic gradient shape was approximated by multi-linear ones LSS parameters have been calculated from two linear gradient runs using a commercially available method

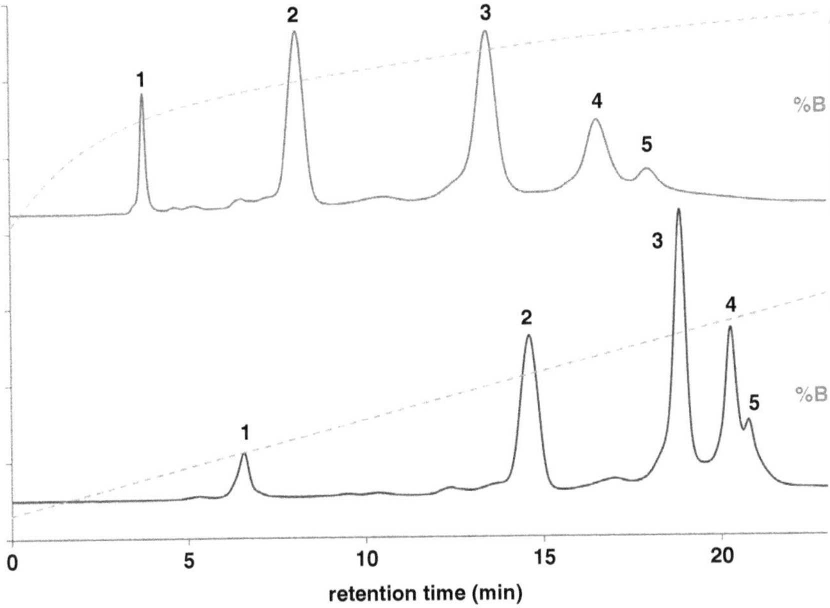

Figure 11.9: Linear (top) and logarithmic (bottom) gradient profiles and chromatograms of brentuximab vedotin. Peaks: DAR0 (1), DAR2 (2), DAR4 (3), DAR6 (4) and DAR8 (6). Mobile phase A: 4 M NaCl with 10 mM phosphate buffer (pH = 7), mobile phase B: 10 mM phosphate buffer (pH = 7) with 8% IPA. Column: Thermo Fisher Scientific MAbPac HIC-10 (100 × 4.6 mm, 5 μm, 1000 Å), *T*: 25°C, gradient program: 0–100% B in 20 min, flow: 0.6 mL/min (with permission from Ref. [27]).

development software, and the logarithmic gradient was approximated with linear segments. Elution profiles of the linear and the logarithmic type gradient are shown in Fig. 11.9.

In the final method, only four linear segments seemed to appropriately approach the logarithmic gradient shape, enabling the use of such profiles in routine laboratories. Again, predicted and experimental chromatograms were in good agreement with an average retention time error of less than $\sim$2% (Fig. 11.10).

The logarithmic profile provided more equidistant peak spacing and shorter analysis time. Another important advantage of the logarithmic gradient against the linear one is its peak focusing effect for the unconjugated mAb. This is particularly useful because the concentration of DAR0

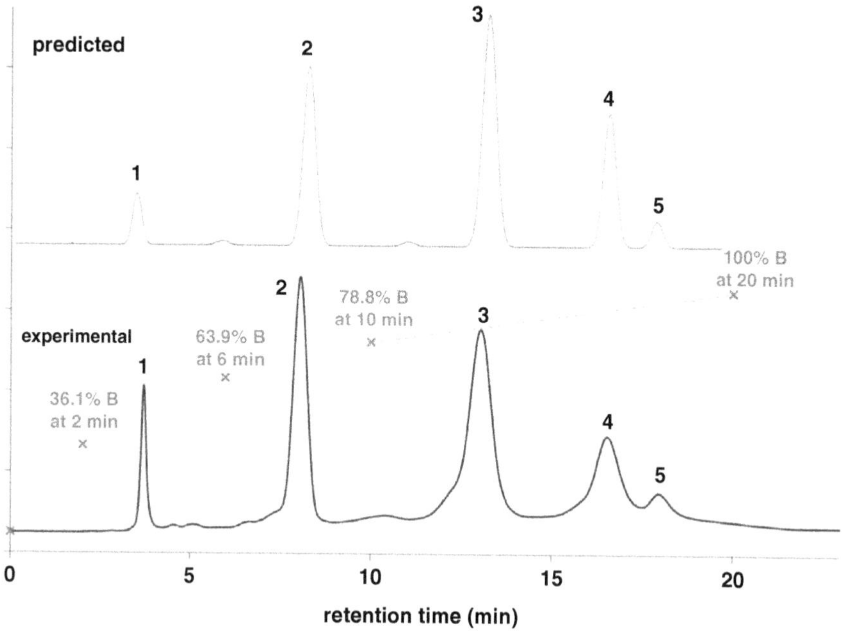

Figure 11.10: Approximation of logarithmic gradient by a 4-segment multi-linear gradient and experimental verification. Mobile phase A: 4 M NaCl with 10 mM phosphate buffer (pH = 7), mobile phase B: 10 mM phosphate buffer (pH = 7) with 8% IPA. Column: Thermo Fisher Scientific MAbPac HIC-10 (100 × 4.6 mm, 5 μm, 1000 Å), *T*: 25°C, gradient program: 0–100% B in 20 min, flow: 0.6 mL/min. Peaks: DAR0 (1), DAR2 (2), DAR4 (3), DAR6 (4) and DAR8 (6) species of brentuximab vedotin (with permission from Ref. [27]).

is often low (the naked mAb is considered as an impurity of the ADC). By utilizing the peak focusing effect, the quantitation limit of DAR0 can be improved.

References

[1] A. Tiselius, Adsorption separation by salting out, *Mineral Geol.* 26B (1948) 1–5.
[2] S. Hjertén, Some general aspect of hydrophobic interaction chromatography, *J. Chromatogr.* 87 (1973) 325–331.
[3] J. Vajda, E. Mueller, Hydrophobic Interaction Chromatography for the Purification of Antibodies, Chapter 7 in *Process Scale Purification of Antibodies*, ed. Uwe Gottschalk, 2nd edition, Wiley, 2017, Hoboken, NJ, USA.
[4] J.A. Queiroz, C.T. Tomaz, J.M.S. Cabral, Hydrophobic interaction chromatography of proteins, *J. Biotechnol.* 87 (2001) 143–159.

[5] J.T. McCue, Theory and use of hydrophobic interaction chromatography in protein purification applications, *Meth. Enzymol.* 463 (2009) 405–414.

[6] B. F. Roettger, M. R. Landisch, Hydrophobic interaction chromatography, *Biotech. Adv.* 7 (1989) 15–29.

[7] S. Fekete, J.-L. Veuthey, A. Beck, D. Guillarme, Hydrophobic interaction chromatography for the characterization of monoclonal antibodies and related products, *J. Pharm. Boimed. Anal.* 130 (2016) 3–18.

[8] M. Haverick, S. Mengisen, M. Shameem, A. Ambrogelly, Separation of mAbs molecular variants by analytical hydrophobic interaction chromatography HPLC: Overview and applications, *mAbs* 6 (2014) 852–858.

[9] A. Wakankar, Y. Chen, Y. Gokarn, F.S. Jacobson, Analytical methods for physicochemical characterization of antibody drug conjugates, *mAbs* 3 (2011) 161–172.

[10] C. Spiess, M. Merchant, A. Huang, Z. Zheng, N.-Y. Yang, J. Peng, D. Ellerman, W. Shatz, D. Reilly, D. G. Yansura J. M. Scheer, Bispecific antibodies with natural architecture produced by co-culture of bacteria expressing two distinct half-antibodies, *Nat. Biotechnol.* 31 (2013) 753–758.

[11] S. Shalitel, Z. Er-el, Hydrophobic chromatography. Use for purification of glycogen synthetase, *Proc. Natl. Acad. Sci. U.S.A.* 70 (1973) 778–781.

[12] J. Porath, Salt-promoted adsorption: recent developments, *J. Chromatogr.* 376 (1986) 331–341.

[13] L. Szepesy, Cs. Horváth, Specific salt effects in hydrophobic interaction chromatography of proteins, *Chromatographia* 26 (1988) 13–18.

[14] A. Vailaya, Cs. Horváth, Retention thermodynamics in hydrophobic interaction chromatography, *Ind. Eng. Chem. Res.* 35 (1996) 2964–2981.

[15] T. Arakawa, S.N. Timasheff, Mechanism of protein salting in and salting out by divalent cation salts: balance between hydration and salt binding, *Biochemistry* 23 (1984) 5912–5923.

[16] C. Tanford, The hydrophobic effect and the organization of living matter, *Science* 200 (1978) 1012–1018.

[17] W. Melander, Cs. Horvath, Salt effects on hydrophobic interactions in precipitation and chromatography of proteins: An interpretation of the lyotropic series, *Arch. Biochem. Biophys.* 183 (1977) 200–215.

[18] J. L. Ochoa, Hydrophobic (interaction) chromatography, *Biochemie* 60 (1978) 1–15.

[19] R.L. Baldwin, Temperature dependence of the hydrophobic interaction inprotein folding, *Proc. Natl. Acad. Sci. U.S.A.* 83 (1986) 8069–8072.

[20] Cs. Horváth, W. Melander, I. Molnár, Solvophobic interactions in liquid chromatography with non-polar stationary phases, *J. Chromatogr.* 125 (1976) 129–156.

[21] I. Molnár, Searching for robust HPLC methods — Csaba Horváth and the solvophobic theory, *Chromatographia* 62 (2005) S7–S17.

[22] L.R. Snyder, *Gradient elution in HPLC: Advances and Perspectives*, Ed. C. Horvath, vol. 1, Academic Press, New York, 1980, pp. 208–316.

[23] J.W. Dolan, L.R. Snyder, Developing a gradient elution method for reversed-phase HPLC, *LC–GC*, 5 (1988) 970–978.

[24] L.R. Snyder, J.W. Dolan, *High-Performance Gradient Elution: The Practical Application of The Linear-Solvent-Strength Model*, John Wiley & Sons, Inc, Hoboken, New Jersey, USA, 2007.

[25] M. Rodriguez-Aller, D. Guillarme, A. Beck, S. Fekete, Practical method development for the separation of monoclonal antibodies and antibody-drug-conjugate species in hydrophobic interaction chromatography, part 1: Optimization of the mobile phase, *J. Pharm. Biomed. Anal.* 118 (2016) 393–403.
[26] A. Cusumano, D. Guillarme, A. Beck, S. Fekete, Practical method development for the separation of monoclonal antibodies and antibody-drug-conjugate species in hydrophobic interaction chromatography, part 2: Optimization of the phase system, *J. Pharm. Biomed. Anal.* 121 (2016) 161–173.
[27] B. Bobály, G. M. Randazzo, S. Rudaz, D. Guillarme, S. Fekete, Optimization of non-linear gradient in hydrophobic interaction chromatography for the analytical characterization of antibody-drug conjugates, *J. Chromatogr. A* 1481 (2017) 82–91.
[28] S. Ihlman, J. Rosengren, S. Hjertén, Hydrophobic interaction chromatography on uncharged Sepharose derivatives. Effects of neutral salts on the adsorption of proteins, *J. Chromatogr.* 131 (1977) 99–108.
[29] G. Rippel, L. Szepesy, Hydrophobic interaction chromatography of proteins on an Alkyl-Superose column, *J. Chromatogr. A* 664 (1994) 27–32.
[30] G. Rippel, A. Bede, L. Szepesy, Systematic method development in hydrophobic interaction chromatography I. Characterization of the phase system and modelling retention, *J. Chromatogr. A* 679 (1995) 17–29.
[31] S. Fekete, J.L. Veuthey, D. Guillarme, Modern column technologies for the analytical characterization of biopharmaceuticals in various liquid chromatographic modes, *LC GC Eur.* (Suppl.: S) (2015) 8–15.
[32] F.Y. Lin, W.Y. Chen, R.C. Ruaan, H.M. Huang, Microcalorimetric studies of the interactions between proteins and hydrophobic ligands in hydrophobic interaction chromatography: effects of chain length, density and the amount of bound protein, *J. Chromatogr. A* 872 (2000) 37–47.
[33] B. Bobaly, A. Beck, J.-L. Veuthey, D. Guillarme, S. Fekete, Impact of organic modifier and temperature on protein denaturation in hydrophobic interaction chromatography, *J. Pharm. Biomed. Anal.* 131 (2016) 124–132.
[34] S.L. Wu, K. Benedek, B.L. Karger, Thermal behavior of proteins in high-performance hydrophobic-interaction chromatography. On-line spectroscopic and chromatographic characterization, *J. Chromatogr.* 359 (1986) 3–17.
[35] S.L. Wu, A. Figueroa, B.L. Karger, Protein conformational effects in hydrophobic interaction chromatography. Retention characterization and the role of mobile phase additives and stationary phase hydrophobicity, *J.Chromatogr.* 371 (1986) 3–27.
[36] A. Vailaya, Cs. Horváth, Retention thermodynamics in hydrophobicinteraction chromatography, *Ind. Eng. Chem. Res.* 35 (1996) 2964–2981.
[37] D. Haidacher, A. Vailaya, Cs. Horváth, Temperature effects in hydrophobicinteraction chromatography, *Proc. Natl. Acad. Sci. U.S.A.* 93 (1996) 2290–2295.
[38] O'Farrell, P.A., Hydrophobic Interaction Chromatography, in *Molecular Biomethods Handbook*, Eds. R.R. John M. Walker, Humana Press, 2008, pp. 731–739.
[39] S. Hjertén, K. Yao, K.O. Eriksson, B. Johansson, Gradient and isocratic high performance hydrophobic interaction chromatography of proteins on agarose columns, *J. Chromatogr.* 359 (1986) 99–109.

[40] A. Goyon, V. D'Atri, B. Bobaly, E. Wagner-Rousset, A. Beck, S. Fekete, D. Guillarme, Protocols for the characterization of therapeutic monoclonal antibodies. I – Non-denaturing chromatographic techniques, *J. Chromatogr. B* 1058 (2017) 73–84.

[41] E. Tyteca, J.-L. Veuthey G. Desmet D. Guillarme, S. Fekete, Computer assisted liquid chromatographic method development for the separation of therapeutic proteins, *Analyst*, 141 (2016) 5488–5501.

[42] B. Bobaly, V. D'Atri, A. Beck, D. Guillarme, S. Fekete, Analysis of recombinant monoclonal antibodies in hydrophilic interaction chromatography: A generic method development approach, *J. Pharm. Biomed. Anal.* 145 (2017) 24–32.

Chapter 12

Computer-assisted Method Development in Characterization of Therapeutic Proteins by Hydrophilic Interaction Liquid Chromatography

Szabolcs Fekete* and Balazs Bobaly

School of Pharmaceutical Sciences,
University of Geneva, University of Lausanne,
CMU — Rue Michel Servet 1, 1211 Geneva 4, Switzerland

**szabolcs.fekete@unige.ch*

12.1 Introduction

Hydrophilic interaction liquid chromatography (HILIC) is a well-established technique for the separation and analysis of small polar compounds. Thanks to recent developments in column technology, wide-pore HILIC phases are now commercially available and enable the separation of peptides, protein fragments and intact proteins with high efficiency [1–3]. It was shown that a mobile phase composition between 80 and 65% acetonitrile in the presence of 0.1% trifluoroacetic acid (TFA) provided optimal conditions to retain proteins and obtain appropriate peak shapes. The selectivity of these HILIC separations have proven to be highly orthogonal to reversed-phase liquid chromatography (RPLC), and some hydrophilic protein-variants (mAb glycoforms) were better resolved in HILIC than in RPLC.

HILIC has already been applied in the past to the field of biopharmaceuticals for released glycan profiling and glycopeptide separations [4, 5]. Wide-pore HILIC phases offer new possibilities in glycan analysis at intact or middle-up levels of analysis [3, 6]. This approach also allows the qualitative comparison of the glycosylation profiles between originator and biosimilar products.

HILIC offers several additional benefits for biopharmaceutical characterization, as inherent compatibility with mass spectrometry (MS), the use of moderate mobile phase temperature for several proteins that are poorly recovered in RPLC and the possibility to couple several columns in series to improve resolving power (peak capacity), thanks to comparatively low mobile phase viscosity [2].

The retention mechanism in HILIC is more complex than in RPLC, and very sophisticated retention models are often required for method development. Trial and error method development approach is still usually performed for HILIC separations. HILIC retention can be considered as a mixed-mode mechanism, combining hydrophilic partitioning, adsorption through hydrogen bonds and various types of possible electrostatic and ionic interactions [7] which may be attractive as well as repulsive [8]. Therefore, HILIC retention models do not follow a perfect linear relationship in most cases [9]. Polynomial, empirical and mixed retention models are often applied for HILIC method development [7, 9–11]. Some quantitative structure-retention relationship (QSRR)-based approaches were also reported for HILIC method development and retention prediction [12–14].

This chapter provides a generic method development approach for the HILIC separation of monoclonal antibody (mAb) sub-units and their hydrophilic variants. A generic approach based on linear relationships and four initial experimental runs is suggested to provide good accuracy and a fast procedure in the practically useful range of the method variables.

12.2 General Considerations for Therapeutic Protein Separations in HILIC

The requirement of an unbiased separation relies on obtaining an acceptable recovery for all the species. Higher temperature generally enhances mass transfer of large molecules, leading to higher separation performance. In addition, temperature is a common parameter in chromatographic method development since selectivity and resolution can be tuned by adjusting this variable [15–18]. In RPLC conditions, it is well known that temperature strongly affects solute adsorption on the stationary phase.

Previous studies showed that intact mAbs, as well as mAb sub-units, show poor recovery in RPLC conditions when working at moderate temperature (e.g. $\leq$ 70°C), thus demonstrating the need to work at 80–90°C to avoid adsorption issues and reach acceptable recovery (e.g. above 90% of the injected protein amount). However in HILIC, it was also reported that lower temperature (e.g. 50–60°C) might result in appropriate recovery for some proteins [2]. Adsorption of digested and reduced NISTmAb, cetuximab, and brentuximab vedotin sub-units were monitored and relative recoveries of the main peaks were reported recently [1]. Typically higher than 90% recovery was observed above 70°C for mAb sub-units. The most critical sub-units were the light chain and the Fd glycovariants. For antibody drug conjugate (ADC) brentuximab vedotin, at least 80°C is required to achieve 90% recovery of the loaded sub-units, whereas some of the sub-units were completely adsorbed at 40°C. Interestingly, the adsorption behavior of the three different categories of sub-units (glycosylated, loaded and naked) can be differentiated, with the most critical group represented by the loaded species, and the most hydrophilic glycosylated sub-units showing the highest recovery in all cases.

In HILIC, retention times and chromatographic profiles — especially of large molecules — may slightly vary during first injections when using brand new columns. Saturation of the active sites by serial injections of concentrated protein samples might be necessary prior to analysis. This behavior can be monitored by the stabilization of retention times and elution profiles. Carefully equilibrated and properly saturated HILIC columns provide comparable retention time repeatability, as usually observed in reversed-phase conditions.

As a result of inappropriate focusing at the column inlet, injection of aqueous protein samples under HILIC conditions may result in fronting, distorted peaks. This issue can be overcome by various ways [2, 19–21]. The simplest procedures incorporate the decrease of the injection volume, the dilution of the sample by organic solvents (preferably by acetonitrile containing 0.1% TFA) and/or the use of an initial fast, steep gradient starting from lower eluent strength (e.g. focusing step) at the beginning of the separation.

12.3 Retention Properties of Protein Sub-units in HILIC, Selecting Method Variables

Working in a relatively limited, practically relevant gradient steepness range, a linear or nearly linear correlation between the gradient time and retention can be observed (linear solvent strength (LSS) model like behavior for k_{app} against tG) [1]. Deviation from linear behavior seems to decrease with increased solute retention. Selectivity between protein sub-units slightly changes, but elution order remains the same whatever the gradient steepness Peak width and therefore retention however strongly depends on tG, thus gradient time can be an important method variable for HILIC method development.

Similarly to gradient steepness, temperature also has a regular effect on solute retention when working in the practically relevant temperature range ($T > 70$°C). Linear fits properly approach the experimental points when constructing a van't Hoff plot ($\log k - 1/T$) [1]. With respect to recovery, practical temperature range is typically restricted to 70–90°C for protein fragments, but is sample dependent.

Other organic modifier than acetonitrile does not make much sense as the use of aprotic organic solvent is mandatory in HILIC. Therefore, ternary mobile phase composition is not worth studying as method variable. Similarly, it does not make sense to try other mobile phase additives, apart from TFA, as TFA provides the best peak shape and appropriate retention for multiply charged large proteins. Only if MS sensitivity needs to be improved, some other additives can be tried; however, chromatographic efficiency will probably decrease.

As a consequence, clearly the two most important method variables are the gradient steepness (tG) and mobile phase temperature (T). A linear retention model based on four initial experiments (tG-T model) should be tried first for HILIC method optimization.

12.4 2D Method Optimization

The impact of tG and T should be studied first at two levels, as linear behavior is expected in the relatively small design space (DS). Preferably a 150×2.1 mm column is suggested to work with as it is a good compromise between peak capacity and analysis time. Currently, only a very limited number of wide-pore phases are available; a good starting point can be the Glycoprotein BEH Amide 1.7 μm material. A flow rate between 0.3 and 0.6 mL/min is recommended, depending on the needs. (Higher separation efficiency but longer analysis time are expected at lower flow rate while faster separation with moderate efficiency is expected at higher flow rate.) As starting point, a flow rate of 0.45 mL/min is a good compromise. Suggested mobile phase A is 0.1% TFA in water, while mobile phase B is 0.1% TFA in acetonitrile. A linear gradient from 75% to 60% B generally elutes all the compounds of therapeutic proteins (mAb units, ADC species, fusion proteins, etc.) Gradient can be run between 75% and 60% B, and temperature can be set at 70 and 90°C.

Figure 12.1(a) shows the resolution map, obtained for IdeS-digested and reduced NISTmAb. As can be seen, lower temperature and longer gradient time are advantageous to improve resolution. However, a mobile phase temperature lower than 70°C is not suggested in order to avoid recovery issues. A good working point occurs at $tG = 15$ min and $T = 70$°C. Figure 12.1(b) shows the corresponding chromatogram. The glycovariants of the Fc/2 unit can be resolved and identified.

Another example is shown in Fig. 12.2. Here, the separation of cetuximab sub-units have been optimized. Cetuximab is a special case of mAbs as it possesses glycolization sites in both the Fc part and Fd arms. Based on the resolution map, an elution order change is observed with temperature (blue curves correspond to co-elution). The Fc/2 species are quite sensitive for temperature, and therefore the selectivity between them can be adjusted significantly by changing the mobile phase temperature.

The final example represents the optimization of an ADC separation. Figure 12.3 illustrates the resolution map and experimentally observed chromatogram for cysteine-linked IgG1 conjugation (brentuximab-vedotin). The loaded (conjugated) and the non-conjugated sub-units show different retention behavior. For brentuximab–vedotin, the retention of loaded species decreases with temperature, while the non-conjugated species (LC and Fd) shows the opposite behavior. Therefore, it may happen that elution order can be changed by temperature. Figure 12.4 shows the elution order change between the LC I and Fd species.

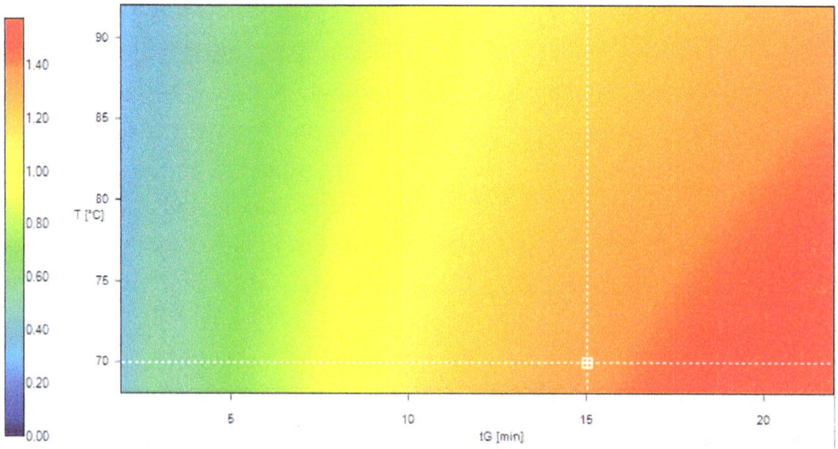

(a)

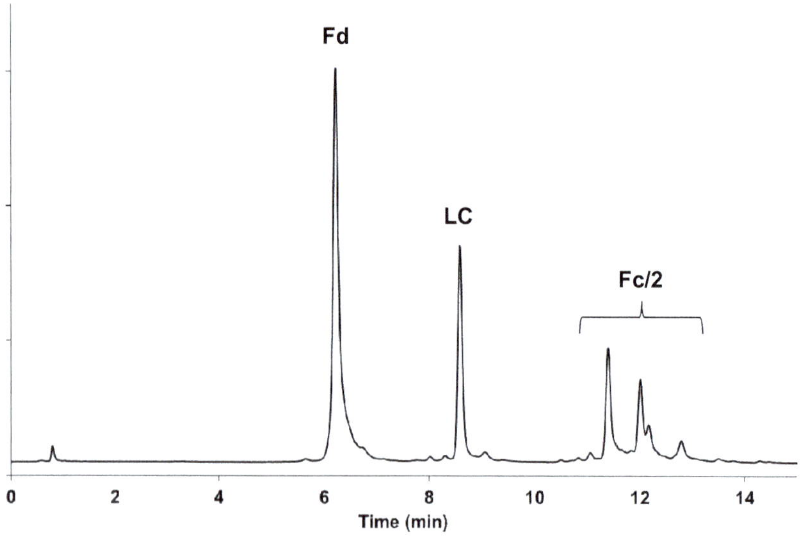

(b)

Figure 12.1: $tG - T$ resolution map in HILIC, obtained for partially digested and reduced NISTmAb (a) and an experimentally measured chromatogram at the working point (b).

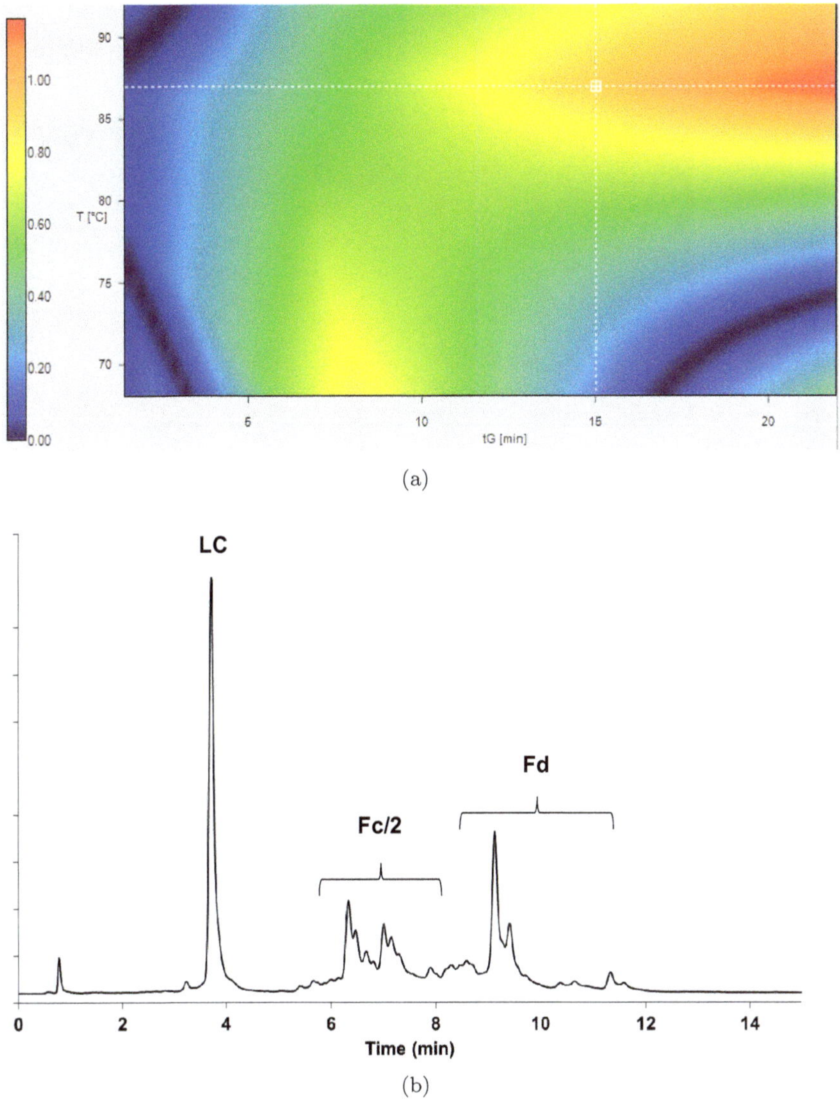

Figure 12.2: $tG - T$ resolution map in HILIC, obtained for partially digested and reduced cetuximab (a) and an experimentally measured chromatogram at the working point (b).

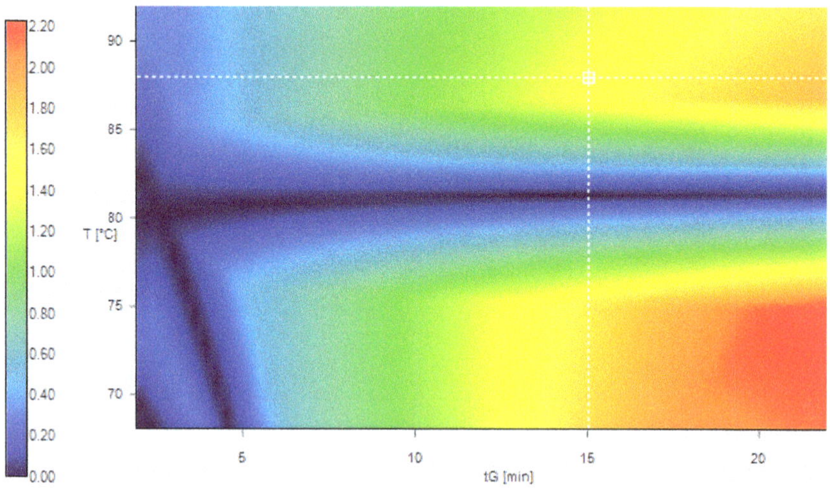

(a)

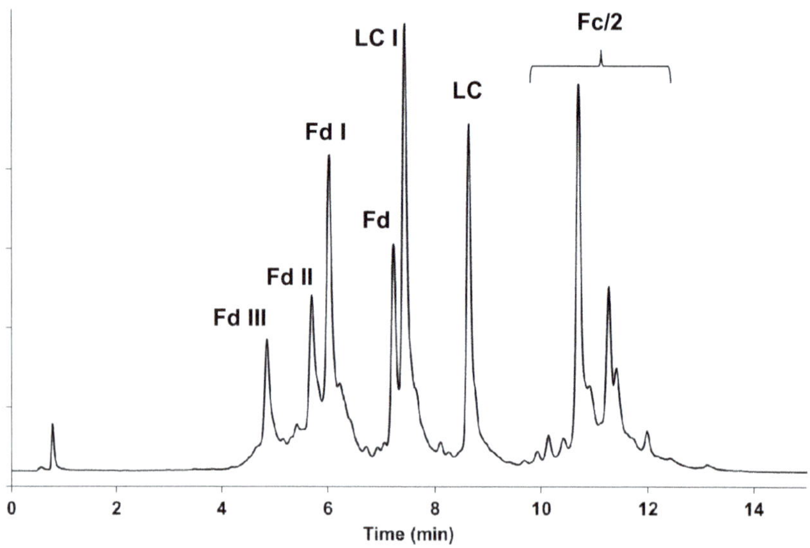

(b)

Figure 12.3: *tG* − *T* resolution map in HILIC, obtained for partially digested and reduced ADC (brentuximab–vedotin) (a) and an experimentally measured chromatogram at the working point (b).

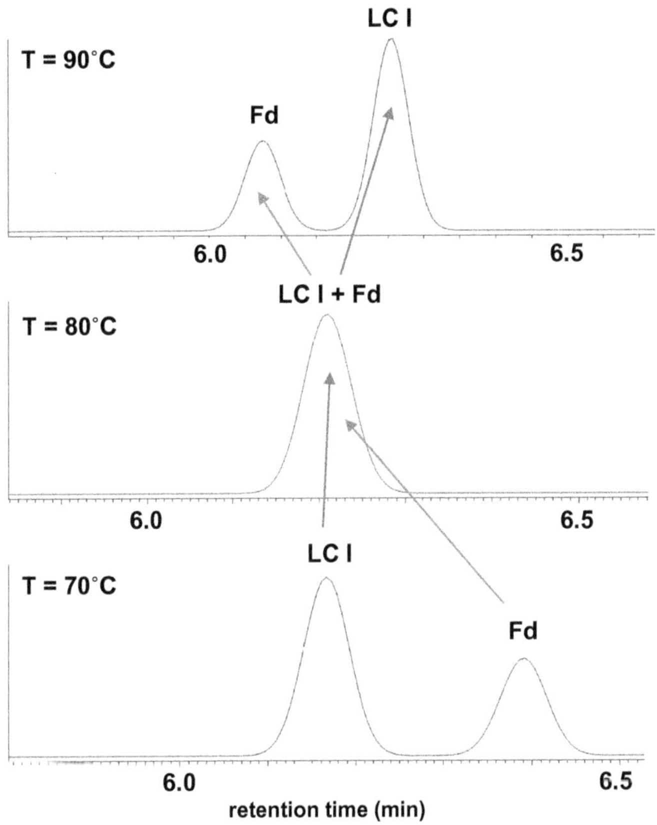

Figure 12.4: Change in elution order by temperature between ADC's LC I and Fd peaks.

References

[1] B. Bobaly, V. D'Atri, A. Beck, D. Guillarme, S. Fekete, Analysis of recombinant monoclonal antibodies in hydrophilic interaction chromatography: A generic method development approach, *J. Pharm. Biomed. Anal.* 145 (2017) 24–32.

[2] A. Periat, S. Fekete, A. Cusumano, J.-L. Veuthey, A. Beck, M. Lauber, D. Guillarme, Potential of hydrophilic interaction chromatography for the analytical characterization of protein biopharmaceuticals, *J. Chromatogr. A* 1448 (2016) 81–92.

[3] V. D'Atri, S. Fekete, A. Beck, M. Lauber, D. Guillarme, Hydrophilic interaction chromatography hyphenated with mass spectrometry: A powerful analytical tool for the comparison of originator and biosimilar therapeutic monoclonal antibodies at the middle-up level of analysis, *Anal. Chem.* 89 (2017) 2086–2092.

[4] M. Mancera-Arteu, E. Gimenez, J. Barbosa, V. Sanz-Nebot, Identification and characterization of isomeric N-glycans of human alfa-acid-glycoprotein by stable isotope labelling and ZIC-HILIC-MS in combination with exoglycosidase digestion, *Anal. Chim. Acta* 940 (2016) 92–103.
[5] J. Ahn, J. Bones, Y.Q. Yu, P.M. Rudd, M. Gilar, Separation of 2-aminobenzamide labeled glycans using hydrophilic interaction chromatography columns packed with 1.7 μm sorbent, *J. Chromatogr. A* 878 (2010) 403–408.
[6] M.A. Lauber, S.M. Koza, Mapping IgG subunit glycoforms using HILIC and a wide-pore amide stationary phase, 2015, Waters application note 720005385EN.
[7] G. Greco, S. Grosse, T. Letzel, Study of the retention behavior in zwitterionic hydrophilic interaction chromatography of isomeric hydroxy- and aminobenzoic acids, *J. Chromatogr. A* 1235 (2012) 60–67.
[8] A.J. Alpert, Electrostatic repulsion hydrophilic interaction chromatography for isocratic separation of charged solutes and selective isolation of phosphopeptides, *Anal. Chem.* 80 (2008) 62–76.
[9] E. Tyteca, A. Périat, S. Rudaz, G. Desmet, D. Guillarme, Retention modeling and method development in hydrophilic interaction chromatography, *J. Chromatogr. A* 1337 (2014) 116–127.
[10] U.D. Neue, H.J. Kuss, Improved reversed-phase gradient retention modeling, *J. Chromatogr. A* 1217 (2010) 3794–3803.
[11] G. Jin, Z. Guo, F. Zhang, X. Xue, Y. Jin, X. Liang, Study on the retention equation in hydrophilic interaction liquid chromatography, *Talanta* 76 (2008) 522–527.
[12] M. Taraji, P.R. Haddad, R.I.J. Amos, M. Talebi, R. Szucs, J.W. Dolan, C.A. Pohl, Prediction of retention in hydrophilic interaction liquid chromatography using solute molecular descriptors based on chemical structures, *J. Chromatogr. A*, 1486 (2017) 59–67.
[13] S.L. Maux, A.B. Nongonierma, R.J. FitzGerald, Improved short peptide identification using HILIC–MS/MS: Retention time prediction model based on the impact of amino acid position in the peptide sequence, *Food Chem.* 173 (2015) 847–854.
[14] E. Tyteca, M. Talebi, R. Amos, S.H. Park, M. Taraji, Y. Wen, R. Szucs, C.A. Pohl, J.W. Dolan, P.R. Haddad, Towards a chromatographic similarity index to establish localized quantitative structure-retention models for retention prediction: Use of retention factor ratio, *J. Chromatogr. A* 1486 (2017) 50–58.
[15] S. Fekete, S. Rudaz, J. Fekete, D. Guillarme, Analysis of recombinant monoclonal antibodies by RPLC: Toward a generic method development approach, *J. Pharm. Biomed. Anal.* 70 (2012) 158–168.
[16] S. Fekete, A. Beck, J. Fekete, D. Guillarme, Method development for the separation of monoclonal antibody charge variants in cation exchange chromatography, Part II: pH gradient approach, *J. Pharm. Biomed. Anal.* 102 (2015) 282–289.
[17] A. Cusumano, D. Guillarme, A. Beck, S. Fekete, Practical method development for the separation of monoclonal antibodies and antibody-drug-conjugate species in hydrophobic interaction chromatography, part 2: Optimization of the phase system, *J. Pharm. Biomed. Anal.* 121 (2016) 161–173.
[18] S. Fekete, I. Molnar, D. Guillarme, Separation of antibody drug conjugate species by RPLC: A generic method development approach, *J. Pharm. Biomed. Anal.* 137 (2017) 60–69.

[19] F. Gritti, J. Sehajpal, J. Fairchild, Using the fundamentals of adsorption to understand peak distortion due to strong solvent effect in hydrophilic interaction chromatography, *J. Chromatogr. A* 1489 (2017) 95–106.
[20] J.C. Heaton, D.V. McCalley, Some factors that can lead to poor peak shape in hydrophilic interaction chromatography, and possibilities for their remediation, *J. Chromatogr. A* 1427 (2016) 37–44.
[21] V. D'Atri, E. Dumont, I. Vandenheede, D. Guillarme, P. Sandra, K. Sandra, Hydrophilic interaction chromatography for the characterization of therapeutic monoclonal antibodies at protein, peptide and glycan levels, *LCGC Europe* 8 (2017) 424–434.

Index

D

E

F

G

N

O

P

Q

U

V

W

www.ingramcontent.com/pod-product-compliance
Ingram Content Group UK Ltd.
Pitfield, Milton Keynes, MK11 3LW, UK
UKHW020350150926
8624IPUK00004B/92